# Urban Transportation Systems

**Scrivener Publishing**
100 Cummings Center, Suite 541J
Beverly, MA 01915-6106

*Publishers at Scrivener*
Martin Scrivener (martin@scrivenerpublishing.com)
Phillip Carmical (pcarmical@scrivenerpublishing.com)

# Urban Transportation Systems

## Sustainability and Future Development

Kundan Meshram

Scrivener Publishing

WILEY

This edition first published 2024 by John Wiley & Sons, Inc., 111 River Street, Hoboken, NJ 07030, USA and Scrivener Publishing LLC, 100 Cummings Center, Suite 541J, Beverly, MA 01915, USA

For more information about Scrivener publications please visit www.scrivenerpublishing.com.



**Wiley Global Headquarters**
111 River Street, Hoboken, NJ 07030, USA

For details of our global editorial offices, customer services, and more information about Wiley products visit us at www.wiley.com.



***Library of Congress Cataloging-in-Publication Data***

ISBN 9781394228249

Cover image: Urban Transportation, Md Nazim Uddin | Dreamstime.com
Cover design by Kris Hackerott

Set in size of 11pt and Minion Pro by Manila Typesetting Company, Makati, Philippines

Printed in the USA

10 9 8 7 6 5 4 3 2 1

# Contents

# Preface

In an era defined by rapid urbanization and burgeoning transportation needs, the quest for sustainable solutions has never been more critical. The book you are about to embark on, *Urban Transportation Systems*, offers a comprehensive exploration of cutting-edge concepts, technologies, and strategies aimed at reshaping the landscape of urban transportation. With an astute focus on sustainability, efficiency, and resilience, this volume unfolds a multifaceted journey through the diverse facets of modern transportation systems.

From the foundational chapters introducing the core principles of sustainable transportation to the intricacies of intelligent transportation systems (ITS), artificial intelligence (AI) applications, geographic information systems (GIS), machine learning (ML), blockchain technology and the promotion of non-motorized transportation (NMT) and pedestrian safety, this book offers a holistic perspective on the challenges and opportunities that urban transportation presents. By examining environmental impact assessments and delving into the potential of advanced technologies and methodologies, it equips readers with the tools to create more sustainable and livable urban environments.

With each chapter, readers will gain a deeper understanding of the intricate interplay between transportation, technology, and sustainability. As the contributors traverse the complex terrain of urban transportation, they shed light on how innovative solutions, data-driven approaches, and visionary thinking can pave the way for a future where cities are more accessible, environmentally friendly, and inclusive characteristics for different use cases. We hope this book serves as a valuable resource for researchers, practitioners, and policymakers committed to shaping a brighter, more sustainable future for urban transportation systems worldwide for different scenarios.

**Kundan Meshram**
Bilaspur, India

# 1

# Introduction to Sustainable Transportation


***Abstract***

This book chapter provides an overview of the traffic and transport problems that arise in urban cities, and the role of mass transportation systems in addressing them in detail for different use cases. The chapter discusses the different modes of transportation and their characteristics, including the public and private transport systems. The advantages and disadvantages of public transport systems are also explored for real-time scenarios. The role of transportation infrastructure in mass transportation systems, including green bays, control stations, mitigation buildings, separator lanes, and safety islands, is also discussed for different use cases. The chapter then introduces sustainable transportation as a solution to various transport issues and explores its benefits. Overall, this chapter provides a comprehensive understanding of the transportation issues faced by urban cities and the potential solutions offered by sustainable transportation characteristics.



***Keywords:*** Sustainable transportation, mass transportation system, public transport system


## 1.1 Introduction

Transportation has been one of the cornerstones of human development, it includes categories like road transportation, air transportation, water transportation, and deep-sea transportation. While road transportation accumulates to nearly 70% of all transportation media, and thereby is considered as one of the most useful modes of transportation, other modes of transportation are also popular when context-sensitive transport is needed. Each of these modes has their own characteristics, including,

- Movement options, which indicates the number and scale of vehicles which the mode of transport can be used for carrying.
- Speed and accessibility, which represents the delay needed, and the number of people to which the medium is capable of reaching.


Kundan Meshram. Urban Transportation Systems: Sustainability and Future Development, (1–18) © 2024 Scrivener Publishing LLC

**Table 1.1** Transportation modes and their characteristics.

| Mode | Movement options | Speed | Accessibility | Cost | Capacity | Intermodal capability |
|---|---|---|---|---|---|---|
| Roadways | Very broad | Moderate | High | Moderate | Low | Very high |
| Railways | Broad | Slow | Moderate | Low | Moderate | Very high |
| Airlines | Narrow | Fast | Low | Very high | Very low | Moderate |
| Waterways | Broad | Very slow | Moderate | Very low | Very high | Very high |
| Pipeline | Very narrow | Very slow | Low | Low | Very high | Very low |

- Cost and capacity, which represents the cost of moving single units, and the number of units which can be moved.
- Intermodal capability, which indicates the capability of the system for interacting with other modes of transportation.

A brief survey of these modes is tabulated in Table 1.1, wherein each of the modes and their characteristics are segregated and compared.

Based on this comparison, various transportation modes and their usage can be estimated for smart cities. In this chapter, a wide variety of transportation problems, their characteristics, advantages, drawbacks, and infrastructure details are discussed in detail.

## 1.2 Traffic and Transport Problems of an Urban City

As the population of any city grows, the number of vehicles also grow exponentially. An example of number of vehicles with regards to number of people in an urban area can be observed from Figure 1.1, which indicates that for each new individual, there is at least one vehicle on the road. In fact, it can be observed that the number of vehicles on road follow Equation 1.1 for any urban area,

$$V_{num} = k^{P_{num}} \tag{1.1}$$

Where, $V_{num}$ *and* $P_{num}$ represent number of vehicles, and number of persons in the area, while $k$ is the area constant, which depends upon geographical size of the area.

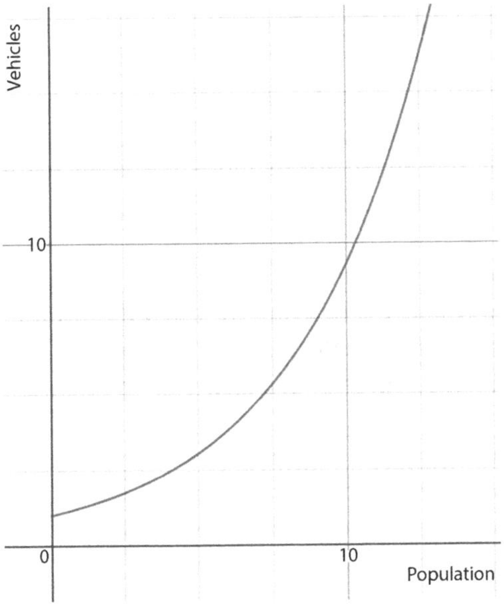


**Figure 1.1** Population v/s number of vehicles.

Due to increase in population size, the following issues arise during transportation:

- Traffic movement and congestion
- Off-peak inadequacy of public transport
- Public transport crowding
- Parking difficulties
- Difficulties for pedestrians
- Traffic noise
- Environmental impact
- Atmospheric pollution

Each of these issues result in increased congestion, and have a multiplicative effect on others. For instance, due to increased traffic movement and congestion, there is an increase in atmospheric pollution, which causes environmental impacts like global warming. Furthermore, due to larger traffic movements, there are parking issues, which further lead to congestion and overcrowding of public transports, which results into difficulties for pedestrians, and causes deficiencies in public transportation systems. Thus, these issues must be tackled during town planning, and mass transportation systems must be designed as discussed in the next sub-section.

## 1.3 Mass Transport System

Due to increase in overall traffic, governmental agencies have deployed mass transportation systems. These systems assist in transporting large number of people from one place to another, thereby reducing the need of multiple vehicle types. Examples of mass transportation systems are:

- City buses, which have moderate level of capacity, and move at moderate speed. These are the second most preferred way of transport, and usually follow a schedule.
- Trolley buses, require dedicated power lines, and thus are used in legacy mode.
- Trams (or light rail) are similar to trolley buses, but have larger capacities. Smart cities have replaced trams with underground and overground metro railways.
- Passenger trains are the most preferred way of communication in metropolitan cities, and are responsible for reducing traffic via maintaining fixed departure and arrival schedules.
- Rapid transit (metros, subways, underground railways, etc.) is one of the most upcoming modern day transportation systems, which are used for high efficiency, low cost and large-scale transportations.
- Ferries and other water taxis are also used in geographies with water bodies. They are useful for low speed, but high-capacity transportation.

All these modes of transportation are categorized under public transportation system, which can be observed from Figure 1.2, and are discussed in the next sub-section of this text. This will assist in analysis of different transportation models, and thereby improving their feasibility for real time deployment.

## 1.4 Modes of Transportation and Characteristics

The transportation industry has a significant impact on how our culture and economy are shaped. However, the traditional forms of transportation, which are mostly dependent on fossil fuels, have had detrimental effects on the environment, people's health, and sustainability in general. It is crucial to investigate alternate forms of transportation that lower greenhouse gas

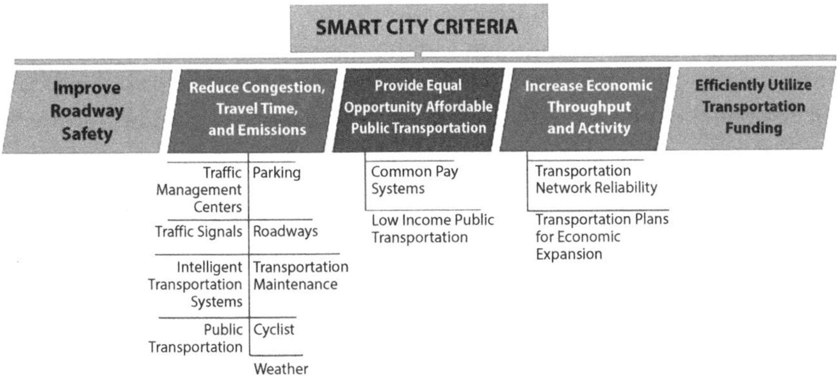


**Figure 1.2** Mass transportation system for smart city.

emissions, advance energy efficiency, and improve livability as the work toward a more sustainable future. This chapter seeks to provide a general overview of several sustainable transportation options and their features.

## 1. Cycling and Walking

The most basic and sustainable forms of transportation are cycling and walking. These modes have a number of benefits, including the promotion of physical activity, a decrease in air pollution and traffic, and a need for less infrastructure. They are best suited for quick travels inside cities or getting to transportation hubs. In order to promote these forms of transportation, walkability, and the availability of bicycle infrastructure, such as designated lanes and bike-sharing programs, are essential.

Characteristics: following characteristics of cycling and walking are given below:

i. High sustainability and zero emissions.
ii. Improve public health by encouraging physical exercise.
iii. Suitable for short journeys.
iv. Demand bicycle facilities and pedestrian-friendly infrastructure.
v. Encourage engagement with the environment and a feeling of community.

## 2. Public Transportation

For a huge number of people, shared mobility alternatives are provided via public transportation systems, which include buses, trams, light rail, subways, and trains. By lowering energy use, emissions per passenger kilometer, and traffic congestion, public transportation has a substantial positive impact on sustainability. By influencing the patterns of land use, effective public transportation systems may improve accessibility, promote social fairness, and aid in urban growth.

Characteristics: the characteristics of public transportation are given as follow:

i. Efficient and cooperative resource usage.
ii. Reduce emissions and traffic congestion.
iii. Accessibility to a variety of locations is provided.
iv. Encourage dense construction and communities that are close to transportation.
v. Need well thought-out infrastructure and encouraging policies.

**3. Carpooling and Ridesharing**

These activities include using private automobiles to transport a number of people in the same direction. These means of transportation may minimize per capita emissions, reduce the number of automobiles on the road, and improve traffic congestion. Carpooling and ridesharing have grown more practical and available with the development of digital platforms and smartphone apps.

Characteristics: the characteristics of carpooling and ridesharing are mentioned below:

i. Improve vehicle occupancy while minimizing the number of vehicles on the road.
ii. Reduced expenses for individual travel.
iii. Enhance community development and social engagement.
iv. Demand supporting policies, platforms, and incentives.
v. Encourage collaboration and shared responsibility among participants.

**4. Electric Vehicles**

Especially for private automobile travel, electric vehicles (EVs) offer a viable sustainable means of transportation. Electric vehicles (EVs) have the potential to drastically cut greenhouse gas emissions and reliance on fossil fuels by swapping out internal combustion engines for electric motors. The use of EVs increases as battery technology advances and charging infrastructure develops.

Characteristics: the following characteristics of electric vehicle are given:

i. No emissions from the exhaust, lowering greenhouse gas emissions and air pollution.
ii. Possibility of charging using renewable energy sources.
iii. Construct the infrastructure for charging.
iv. Continuous advancements in battery technology and driving distance.
v. Aid in energy source diversification and lessen reliance on fossil fuels.

**5. Technologies and Alternative Fuels**

Several alternative fuels and technologies are being investigated in addition to electric cars to decarbonize transportation. These include automobiles that run on biofuels, compressed natural gas (CNG), liquefied petroleum gas (LPG), and hydrogen fuel cell cars. With these options, transportation's carbon footprint will be reduced, and the energy sources used to power cars will be more varied.

Characteristics: the characteristics of technologies and alternative fuels are shown below:

i. Potential for decreased reliance on fossil fuels and greenhouse gas emissions.
ii. Construct the necessary infrastructure to produce and distribute gasoline.
iii. Ongoing development and research aimed at enhancing availability and efficiency.
iv. Taking into account feedstock sustainability and life cycle emissions.

In order to make the transition to sustainable mobility, it is necessary to combine a number of different modes and technologies that all work to improve air quality, reduce emissions, and improve quality of life process. Walking and cycling, public transit, carpooling and ridesharing, electric automobiles, and alternative fuels are just a few of the modes covered in this chapter that show how a more sustainable and effective transportation system might be developed for real-time scenarios. To fully reap the benefits of sustainable mobility, however, a complete strategy comprising encouraging legislation, investments in infrastructure, and modifications in personal behavior is required for different use cases.

## 1.5 Public Transport System

Public transport systems are schemes initiated by governments, and includes buses, railways, and taxis. These transportation systems are designed by governments via geographical and population data analysis, which can be observed from Table 1.2 as follows.

Thus, from Table 1.2 and Figure 1.3, it is observed that 22% people walk on foot, 8% use their cycles, 14% use their own motorbikes, 10% use personal cars, only 2% use autorickshaws, 21% of people depend on buses, only 3% utilize trains, and 0.4% utilize water transportation. Based on this data, governmental agencies will focus on brining more people on buses, and reduce autorickshaws as they are underutilized. Using such data analysis, public transportation systems are planned, and continuous evaluation is done in order to estimate the number of vehicle types needed for a given geographical area. A wide number of systems are developed for optimizing public transportation, for instance, the work in References [1] and [2] proposes models for intelligent transportation during COVID-19 pandemic, and define use of modern mediums of communication including

**Table 1.2** Data analysis for government use taken from the city of Bhopal during the financial year 2015–16.

| Mode of travel | Total - persons | Total - male | Total - female | Male traveler ratio (in %) | Female traveler ratio (in %) |
|---|---|---|---|---|---|
| All modes | 4,704,771 | 4,013,162 | 691,609 | 77 | 15 |
| On foot | 1,042,224 | 873,113 | 169,111 | 76 | 24 |
| Moped/scooter/ motor cycle | 663,546 | 639,014 | 24,531 | 86 | 14 |
| Bicycle | 420,854 | 411,326 | 9528 | 88 | 12 |
| Bus | 1,015,380 | 843,516 | 171,864 | 75 | 25 |
| Car/jeep/van | 514,878 | 416,717 | 98,161 | 73 | 27 |
| Train | 138,263 | 109,250 | 29,012 | 71 | 29 |
| Tempo/autorickshaw/ taxi | 108,869 | 83,960 | 24,909 | 69 | 31 |
| Water transport | 2027 | 1687 | 340 | 75 | 25 |
| Any other | 53,807 | 41,958 | 11,849 | 70 | 30 |
| No travel | 744,923 | 592,620 | 152,303 | 72 | 28 |

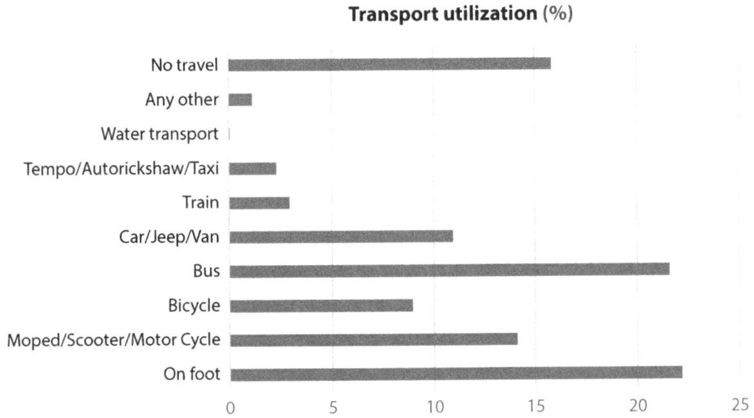


**Figure 1.3** Percentage of transport utilization for Bhopal in 2016–17.

Bluetooth low energy (BLE). These models inform the Urban Traffic Data Management Centre for effective control and service usage. The performance of these systems can be improved using the work in References [3–5] where models for real-time rerouting, total emergency prevention, and optimization of passenger energy cost are defined. These models assist in reducing cost of deployment, cost of maintenance, and running cost of the system via exploration of signal processing and filtering methods on real-time traffic data. Internet of things (IoT) systems are also used for public transport optimization. The work in Reference [6] describes design of a novel context sensitive intelligent public transport system that utilizes software defined networks (SDNs), for flexible control of public transportation resources, resource allocation, suggestion of alternate routes, and calibration of transportation modes. This assists in improving speed and efficiency of public transport.

Furthermore, the work in Reference [7] suggest use of be-in and be-out (BIBO) method for making payments on highways, thereby improving traffic speeds. The model utilizes BLE beacons with reduced delay, and low energy for secure and highly effective payments. Similarly, models proposed in References [8–10] discuss use of intelligent public transport for enhancing passengers' safety, reducing congestion via on-the-fly analysis, and sustainable transport using IoT devices. Work proposed in References [11–13] also directs use of machine learning models like optimal least square (OLS), support vector regression, convolutional neural networks (CNNs) along with service driven middleware approaches, and fare evasion models for public transport. These models assist in reducing the delay, and cost of public transport management, while improving the tax collection, fare collection, and maintaining road quality. Geography specific studies are mentioned in References [14] and [15], which discuss use of intelligent public transport in Asian countries, and suggest methods to improve their overall performance.

## 1.6 Advantages and Disadvantages of Public Transport System

Based on previous studies, following are the advantages of public transport systems:

1. Improves community health by allowing people to socialize with each other.
2. Economic benefits to the community by creating a greater number of jobs, and reducing dependency on private vehicles.
3. Improves fuel efficiency via reduction of number of vehicles on the road, and increase efficiency of per km travel cost per passenger.
4. Public transportation reduces air pollution by reducing number of private vehicles on the road, thereby reducing dependency on private transportation.
5. Reduce road congestion via increasing the number of persons per unit carrying area.
6. Improves community mobility via improved socialization during travel.
7. Provides an equitable transportation system.
8. Public transportation improves commuters' productivity by fixing a schedule for travel, thereby inculcating a schedule in day-to-day activities.

## 1.7 Role of Transportation in Mass Transportation Systems

Public transportation is capable of providing jobs, resources, medical and healthcare, opportunities for recreation, etc. for a wide variety of individuals. A large number of people travelling to far distances do not possess personal vehicles, and thus rely on public transportation. Due to mass transportation, people are able to travel in groups, which assists them in socializing, reducing fuel costs, thereby increasing their financial status.

Faster and frequent transportation systems lead to a more secure environment for public, which can be described in terms of medical security, personnel safety, etc. The areas that have good public transport systems automatically have better economy thereby offering better opportunities to people and businesses for settling, thus inculcating growth. Moreover, when emergencies arise, these transport systems assist in providing critical services, and thus improving trust level in the system, which attracts more people for achieving better living opportunities. These systems also assist in reducing pollution, congestion, accidents, sound levels, etc. when compared with individual private transport systems, which makes them beneficial for riders and their counterparts.

## 1.8 Public-Private Transport System

Based on the analysis mentioned in the previous section, government makes plans for public- private transportation systems. Such a plan can be observed from Figure 1.4, which indicates if government wants to reduce number of buses from road to boost car sales, then they can reduce bus frequency, and reduce car costs. As a result, there is a greater number of cars

**Figure 1.4** Modes of transportation.

on the street, which indicates higher probability of congestion, that might reduce demand for cars.

But at this point, the government can reduce frequency of buses, and increase operational costs of public transport, which reduces the number of buses from the street, thereby reducing dependency on public transport. This increases car sales, and reduces the number of trips via public transport options.

## 1.9 Transportation Infrastructure

In order to design transportation systems, efficient infrastructure is needed, which involves design of the following entities,

1. Green Bays
   These are mini transport systems, which allow vehicles to stop and recharge for fuel, check their air pressure, etc. Green bays are suited for vehicles which have to travel long routes, thereby assisting them in continuing their travel with minimum disturbance.
2. Control Stations
   These are mini stations, which are used for controlling transportation routes, public onboarding and deboarding, vehicle maintenance, route upgradation, etc. Due to these stations,

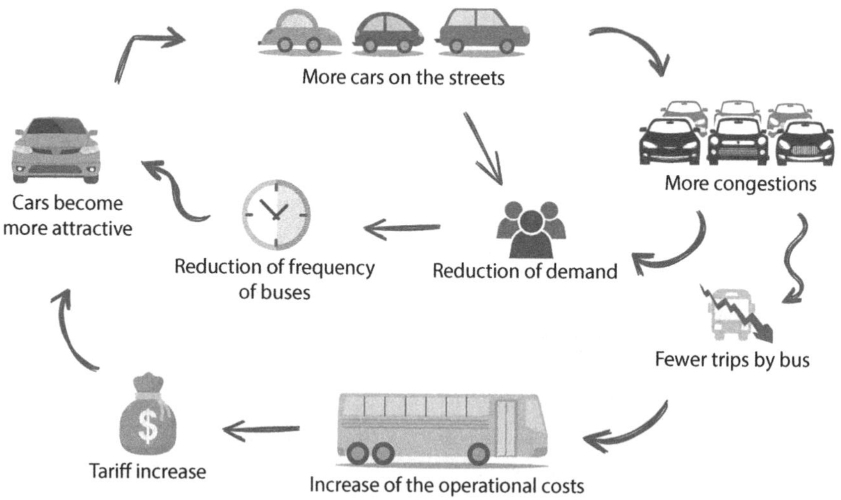


**Figure 1.5** Government planning process to boost car sales.

optimization in vehicle technology is possible. These stations also assist in improving overall efficiency of vehicle routing using on-site upgrades, which assists in optimizing point-to-point distance, thereby reducing fuel consumption as shown in Figure 1.5.

3. Mitigation Buildings
   These buildings are like train stations, wherein different businesses can be setup. These buildings assist in passenger boarding/deboarding, goods boarding/deboarding, route upgradation, maintenance, etc. Furthermore, these buildings assist in improving overall efficiency of routing via connecting various junctions with each other at common points of interest.
4. Separator Lanes
   Separator lanes assist in dividing entire area into multiple parts, where each part is responsible for serving a particular geography of people. These lanes also assist in separating areas with different needs, for instance, industrial lines are always different from commercial and residential lines. This division is useful for segregating costs for different points of interest, and thereby assisting in price fixation, and tax collection.
5. Safety Islands
   Safety islands are used for storing bulk transport vehicles, which includes both old, and new ones. These islands are extremely useful for auctioning, leasing out new vehicles, and reducing dependency on external agencies during upgrades.

A brief summary of different infrastructure modes is tabulated in Table 1.3, which assists in finding the consistency of these modes for smart city planning.

These metrics assist in the evaluation of infrastructure placements in the city of Bhopal, which can be used for estimation of different key-locations in the city where control stations, mitigation buildings, etc. can be placed for efficient access.

## 1.10 Introduction to Sustainable Transportation, and How It Can Solve Various Issues

In today's world, transportation is crucial since it links people, products, and services. Traditional forms of transportation, on the other hand,

**Table 1.3** Summary of different infrastructure components for the city of Bhopal during year 2016–17.

| Metric | Average | Std. deviation | Loading factor | Variance (%) | Consistency |
|---|---|---|---|---|---|
| Land transportation | | | | | |
| Density of railway (= railway length/land area) | 0.0144 | 0.0135 | 0.7902 | 66.8493 | 0.7443 |
| Density of road (= roadway length/ land area) | 0.3843 | 0.2925 | 0.7686 | N/A | N/A |
| Quality of road (= highway length/ total roadway length) | 0.0144 | 0.0117 | 0.7677 | N/A | N/A |
| Air transportation | N/A | N/A | N/A | N/A | N/A |
| Area of airport lounge | 51093.65 | 66021.68 | 0.7785 | 61.8219 | 0.7605 |
| Length of runway | 2975.516 | 337.0698 | 0.6561 | N/A | N/A |
| Project cargo network | 60.8373 | 15.7644 | 0.8172 | N/A | N/A |
| Density of airport (= the number of airports/ land area in a province) | 0.0333 | 0.0495 | 0.7218 | N/A | N/A |
| Water transportation | N/A | N/A | N/A | N/A | N/A |
| Sum of number of berth/the distance to seaport | 1.008 | 3.3156 | 0.846 | 79.4907 | N/A |
| Sum of number of deep- berth/ the distance to seaport | 0.297 | 0.7443 | 0.846 | N/A | N/A |

mainly depend on fossil fuels, which have several negative effects on the environment, society, and the economy. On the other hand, sustainable transportation provides a comprehensive strategy that tries to solve these problems by lessening environmental effects, advancing social equality, and improving economic efficiency. This chapter gives a general overview of sustainable transportation and considers how it may successfully address a number of urgent problems.

1. Environmental Concerns
   a) Climate change and greenhouse emissions: traditional transportation systems, principally via the burning of fossil fuels, are significant contributors to greenhouse gas emissions. Electric cars, mass transit, and alternative fuels are just a few examples of sustainable transportation options that may drastically cut emissions and lessen the effects of climate change.
   b) Air pollution: vehicle emissions, particularly those from internal combustion engines, are a major cause of poor air quality, which in turn causes respiratory illnesses and other health problems. Alternatives to conventional transportation that create zero or little emissions, such electric cars and active modes like walking and cycling, help to improve public health and reduce air pollution.
   c) Resource depletion: transportation cannot continue to be powered by finite fossil fuel reserves. By promoting energy efficiency, alternative fuels, and renewable energy sources, sustainable transportation lessens reliance on non-renewable resources and ensures that there will always be enough energy for transportation requirements.

   Traffic congestion has developed as a consequence of growing urbanization, which results in lost productivity, frustration, and wasted time. By maximizing road space and encouraging shared mobility, sustainable transportation options including effective public transit, carpooling, and active modes may help minimize congestion.
2. Social and Health Concerns
   a) Social equity: marginalized groups are disproportionately affected by limited transportation alternatives, which limits their mobility and possibilities. By linking individuals to necessary services, education, work, and leisure pursuits, sustainable transportation, with an emphasis on

inexpensive and accessible public transit, may alleviate these imbalances and enhance social inclusion.

b) Health and wellbeing: transportation systems that prioritize cars discourage physical exercise, which leads to sedentary lifestyles and related health issues. Walking, bicycling, and using public transportation are examples of sustainable modes of transportation that promote active lives, improve physical and mental health, and lower the incidence of chronic illnesses.

3. Economic Concerns

a) Rising fuel prices: varying oil prices have a big influence on transportation expenses and the economy as a whole. Sustainable transportation helps lower fuel consumption and offers prospects for cost savings and financial stability via the use of alternative fuels and energy- efficient technology.

Investments in sustainable transportation infrastructure, such as active transportation infrastructure and public transit networks, may promote economic growth. As a result of these investments, metropolitan areas are revitalized, companies are attracted, and jobs are created, increasing economic prospects and enhancing quality of life sets.

b) Productivity and efficiency: sustainable transportation places an emphasis on multimodal connectivity and resource-efficient resource usage. It increases overall system efficiency, decreases travel times, and increases productivity for both consumers and enterprises by offering seamless and integrated transportation alternatives.

A broad variety of urgent concerns, such as climate change, air pollution, traffic congestion, social unfairness, and economic difficulties, may be greatly improved with sustainable transportation. Sustainable transportation provides a route to a more egalitarian, ecologically responsible, and economically viable transportation system by supporting energy-efficient technology, alternative fuels, public transit, and active modes. To reach a future that is cleaner, healthier, and more prosperous for everyone, governments, corporations, communities, and people must work together and invest in sustainable mobility solutions.

## References


1. Jain, S., Jain, S.S. Development of Intelligent Transportation System and Its Applications for an Urban Corridor During COVID-19. *J. Inst. Eng. India Ser. B* (2021). https://doi.org/10.1007/s40031-021-00556-y
2. R. Chawla, M. Dhakate and S. Chaurasia, "System and Method for Smart Public Transportation System," *2020 International Conference Industry 4.0 Technology (I4Tech)*, 2020, pp. 51-54, doi: 10.1109/I4Tech48345.2020.9102635.
3. S. Kalra, S. Momin, T. S. Kulkarni and V. Lohani, "Real Time Re-routing of Public Transportation System," *2019 IEEE Bombay Section Signature Conference (IBSSC)*, 2019, pp. 1-5, doi: 10.1109/IBSSC47189.2019.8973089.
4. Y. Yin, "Thoughts on the Construction of Urban Public Transport Emergency Prevention and Control System in China—learning from the United States," *2020 International Conference Urban Engineering Management Science (ICUEMS)*, 2020, pp. 331- 334, doi: 10.1109/ICUEMS50872.2020.00079.
5. X. Lai, J. Teng and L. Ling, "Evaluating Public Transportation Service in a Transit Hub based on Passengers Energy Cost," *2020 IEEE 23rd International Conference Intelligent Transportation Systems (ITSC)*, 2020, pp. 1-7, doi: 10.1109/ITSC45102.2020.9294662.
6. S. Chavhan, D. Gupta, B. N. Chandana, A. Khanna and J. J. P. C. Rodrigues, "IoT-Based Context-Aware Intelligent Public Transport System in a Metropolitan Area," in *IEEE Internet Things J.*, vol. 7, no. 7, pp. 6023-6034, July 2020, doi: 10.1109/JIOT.2019.2955102.
7. M. A. Bitew, A. Muhammad, D. H. Fandiantoro, H. Boedinoegroho and A. Kurniawan, "E-Payment For Public Transportation Using BIBO Method Based On Bluetooth Low Energy Beacon," *2020 International Conference Computer Engineering, Network, Intelligent Multimedia (CENIM)*, 2020, pp. 199-204, doi: 10.1109/CENIM51130.2020.9297901.
8. W. Agustiono, "Integrated Public Transportation Systems Model for Passengers' Convenience and Safety," *2020 6th Information Technology International Seminar (ITIS)*, 2020, pp. 243-248, doi: 10.1109/ITIS50118.2020.9321066.
9. M. Skhosana and A. E. ezugwu, "Irenbus: A Real-Time Public Transport Management System," *2020 Conference Information Communications Technology Society (ICTAS)*, 2020, pp. 1-7, doi: 10.1109/ICTAS47918.2020.234000.
10. B. V. Kumar, A. Ravishankar, A. Karan, K. Vishal and J. A. Praseeth Kumar, "A Smart Public Transportation System for Reliable and Hassle Free Conveyance in Sustainable Smart Cities," *2020 International Conference Computer Communication Informatics (ICCCI)*, 2020, pp. 1-6, doi: 10.1109/ICCCI48352.2020.9104094.
11. D. Panovski and T. Zaharia, "Real-Time Public Transportation Prediction with Machine Learning Algorithms," *2020 IEEE International Conference Consum. Electronics (ICCE)*, 2020, pp. 1-4, doi: 10.1109/ICCE46568.2020.9043077.
12. Nauman Ahmad Khan, Jean-Christophe Nebel, Souheil Khaddaj, Vesna Brujic-Okretic, "Scalable System for Smart Urban Transport Management",

*J. Adv. Transp.*, vol. 2020, Article ID 8894705, 13 pages, 2020. https://doi.org/10.1155/2020/8894705

13. Barabino, B., Lai, C. and Olivo, A. Fare evasion in public transport systems: a review of the literature. *Public Transp. {b}*12, {/b}27–88 (2020). https://doi.org/10.1007/s12469-019- 00225-w
14. Noman, S.M., Ahmed, A. and Ali, M.S. Comparative analysis of public transport modes available in Karachi, Pakistan. *SN Appl. Sci. {b}*2, {/b}967 (2020). https://doi.org/10.1007/s42452-020-2678-3
15. Elkosantini, Sabeur and Darmoul, Saber. (2013). Intelligent Public Transportation Systems: A Review of Architectures and Enabling Technologies. *2013 International Conference Adv.Logistics Transport, ICALT* 2013. 10.1109/ICAdLT.2013.6568465.

# 2

# Use of Sustainable Transportation for Urban Scenarios

***Abstract***
This book chapter provides an introduction to urban transportation scenarios and explores advanced operational concepts of urban transportation. The chapter focuses on the bus lane system, discussing its advantages and limitations in worldwide scenarios. It also discusses rail systems, including the different types of rail systems such as monorail and metro, and their respective advantages and disadvantages. The chapter also examines the skywalk and under bridge systems and their advantages. Finally, the chapter explores how to measure the performance of transit systems. By providing an overview of these different transportation scenarios and their performance metrics, this chapter offers valuable insights into the transportation planning and management of urban areas.

***Keywords:*** Urban public transport system, bus rapid transit system, electric vehicles

## 2.1 Introduction to Urban Transportation Scenarios

Urban public transport system (UPTS) refers to the various modes of public transportation available in cities and urban areas, such as buses, trains, trams, subways, and light rail. The UPTS is essential for the mobility of urban dwellers, providing a means of transportation that is affordable, safe, and efficient. This document will provide a detailed overview of the UPTS, its importance, and the challenges it faces.

Importance of UPTS. A well-developed UPTS is essential for sustainable urban development as it helps to reduce traffic congestion, air pollution, and carbon emissions. It also promotes social inclusion by providing affordable transportation options for low-income residents, students, and senior citizens. Furthermore, it supports economic growth by improving accessibility to employment centers, commercial areas, and educational institutions [1–3].

---

Kundan Meshram. Urban Transportation Systems: Sustainability and Future Development, (19–40) © 2024 Scrivener Publishing LLC

Modes of UPTS. The UPTS consists of several modes of transportation, as shown in Figure 2.1, including:

1. Buses: buses are the most common mode of transportation in urban areas. They are relatively affordable, flexible, and accessible, and can transport a large number of passengers.
2. Trains: trains are ideal for longer distances and can transport a large number of passengers quickly. They can be further classified into commuter trains, light rail, and metro rail.
3. Subways: subways are underground trains that operate on dedicated tracks. They are typically used in densely populated urban areas and provide a fast and efficient means of transportation.
4. Trams: trams are similar to buses but operate on dedicated tracks, providing a smoother ride. They are commonly used in European cities.

**Figure 2.1** Typical transportation scenarios.

Challenges faced by UPTS. Despite its importance, the UPTS faces several challenges that hinder its effectiveness, including:

1. Insufficient funding: most UPTS systems are underfunded, which limits their ability to expand and upgrade their infrastructure and services.
2. Congestion: traffic congestion can significantly impact the efficiency and reliability of UPTS systems, resulting in delays and reduced capacity.
3. Safety and security: safety and security are major concerns for UPTS systems, particularly in areas with high crime rates.
4. Lack of integration: lack of integration between different modes of transportation, such as buses and trains, can hinder the effectiveness of UPTS systems.

Solutions to improve UPTS. Several solutions can help improve the effectiveness of UPTS systems, including:

1. Increased funding: increased funding can help UPTS systems to expand their infrastructure and services, improve safety and security, and enhance the user experience.
2. Integration of different modes of transportation: integration of different modes of transportation can improve the efficiency and reliability of UPTS systems, making them more attractive to commuters.
3. Use of technology: the use of technology, such as intelligent transportation systems (ITS), can improve the efficiency and reliability of UPTS systems by providing real- time information on schedules, delays, and routes.
4. Promotion of sustainable transportation: the promotion of sustainable transportation, such as cycling and walking, can reduce congestion and improve the effectiveness of UPTS systems.

Thus, the UPTS is essential for sustainable urban development, promoting social inclusion, and supporting economic growth [4–6]. However, it faces several challenges that hinder its effectiveness, including insufficient funding, congestion, safety and security, and lack of integration. Solutions such as increased funding, integration of different modes of transportation, use of technology, and promotion of sustainable transportation can help improve the effectiveness of UPTS systems.

## 2.2 Advanced Operation Concepts of Public Transportation

Advanced operation concepts of public transportation refer to the use of advanced technologies and strategies to improve the efficiency and effectiveness of public transportation systems. These concepts are designed to enhance the user experience, reduce travel time, increase safety, and reduce the environmental impact of public transportation. This document will provide a detailed overview of advanced operation concepts of public transportation and their importance.

Importance of advanced operation concepts of public transportation. Advanced operation concepts of public transportation are essential for enhancing the user experience, improving safety, and reducing the environmental impact of public transportation systems. They help to improve the efficiency and reliability of public transportation systems, making them more attractive to commuters. Furthermore, they can help reduce the cost of operating public transportation systems, making them more sustainable in the long term.

Advanced operation concepts of public transportation. The following are some of the advanced operation concepts of public transportation:

1. Intelligent transportation systems (ITS): ITS involves the use of advanced technologies, such as GPS (global positioning system), sensors, and communication systems, to improve the efficiency and effectiveness of public transportation systems. Intelligent transportation systems can provide real-time information on travel times, schedules, and delays, enabling commuters to plan their journeys more effectively.
2. Mobility as a service (MaaS): MaaS is a concept that involves integrating various modes of transportation, such as buses, trains, and taxis, into a single platform. This platform provides commuters with a range of transportation options and enables them to plan and pay for their journeys using a single app.
3. Demand-responsive transport (DRT): DRT is a concept that involves providing transportation services on demand, rather than on a fixed schedule. Demand-responsive transport can be used in areas with low demand for public transportation, making it more cost-effective than traditional bus services.
4. Automated transit networks (ATN): ATN involves the use of automated vehicles, such as driverless buses or trains,

to transport passengers. Automated transit networks can improve safety, reduce travel time, and increase the capacity of public transportation systems.

5. Transit-oriented development (TOD): TOD involves the development of compact, mixed-use communities around public transportation hubs. This concept encourages the use of public transportation and reduces reliance on private cars, reducing traffic congestion and air pollution sets [7–9].

Challenges faced by advanced operation concepts of public transportation. Advanced operation concepts of public transportation face several challenges that hinder their implementation, including:

1. High costs: implementing advanced operation concepts of public transportation can be expensive, requiring significant investments in technology and infrastructure.
2. Resistance to change: public transportation systems have long been operated in traditional ways, and introducing new concepts can be met with resistance from stakeholders.
3. Technical challenges: implementing advanced technologies, such as driverless vehicles, can present technical challenges, such as ensuring safety and reliability.
4. Data privacy and security: the use of advanced technologies, such as ITS, can raise concerns about data privacy and security.

Solutions to improve advanced operation concepts of public transportation. Several solutions can help improve the implementation of advanced operation concepts of public transportation, including:

1. Collaboration between stakeholders: collaboration between transportation providers, technology companies, and governments can help to identify and overcome challenges.
2. Pilot projects: pilot projects can help to test and refine new concepts before implementing them on a larger scale.
3. Public education and awareness: public education and awareness campaigns can help to increase support for new concepts among commuters and other stakeholders.
4. Data privacy and security measures: implementing strong data privacy and security measures can help to address concerns about the use of advanced technologies.

Thus, advanced operation concepts of public transportation are essential for improving the efficiency and effectiveness of public transportation systems. Intelligent transport systems, MaaS, DRT, ATN, and TOD are some of the advanced operation concepts that can help to achieve these goals. However, implementing these concepts can be challenging, and they require significant investments in technology and infrastructure. Collaborative efforts between stakeholders, pilot projects, public education and awareness, and strong data privacy and security measures can help to address these challenges and improve the implementation of advanced operation concepts of public transportation. By adopting these concepts, public transportation systems can provide a more reliable, efficient, and sustainable mode of transportation for commuters, while also reducing traffic congestion and environmental impacts.

## 2.3 BRTS and Bus Lane System

Bus rapid transit system (BRTS) and bus lane system are two important concepts in urban transportation that are designed to improve the efficiency of public transportation. Bus rapid transit system is a bus-based public transportation system that uses dedicated lanes, stations, and other infrastructure to provide fast, reliable, and efficient service as shown in Figure 2.2. Bus lane system, on the other hand, involves the designation of specific lanes on roads for the exclusive use of buses, allowing them to bypass traffic congestion and provide faster service. This document will provide a detailed overview of BRTS and bus lane system and their importance in urban transportation [10–12].

Importance of BRTS and bus lane system. Bus rapid transit system and bus lane system are important concepts in urban transportation because they provide an efficient, affordable, and environmentally friendly mode of transportation for commuters. They offer several benefits, including:

1. Improved travel time: by providing dedicated lanes or separate corridors, BRTS and bus lane system can significantly reduce travel time, making public transportation more attractive to commuters.
2. Reduced congestion: by providing dedicated lanes or separate corridors, BRTS and bus lane system can reduce traffic congestion and improve traffic flow, reducing travel time for all road users.

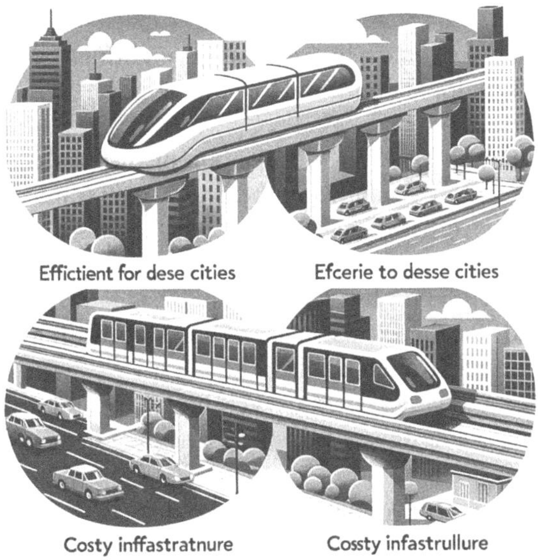


**Figure 2.2** Infrastructure sets.

3. Environmental benefits: by encouraging the use of public transportation, BRTS and bus lane system can reduce the number of cars on the road, reducing air pollution and greenhouse gas emissions.
4. Cost-effective: compared to other modes of transportation, such as private cars, BRTS and bus lane system are more cost-effective, providing an affordable mode of transportation for commuters.

Bus rapid transit system is a high-quality, bus-based transit system that uses dedicated lanes, stations, and other infrastructure to provide fast, reliable, and efficient service. The key features of BRTS include:

1. Dedicated lanes: BRTS systems have dedicated lanes that are separate from regular traffic lanes, allowing buses to bypass traffic congestion and improve travel time.
2. Stations: BRTS systems have dedicated stations that are designed to be efficient and user-friendly, providing passengers with easy access to buses and comfortable waiting areas.
3. Fare collection: BRTS systems use advanced fare collection systems, such as smart card technology or mobile payments, to reduce boarding times and improve efficiency.

4. Bus fleet: BRTS systems use high-capacity buses that are designed for quick boarding and alighting, with features such as multiple doors and low floors.

Bus lane system involves the designation of specific lanes on roads for the exclusive use of buses, allowing them to bypass traffic congestion and provide faster service. The key features of bus lane system include:

1. Designated lanes: bus lane system designates specific lanes on roads for the exclusive use of buses, providing them with a clear path through traffic.
2. Signage: bus lane system uses clear signage and markings to indicate the designated bus lanes, helping drivers to stay in the correct lanes.
3. Enforcement: bus lane system requires strict enforcement to ensure that other vehicles do not use the designated bus lanes, improving the efficiency of the system.
4. Integration with BRTS: bus lane system can be integrated with BRTS systems to provide an even more efficient and reliable mode of public transportation.

Challenges faced by BRTS and bus lane system. Bus rapid transit system and bus lane system face several challenges that can hinder their effectiveness, including:

1. Limited space: the implementation of dedicated lanes or separate corridors for buses can be challenging in areas with limited space.
2. High initial cost: the construction of dedicated lanes or separate corridors can be expensive, requiring significant investments in infrastructure.
3. Resistance to change: BRTS and bus lane system may face resistance from stakeholders who are resistant to change or who may be affected by the changes.
4. Maintenance: dedicated lanes and other infrastructure require ongoing maintenance, which can be costly and time-consuming.
5. Enforcement: enforcement of dedicated bus lanes can be difficult, and violations can reduce the effectiveness of the system.

Strategies to overcome challenges. To overcome the challenges faced by BRTS and bus lane system, several strategies can be implemented, including:

1. Collaborative planning: collaborative planning between stakeholders can help to ensure that the implementation of BRTS and bus lane system is coordinated and effective.
2. Public education and awareness: public education and awareness campaigns can help to inform commuters and other stakeholders about the benefits of BRTS and bus lane system, addressing resistance to change.
3. Pilot projects: pilot projects can be used to test the feasibility of BRTS and bus lane system in specific areas, allowing for adjustments and modifications before full-scale implementation.
4. Data-driven decision-making: data-driven decision-making can help to ensure that BRTS and bus lane system are implemented in areas with the greatest potential for success, reducing the risk of failure.

Thus, it can be observed that BRTS and bus lane system are important concepts in urban transportation that provide an efficient, affordable, and environmentally friendly mode of transportation for commuters. While they face several challenges, including limited space, high initial cost, resistance to change, maintenance, and enforcement, collaborative planning, public education and awareness, pilot projects, and data-driven decision-making can help to address these challenges and improve their effectiveness. By implementing BRTS and bus lane system, urban transportation systems can provide a more reliable, efficient, and sustainable mode of transportation for commuters, while also reducing traffic congestion and environmental impacts for different use cases [13–15].

## 2.4 Advantages and Limitations in Worldwide Transport Scenario

The utilization of sustainable transportation in urban scenarios has gained substantial attention in recent years, primarily due to its potential to mitigate environmental issues, enhance the quality of life, and promote economic development. This chapter delves into the advantages and limitations of sustainable transportation in worldwide scenarios. It explores how sustainable transportation systems offer numerous benefits while also addressing the challenges and constraints they face on a global scale.

Advantages of sustainable transportation worldwide:

1. Environmental benefits: sustainable transportation, such as electric vehicles (EVs), public transit, and cycling infrastructure, significantly reduces greenhouse gas emissions. These systems decrease air pollution, lower the carbon footprint, and help combat climate change. This is particularly crucial as cities worldwide grapple with worsening air quality and the impacts of global warming.
2. Reduced congestion: sustainable transportation modes, like efficient public transit networks and carpooling, contribute to reduced traffic congestion. This leads to smoother traffic flow, shorter commute times, and improved overall mobility, which is especially valuable in densely populated urban areas.
3. Improved public health: active transportation options like walking and cycling promote physical activity, leading to healthier populations. Sustainable transportation systems also lower exposure to air pollutants, reducing the risk of respiratory diseases and enhancing public health.
4. Economic advantages: investment in sustainable transportation infrastructure generates job opportunities and stimulates economic growth. Additionally, it can reduce the economic burden associated with traffic accidents, healthcare costs, and the importation of fossil fuels.
5. Enhanced accessibility: sustainable transportation promotes inclusive mobility by providing accessible options for all, including people with disabilities. Well-designed public transit and pedestrian-friendly infrastructure improve urban accessibility and social equity.
6. Energy efficiency: sustainable transportation systems tend to be more energy-efficient than traditional modes. Electric vehicles, for example, can use renewable energy sources, reducing energy consumption and dependence on fossil fuels.

Limitations of sustainable transportation worldwide:

1. Infrastructure investment: the transition to sustainable transportation requires significant investment in infrastructure. Developing and maintaining public transit systems, bike lanes,

charging stations, and pedestrian-friendly areas can be expensive and time-consuming.
2. Land use challenges: retrofitting existing urban areas to accommodate sustainable transportation can be challenging. Conflicting land uses, limited space, and resistance from stakeholders may hinder progress.
3. Behavior change: encouraging people to shift from personal vehicles to sustainable transportation modes can be difficult. It often requires a change in mindset and habits, which can be a slow and gradual process.
4. Equity concerns: implementing sustainable transportation solutions must consider equity issues. Low-income communities often have limited access to these systems and may face displacement due to gentrification associated with transit-oriented development.
5. Technological barriers: advancements in sustainable transportation technology, like EV charging infrastructure, may face technical challenges, including compatibility issues and scalability concerns.
6. Political and policy hurdles: sustainable transportation initiatives are often subject to political debates and policy inconsistencies. Long-term planning and political will are crucial to overcoming these hurdles.

The adoption of sustainable transportation in urban scenarios worldwide offers a multitude of advantages, ranging from environmental benefits and improved public health to economic growth and reduced congestion. However, it is not without its limitations, including the need for substantial infrastructure investment, behavior change, and addressing equity concerns. To fully realize the potential of sustainable transportation, it is essential for governments, urban planners, and communities to work together in overcoming these challenges and creating more sustainable and livable cities for future generations.

## 2.5 Advantages and Limitations in Indian Transport Scenario

The transport scenario in India is complex, with a diverse range of modes of transport and a rapidly growing economy. While India has made significant progress in developing its transport infrastructure, there are still significant challenges that need to be addressed. This document will provide an overview of the advantages and limitations of the transport scenario in India for different use cases.

Advantages of Indian transport scenario:

1. Diverse modes of transport: India has a diverse range of modes of transport, including road, rail, air, and water transport. This diversity provides commuters and businesses with a range of options for transportation, allowing for greater flexibility and choice.
2. Low cost of transportation: compared to other countries, transportation costs in India are relatively low. This makes transportation more accessible to people with lower incomes, allowing for greater economic mobility.
3. Rapidly growing economy: India has a rapidly growing economy, which is driving the demand for transportation infrastructure. This growth is fueling investment in transportation, with significant investments being made in infrastructure development.
4. Improving infrastructure: India is making significant investments in improving its transportation infrastructure, including the construction of new highways, airports, and ports. This investment is improving the quality and efficiency of transportation in the country.
5. Large workforce: India has a large workforce, which is driving demand for transportation services. This demand is fueling the growth of the transportation industry and creating new opportunities for businesses and entrepreneurs.

Limitations of Indian transport scenario:

1. Poor infrastructure in rural areas: while India is making significant investments in improving its transportation infrastructure, much of this investment is focused on urban areas. Rural areas often have poor transportation infrastructure, limiting economic mobility and access to services.
2. Traffic congestion: India's cities are known for their traffic congestion, which can significantly increase travel times and reduce the efficiency of transportation. This congestion is caused by a range of factors, including poor road design,

inadequate public transportation, and a high number of vehicles on the road.

3. Lack of safety: India's transportation system has a poor safety record, with high rates of accidents and fatalities. This is due to a range of factors, including poor infrastructure, inadequate safety regulations, and a lack of enforcement.
4. Limited public transportation: while India has a range of modes of transport, public transportation is often limited and inadequate, particularly in rural areas. This limits access to transportation for people with lower incomes and can reduce economic mobility.
5. Environmental impact: India's transportation system has a significant environmental impact, contributing to air and water pollution and climate change. This is due to a range of factors, including a high number of vehicles on the road and inadequate regulations and enforcement.

The transport scenario in India has a range of advantages and limitations. While the diversity of modes of transport and low cost of transportation provide advantages, the poor infrastructure in rural areas, traffic congestion, lack of safety, limited public transportation, and environmental impact are significant challenges that need to be addressed. To address these challenges, India needs to make significant investments in improving its transportation infrastructure, particularly in rural areas, and strengthening safety regulations and enforcement. Additionally, India needs to invest in public transportation and sustainable transportation options to reduce the environmental impact of transportation and improve access to transportation for people with lower incomes. By addressing these challenges, India can build a more efficient, safe, and sustainable transportation system that supports economic growth and improves the quality of life for its citizens.

## 2.6 Rail System and Its Types (in the Context of Monorail, Metro, etc.)

Rail transportation is an essential part of the transport system in many countries. It is a safe, efficient, and eco-friendly mode of transportation, with several advantages over other forms of transportation. Rail systems can be broadly classified into two categories: conventional and rapid transit systems. In this section, it will discuss the different types of rail systems, with a particular focus on monorail and metro rail systems.

Conventional rail systems. Conventional rail systems are designed to transport freight and passengers over long distances. These systems use a network of tracks and trains to transport cargo and passengers across the country. Conventional rail systems include diesel-electric locomotives, which are powered by diesel engines and electric motors. Conventional rail systems are used for long-distance travel, and they are efficient in transporting large quantities of goods over long distances.

1. Rapid transit systems: it is designed to provide efficient transportation within urban areas. They are characterized by high speeds, frequent service, and a high capacity for passengers. Rapid transit systems are further classified into two categories: monorail and metro rail systems.
2. Monorail system: A monorail system is a type of rapid transit system that uses a single rail track. Monorail trains run on a narrow track that is elevated above the ground, allowing for easy movement of traffic below. Monorail systems are primarily used in urban areas for short-distance travel. They are known for their speed and efficiency, and they are designed to carry a large number of passengers in a single trip.
3. Metro rail system: The metro rail systems, also known as subway systems, are rapid transit systems that operate in large urban areas. They are designed to provide high-speed transportation within the city, with trains running on tracks that are either elevated or underground. Metro rail systems are characterized by high speeds, frequent service, and a high capacity for passengers. They are designed to carry a large number of passengers in a single trip, making them an efficient mode of transportation in congested urban areas.

Differences between monorail and metro rail systems. While both monorail and metro rail systems are types of rapid transit systems, there are several differences between the two. The main differences are:

1. Track configuration: monorail systems use a single rail track, while metro rail systems use a double track.
2. Route configuration: monorail systems are primarily used for short-distance travel within urban areas, while metro rail systems are used for both short and long-distance travel.

3. Capacity: metro rail systems have a higher passenger capacity than monorail systems.
4. Speed: monorail systems are generally faster than metro rail systems, with trains traveling at speeds of up to 80 km/h.
5. Cost: monorail systems are generally more expensive to build than metro rail systems, due to the need for specialized infrastructure.

In summary, rail transportation is an essential mode of transportation in many countries. Rail systems can be broadly classified into conventional and rapid transit systems, with monorail and metro rail systems being the two main types of rapid transit systems. While both monorail and metro rail systems are efficient modes of transportation, they differ in terms of track configuration, route configuration, capacity, speed, and cost. The choice of rail system depends on several factors, including the distance to be traveled, the capacity required, and the cost of construction. By selecting the appropriate rail system, urban transportation systems can provide a reliable, efficient, and sustainable mode of transportation for commuters.

## 2.7 Advantages and Disadvantages of Rail System

Rail transportation is a mode of transportation that involves the use of trains to transport passengers and goods. The rail system has several advantages and disadvantages, which are discussed in detail below.

Advantages of rail system:

1. Environmentally friendly: rail transportation is an eco-friendly mode of transportation as it produces fewer emissions than other modes of transportation like cars and airplanes. Trains consume less fuel per passenger mile, reducing greenhouse gas emissions and air pollution.
2. Safety: rail transportation is safer than other modes of transportation as it has fewer accidents per passenger mile. Trains are designed to carry large numbers of passengers and goods, reducing the number of vehicles on the road, which also decreases the risk of accidents.
3. Cost-effective: rail transportation is cost-effective as it can transport a large amount of goods and passengers in a single trip. This reduces the cost of transportation per unit and can lead to lower prices for goods and services.

4. Efficient: rail transportation is an efficient mode of transportation as it can transport goods and passengers over long distances quickly. Trains can reach high speeds and have fewer stops, reducing travel time.
5. Reliable: rail transportation is a reliable mode of transportation as trains have a fixed schedule and run on time. This makes it easy for passengers and businesses to plan their activities around train schedules.

Disadvantages of rail system:

1. Limited accessibility: rail transportation is limited in terms of accessibility as it requires infrastructure such as tracks and stations. This can make it difficult for passengers and goods to access remote or rural areas.
2. High capital costs: building and maintaining rail infrastructure is expensive, making it difficult to expand and improve rail systems. This can lead to a lack of investment in rail transportation, limiting its growth and effectiveness.
3. Inflexibility: rail transportation is inflexible as it operates on fixed tracks and schedules. This can make it difficult to adapt to changes in demand or to meet the needs of customers who require customized transportation solutions.
4. Congestion: rail transportation can suffer from congestion during peak travel times, leading to delays and reduced efficiency. This can also impact the reliability of rail transportation as delays can affect schedules and delivery times.
5. Security: rail transportation can be vulnerable to security threats such as terrorism and sabotages. This requires increased security measures, which can be expensive and time- consuming for different scenarios.

The rail system has several advantages and disadvantages, which need to be carefully considered when choosing a mode of transportation. While the rail system is an eco-friendly, safe, cost-effective, efficient, and reliable mode of transportation, it is limited in terms of accessibility, has high capital costs, is inflexible, can suffer from congestion, and is vulnerable to security threats. By understanding the advantages and disadvantages of rail transportation, governments and businesses can make informed decisions about the use and expansion of rail transportation systems.

## 2.8 Skywalk and Under Bridge and Its Advantages

Skywalks and under bridges are pedestrian infrastructure systems that are designed to improve safety and convenience for pedestrians. A skywalk is a pedestrian bridge that connects two or more buildings or areas, while an under bridge is a pedestrian tunnel that runs underneath a road or highway. Both skywalks and under bridges offer several advantages, which are discussed in detail below.

Advantages of skywalks:

1. Pedestrian safety: skywalks improve pedestrian safety by providing a safe and secure route for pedestrians to cross busy roads or intersections. They also reduce the risk of accidents by separating pedestrians from vehicular traffic.
2. Convenience: skywalks are convenient for pedestrians as they provide a direct route from one location to another without having to navigate through heavy traffic or busy intersections.
3. Accessibility: skywalks improve accessibility for people with disabilities as they provide a barrier-free and accessible route for people with mobility impairments, such as wheelchair users or people with crutches.
4. Weather protection: skywalks provide protection from inclement weather conditions such as rain, snow, and extreme heat or cold. This makes it more comfortable for pedestrians to walk and reduces the risk of accidents caused by slippery or icy sidewalks.
5. Promotes walkability: skywalks promote walkability by encouraging people to walk instead of using motorized transportation. This reduces traffic congestion, air pollution, and carbon emissions, leading to a more sustainable and livable city.

Advantages of under bridges:

1. Pedestrian safety: under bridges improve pedestrian safety by providing a safe and secure route for pedestrians to cross busy roads or intersections. They also reduce the risk of accidents by separating pedestrians from vehicular traffic.
2. Convenience: under bridges are convenient for pedestrians as they provide a direct route from one location to another without having to navigate through heavy traffic or busy intersections.

3. Accessibility: under bridges improve accessibility for people with disabilities as they provide a barrier-free and accessible route for people with mobility impairments, such as wheelchair users or people with crutches.
4. Noise reduction: under bridges reduce noise pollution by separating pedestrians from vehicular traffic, making it quieter and more pleasant for pedestrians to walk.
5. Aesthetically pleasing: under bridges can be designed to be aesthetically pleasing, adding to the visual appeal of the surrounding area.

Skywalks and under bridges are important pedestrian infrastructure systems that provide several advantages, including improved pedestrian safety, convenience, accessibility, weather protection, and reduced noise pollution. By promoting walkability and sustainable transportation, these systems can help to create more livable and environmentally friendly cities. However, it is important to carefully consider the costs and benefits of constructing and maintaining these systems, and to ensure that they are designed and implemented in a way that meets the needs of all pedestrians, including those with disabilities.

## 2.9 Measuring Performance of Transit Systems

Transit systems play an important role in facilitating mobility and accessibility for people in urban areas. Measuring the performance of transit systems is crucial for ensuring that they are providing safe, reliable, and efficient services to passengers. There are several metrics that are commonly used to evaluate the performance of transit systems, which are discussed in detail below.

1. Ridership: one of the most common metrics used to measure the performance of transit systems is ridership. This refers to the number of passengers that use the transit system over a specific period of time. Higher ridership levels generally indicate that the transit system is meeting the needs of its users and providing services that are in high demand.
2. On-time performance: on-time performance is another important metric used to evaluate the performance of transit

systems. It refers to the percentage of transit vehicles that arrive at their destination within a specified time frame. Higher on-time performance levels indicate that the transit system is providing reliable services to passengers.

3. Average speed: average speed is a metric that measures the average speed of transit vehicles over a specific distance or route. Higher average speeds generally indicate that the transit system is providing efficient and effective services to passengers.
4. Service frequency: service frequency refers to the number of transit vehicles that operate on a particular route over a specific period of time. Higher service frequencies generally indicate that the transit system is providing more frequent and convenient services to passengers.
5. Capacity utilization: capacity utilization measures the number of passengers that are using the available capacity on transit vehicles. Higher capacity utilization levels indicate that the transit system is effectively using its resources to meet the needs of passengers.
6. Operating costs: operating costs refer to the expenses associated with operating and maintaining the transit system. These include expenses such as fuel costs, labor costs, and maintenance costs. Lower operating costs generally indicate that the transit system is operating in a more cost-effective manner.
7. Customer satisfaction: customer satisfaction is a metric that measures the level of satisfaction of transit system users. Higher levels of customer satisfaction generally indicate that the transit system is meeting the needs and expectations of its users.

Measuring the performance of transit systems is crucial for ensuring that they are providing safe, reliable, and efficient services to passengers. The metrics discussed above are commonly used to evaluate the performance of transit systems. Transit agencies should regularly monitor and analyze these metrics to identify areas where improvements can be made and to ensure that the transit system is meeting the needs of its users. By continuously measuring and improving the performance of transit systems, cities can create more livable and sustainable urban environments.

## References

1. F. -Y. Wang *et al.*, "Transportation 5.0: The DAO to Safe, Secure, and Sustainable Intelligent Transportation Systems," in *IEEE Trans. Intell. Transp. Syst.*, vol. 24, no. 10, pp. 10262-10278, Oct. 2023, doi: 10.1109/TITS.2023.3305380.
2. Z. Liu, C. Lyu, J. Huo, S. Wang and J. Chen, "Gaussian Process Regression for Transportation System Estimation and Prediction Problems: The Deformation and a Hat Kernel," in *IEEE Trans. Intell. Transp. Syst.*, vol. 23, no. 11, pp. 22331-22342, Nov. 2022, doi: 10.1109/TITS.2022.3155527.
3. Q. Wang, J. Zheng, H. Xu, B. Xu and R. Chen, "Roadside Magnetic Sensor System for Vehicle Detection in Urban Environments," in *IEEE Trans. Intell. Transp. Syst.*, vol. 19, no. 5, pp. 1365-1374, May 2018, doi: 10.1109/TITS.2017.2723908.
4. M. Xu, K. An, Z. Ye, Y. Wang, J. Feng and J. Zhao, "A Bi-Level Model to Resolve Conflicting Transit Priority Requests at Urban Arterials," in *IEEE Trans. Intell. Transp. Syst.*, vol. 20, no. 4, pp. 1353-1364, April 2019, doi: 10.1109/TITS.2018.2853102.
5. A. L. Alfeo, M. G. C. A. Cimino, S. Egidi, B. Lepri and G. Vaglini, "A Stigmergy-Based Analysis of City Hotspots to Discover Trends and Anomalies in Urban Transportation Usage," in *IEEE Trans. Intell. Transp. Syst.*, vol. 19, no. 7, pp. 2258-2267, July 2018, doi: 10.1109/TITS.2018.2817558.
6. F. Zhu, Y. Lv, Y. Chen, X. Wang, G. Xiong and F. -Y. Wang, "Parallel Transportation Systems: Toward IoT-Enabled Smart Urban Traffic Control and Management," in *IEEE Trans. Intell. Transp. Syst.*, vol. 21, no. 10, pp. 4063-4071, Oct. 2020, doi: 10.1109/TITS.2019.2934991.
7. C. Park, G. S. Kim, S. Park, S. Jung and J. Kim, "Multi-Agent Reinforcement Learning for Cooperative Air Transportation Services in City-Wide Autonomous Urban Air Mobility," in *IEEE Trans. Intell. Veh.*, vol. 8, no. 8, pp. 4016-4030, Aug. 2023, doi: 10.1109/TIV.2023.3283235.
8. T. Li and W. Xu, "Multiperiod Location Models for Urban System Planning with Fuzzy Intercity Passenger Transportation Demands," in *IEEE Trans. Cybern.*, vol. 52, no. 1, pp. 459-471, Jan. 2022, doi: 10.1109/TCYB.2020.2973358.
9. Y. Zhou and J. Wang, "Critical Link Analysis for Urban Transportation Systems," in *IEEE Trans. Intell. Transp. Syst.*, vol. 19, no. 2, pp. 402-415, Feb. 2018, doi: 10.1109/TITS.2017.2700080.
10. K. Manchella, A. K. Umrawal and V. Aggarwal, "FlexPool: A Distributed Model-Free Deep Reinforcement Learning Algorithm for Joint Passengers and Goods Transportation," in *IEEE Trans. Intell. Transp. Syst.*, vol. 22, no. 4, pp. 2035-2047, April 2021, doi: 10.1109/TITS.2020.3048361.
11. G. Zhong, X. Wan, J. Zhang, T. Yin and B. Ran, "Characterizing Passenger Flow for a Transportation Hub Based on Mobile Phone Data," in *IEEE Trans. Intell. Transp. Syst.*, vol. 18, no. 6, pp. 1507-1518, June 2017, doi: 10.1109/TITS.2016.2607760.

12. C. Wang, Z. Ye and W. Wang, "A Multi-Objective Optimization and Hybrid Heuristic Approach for Urban Bus Route Network Design," in *IEEE Access*, vol. 8, pp. 12154- 12167, 2020, doi: 10.1109/ACCESS.2020.2966008.
13. D. Li, M. Yang, C. -J. Jin, G. Ren, X. Liu and H. Liu, "Multi-Modal Combined Route Choice Modeling in the MaaS Age Considering Generalized Path Overlapping Problem," in *IEEE Trans. Intell. Transp. Syst.*, vol. 22, no. 4, pp. 2430-2441, April 2021, doi: 10.1109/TITS.2020.3030707.
14. C. Zheng, Y. Gu, J. Shen, M. Du, G. Ma and K. Sun, "Comprehensive Evaluation of Urban Integrated Transportation Network With Multivariate Statistical Analysis: A Case Study of Haimen," in *IEEE Access*, vol. 9, pp. 165034-165048, 2021, doi: 10.1109/ACCESS.2021.3132169.
15. T. W. Ching and W. Li, "A Superconducting Linear Variable Reluctance Machine for Urban Transportation Systems," in *IEEE Trans. Appl. Supercond.*, vol. 28, no. 3, pp. 1-5, April 2018, Art no. 0601805, doi: 10.1109/TASC.2018.2801853.

3

# Background on Intelligent Transportation and Telemetric Systems


***Abstract***

In recent years, intelligent transportation systems (ITS) and telematics systems have transformed the transportation business by delivering new solutions to problems within the industry such as traffic congestion, safety concerns, and environmental issues. These technologies have made it possible for cars and the infrastructure that supports transportation to communicate with one another in real time. This has made it easier for drivers, vehicles, and the systems that control traffic to share information with one another.

Intelligent transportation systems include a wide variety of technologies, such as sophisticated traffic management systems, smart parking, linked and autonomous cars, and intelligent transportation infrastructure. These technologies are grouped together under the umbrella term "intelligent transportation infrastructure." These systems gather and analyze data on traffic patterns, road conditions, and vehicle performance by making use of a variety of data sources. Some examples of these data sources are global positioning systems (GPS), sensors, and cameras. Telematics systems, on the other hand, are primarily concerned with the gathering and transmission of data from individual cars. This data may include information on the location of the vehicle, its speed, its fuel consumption, and a variety of other important parameters. This data is used to improve driver safety, enhance vehicle performance, and offer fleet management with useful insights. In this chapter, we not only go over the principles in great depth but also analyze their applications and use cases.



***Keywords:*** Intelligent transportation systems, global positioning systems, telematics system, traffic management


## 3.1 Introduction

The acronym ITS stands for "Intelligent transportation system," and it refers to sophisticated technical systems that make use of cutting-edge technology to increase the effectiveness, safety, and sustainability of transportation. These systems include a wide variety of applications, such as traffic


Kundan Meshram. Urban Transportation Systems: Sustainability and Future Development, (41–62) © 2024 Scrivener Publishing LLC

management, linked and autonomous vehicle transportation, car-to-infrastructure communication, and smart parking [1, 2].

Telematics systems are an essential part of the ITS. Telematics systems are used to gather and send data from individual cars in order to enhance the performance of vehicles, as well as the safety of drivers and the administration of fleets. These systems make use of a wide variety of sensors, including global positioning systems (GPS), accelerometers, and gyroscopes, in order to gather information on the location of the vehicle, its speed, its fuel consumption, and several other crucial metrics.

The transportation sector makes considerable use of telematics systems in order to improve the performance of vehicles and lower their overall fuel consumption. Telematics data, for instance, enables fleet managers to monitor vehicle routes, improve driver behavior, and cut down on idle time, all of which contribute to considerable cost savings and better operational efficiency [3, 4].

The control of traffic is yet another important use of ITS. Real-time traffic data may be collected and analyzed by traffic management systems by using a variety of data sources. Some examples of these data sources are traffic cameras, sensors, and GPS. The use of this data allows for the management of traffic flow, the optimization of signal timing, and the reduction of congestion. Intelligent traffic management systems, for instance, are able to make real-time adjustments to the timing of traffic signals depending on the flow of traffic. This helps to alleviate congestion and improve travel times.

Connected and autonomous cars, often known as CAVs, are an additional important use of ITS. Connected and autonomous cars (CAVs) are able to drive on roadways in an autonomous manner while maintaining a high level of safety thanks to telematics systems and vehicle-to-infrastructure connection. These cars make judgments and make adjustments to driving behavior based on data received in real time, which lowers the likelihood of being involved in an accident and improves road safety [5, 6].

## 3.2 Definitions

Despite the fact that ITS is a vast area of research, the following are some definitions that are particularly pertinent to intelligent transportation and telematics systems:

1. Intelligent transportation networks, often known as ITS, are a collection of cutting-edge technology and applications with the overarching goal of enhancing the effectiveness,

safety, and long-term viability of transportation networks. Applications such as traffic management, vehicle-to-infrastructure communication, linked and autonomous cars, and smart parking are all included under the umbrella of ITS.
2. Telematics systems are a kind of technology that permits the transfer of data from individual cars to a central server. This data can then be utilized to enhance the operation of the vehicle, as well as the behavior of the driver and the administration of the fleet. Telematics systems gather data on a vehicle's location, speed, fuel consumption, and other important parameters by using a variety of sensors such as GPS, accelerometers, and gyroscopes.
3. Vehicle-to-infrastructure communication, often known as V2I communication, is a technology that facilitates communication between automobiles and transportation infrastructure, such as traffic lights, road sensors, and toll booths. Vehicle-to-infrastructure communication, also known as V2I communication, enables cars to receive real-time information on road conditions, traffic patterns, and other crucial data, allowing for the optimization of travel routes, reduction of congestion, and improvement of safety.
4. Vehicle-to-vehicle (V2V) communication is a technology that facilitates communication between automobiles, enabling them to communicate information such as position, speed, and road conditions. V2V stands for "vehicle-to-vehicle." Communication between cars (vehicle-to-vehicle, or V2V) is an essential component of both connected and autonomous vehicles. It enables the former to make choices in real time and modify their driving behavior in response to the information they get from the latter.
5. Autonomous vehicles (AVs) are vehicles that are able to function without the assistance of a human driver. These vehicles use a variety of sensors, cameras, and sophisticated algorithms to make choices and change their driving behavior. Autonomous vehicles have the potential to vastly decrease the number of accidents that occur on the road and to considerably enhance the effectiveness of transportation networks.
6. "Big data" is a phrase that is used to represent big and complicated data sets that are not able to be handled using typical techniques of data processing because they are too massive. Big data is a term that is used in the context of ITS and

telematics systems. It refers to the huge quantity of data that is gathered from a variety of sources, such as weather sensors, transportation infrastructure, and cars. The processing of big data using modern analytics tools allows for the extraction of useful insights and the generation of data-driven choices.

These will be used at various points throughout this chapter to explain a variety of different ideas.

## 3.3 Features and Objectives of ITS

Intelligent transportation systems, often known as ITS, are a collection of cutting-edge technology and applications with the overarching goal of enhancing the effectiveness, safety, and sustainability of existing transportation networks. The term intelligent transportation systems (ITS) refers to a broad variety of applications, some of which include traffic management, V2I communication, linked and autonomous cars, and smart parking. In the next part, we shall delve more into the characteristics and goals of ITS [7–9]. ITS characteristics include:

1. Real-time data gathering and analysis: in order to collect and analyze real-time data on traffic patterns, road conditions, and vehicle performance, ITS makes use of a wide variety of sensors, cameras, and other data collecting equipment.
2. Intelligent traffic management systems: ITS makes it possible for traffic management systems to make real-time adjustments to signal timings in order to improve traffic flow, optimize traffic flow, and decrease congestion.
3. Communication between cars and transportation infrastructure: ITS makes it possible for automobiles and transportation infrastructure, such as traffic lights, road sensors, and toll booths, to communicate with one another. This makes it possible for cars to get up-to-the- minute information on the state of the roads, the flow of traffic, and other crucial data, allowing for the optimization of travel routes, the reduction of congestion, and the improvement of safety.
4. Telematics systems: the Internet of Things (ITS) permits the transfer of data from individual cars to a central computer. This data may subsequently be utilized to enhance vehicle performance, driver behavior, and fleet management.

5. Connected and self-driving cars: ITS makes it possible for vehicles to interact with one another and with the transportation infrastructure, which allows them to drive themselves and do so in a safe manner while on the road.

The following are the goals of ITS [10–12]:

1. Enhance road safety: this is one of the key goals of ITS, which aims to enhance road safety. Real-time monitoring of road conditions and vehicle performance is made possible by ITS, which enables the early detection of potential hazards and accidents. Intelligent transportation systems also aim to improve travel times and optimize traffic flow, which will make transportation more efficient and reduce the economic and environmental costs of traffic congestion.
2. Improve environmental friendliness: the goal of ITS is to lessen the effect that transportation has on the environment by encouraging the use of cleaner and more environmentally friendly forms of transportation, such as electric automobiles and public transit.
3. Increase mobility: the goal of ITS is to both increase access to transportation and improve mobility for all users, including walkers, cyclists, and individuals with impairments.

Therefore, ITS has the potential to completely revolutionize the transportation business by making it much more secure, productive, and environmentally friendly. Intelligent transportation system was designed with the goals of boosting mobility, reducing environmental impact, increasing efficiency, and improving safety. These are the primary characteristics and aims of ITS. It is anticipated that there will be a fast increase in the use of ITS in the future years, as more cities and nations all over the globe become aware of the advantages offered by cutting-edge technology and applications.

## 3.4 History of ITS and Its Development Over the World

Intelligent transportation systems (ITS) have a lengthy and eventful history that can be traced all the way back to the 1960s. Increasing traffic congestion, worries about personal safety, and the need for transportation systems that are more environmentally friendly and efficient all played a role in driving the development of ITS. In the following part, we shall discuss the origins of ITS as well as its growth in various parts of the globe.

1960s and 1970s: the desire to enhance traffic flow and minimize congestion in metropolitan areas was a driving force in the early development of ITS. The installation of traffic lights that could modify their timing in response to real-time traffic data was one of the first uses of ITS. After this, real-time traffic monitoring systems and electronic toll collecting systems were put into place [13, 14].

Decades of the 1980s and 1990s: these decades witnessed a rise in the widespread use of ITS technology in both the United States and Europe. The Intelligent vehicle highway systems (IVHS) initiative was launched in the United States by the Federal Highway Administration (FHWA) with the goal of encouraging the development of new ITS technologies and their incorporation into existing infrastructure. The IVHS program provided funding for a wide variety of research studies and pilot initiatives that were centered on enhancing traveler information, traffic management, and vehicle safety [15, 16]. The traffic scenarios shown in Figure 3.1.

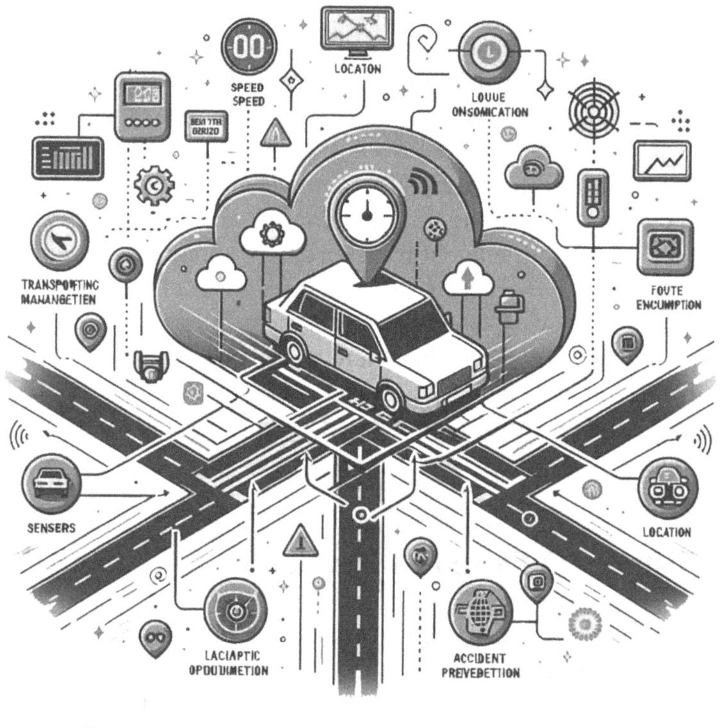


**Figure 3.1** Traffic scenarios.

In Europe, the necessity to handle growing traffic congestion as well as rising worries about the environment drove the development of ITS. The DRIVE initiative was initiated by the European Union with the purpose of encouraging the development and use of ITS systems. The DRIVE program provided financial support for a variety of research initiatives with the goals of improving sustainable transportation, traffic management, and vehicle safety.

The decade of the 2000s and continuing into the present day. The decade of the 2000s witnessed the widespread implementation of ITS technology across the globe. Applications of intelligent transportation systems, such as sophisticated traffic management systems, electronic toll collection, and traveler information systems, have become widespread throughout many nations. Additionally, the development of telematics systems and technology for linked vehicles brought up new prospects for increasing the safety, efficiency, and sustainability of transportation.

The technologies that make up ITS are constantly developing and becoming more advanced. It is anticipated that the transportation sector would undergo a sea change as a result of the widespread adoption of CAVs. Connected and autonomous cars have the potential to greatly decrease the number of accidents and deaths that occur on the road, in addition to improving the flow of traffic and lowering the effect that transportation has on the environment [17, 18].

Therefore, the history of ITS covers many decades and has been pushed by a variety of causes, such as increased traffic congestion, worries about safety, and the desire for transportation systems that are more efficient and sustainable. The development of ITS has been pushed forward by a variety of organizations, including government agencies, business enterprises, and academic institutes located all over the globe. It is projected that the widespread deployment of CAVs will have a profoundly transformational effect on the transportation sector in the years to come [19, 20]. In the present day, ITS technologies are continuously developing and improving for real-time scenarios.

## 3.5 Telemetric Concept

Telematics is an essential part of ITS, which makes it possible to gather and send data from individual cars to a centralized server. These data may be put to use to improve the operation of vehicles, the conduct of drivers, and the administration of fleets. In the next part, it is going to go more into the telemetric idea that is central to ITS. The role of Telematics in ITS: within the context of ITS, "telematics" refers to the transfer of data from individual cars to a centralized server. These data may include information on the position of the vehicle, its speed, acceleration, and braking, as well as the amount of gasoline used and other performance parameters. These transmissions take place through cellular networks or communication systems that are based on satellites [21–23].

Telematics in ITS may be put to use for a variety of purposes, including the following:

1. Optimization of fleet management telematics: this data may be utilized to improve fleet management, which enables fleet managers to track vehicle whereabouts, monitor driver conduct, and plan repairs in a more effective manner.
2. Insurance: telematics data may be utilized by insurance firms to offer usage-based insurance, which enables drivers to pay for insurance that is dependent on how they really drive rather than a predetermined amount.
3. Traffic management: data from telematics may be utilized by traffic management systems to monitor the flow of traffic, improve the timing of signal timings, and lessen the amount of congestion.
4. Telematics: this data may be utilized to allow connected and autonomous cars to function in a manner that is both safe and efficient when they are driving on public roads.

The following are some advantages of telematics in ITS:

1. Telematics: this data may be used to monitor driving behavior, identify possible risks, and send real-time notifications to drivers, all of which contribute to a reduction in the likelihood that drivers would be involved in an accident.
2. An increase in productivity: the data gathered by telematics may be utilized to enhance vehicle performance, therefore lowering fuel consumption, and making it easier to schedule maintenance.
3. Cost reductions: the data provided by telematics may be utilized to improve fleet management, which can lead to a decrease in operating expenses and an increase in profitability.
4. Benefits to the environment: the data gathered by telematics may be used to encourage the adoption of cleaner and more sustainable forms of transportation, hence lowering the effect that transportation has on the environment.

Problems that telematics presents in ITS:

1. Data privacy: telematics data may include sensitive information, such as the position of the car and the driving behavior of the driver. This fact raises worries about the data's privacy and security.
2. Telematics systems: restrictions imposed by technology on telematics systems need a solid communication infrastructure and dependable data transmission, both of which might be difficult to provide in some regions.
3. Standardization: because there are presently no uniform standards for telematics data, it is difficult to transfer data across various systems and applications. This is a barrier to widespread use of telematics technology.

Because of this, telematics is an essential component of ITS since it enables the gathering and transfer of data from individual cars to a centralized server. Data collected via telematics may be put to use in a variety of applications, such as managing fleets, providing insurance, directing traffic, and operating autonomous cars. Telematics in ITS has a number of benefits, some of which include enhanced efficiency, decreased costs, better safety, and reduced environmental impact. Nevertheless, telematics in ITS offers a number of obstacles, including those pertaining to data protection, limits in technology, and standardization. In general, telematics is a crucial technology for the future of transportation since it will make transportation systems safer, more efficient, and more environmentally friendly [24, 25].

## 3.6 Transport Telemetric

The transportation business may be made more efficient by the use of a technology known as transport telematics, which makes use of telecommunications and information technologies. This technology makes it possible to collect, analyze, and transmit data pertaining to the movement of people and products. As a result, it offers insights on traffic patterns, vehicle performance, and driver behavior. In the following sub-section, it shall go more into the idea of telematics in the transportation industry.

Telematics in the transport sector. Components: The term "transport telematics" refers to a number of different components, including:

1. Telemetry systems: telemetry systems are used to gather and send data from individual cars to a central server. Telemetry systems are utilized in a variety of applications. These data may include information on the position of the vehicle, its speed, acceleration, and braking, as well as the amount of gasoline used and other performance parameters.
2. Geographical information networks (GIS): the technology behind GIS is used to organize and analyze geographical data that is associated with transportation networks. The use of this technology allows the mapping of traffic patterns, the identification of congested locations, and the enhancement of traffic flows.
3. Intelligent transport systems (ITS): The technology behind ITS is utilized to improve traffic management systems. This improves the ability to monitor traffic flow in real time, optimizes signal timings, and decreases congestion.
4. Vehicle-to-vehicle (V2V) and vehicle-to-infrastructure (V2I) communication: V2V and V2I communication makes it possible for connected cars to interact with one another as well as with the systems that make up the infrastructure. This makes it possible to get real-time information on the state of the traffic and improves the efficiency of route planning.

Uses of transport telematics [22, 23]: There are many different uses for transport telematics, including the following:

1. Traffic management: the application of transport telematics allows for the monitoring of traffic flow, the optimization of signal timing, and the reduction of congestion.
2. Fleet management: transport telematics may be used to improve fleet management, which enables fleet owners to track the whereabouts of their vehicles, monitor the conduct of their drivers, and plan repairs in a more effective manner.
3. Intelligent parking systems: transport telematics may be utilized to improve parking systems by giving real-time information on the availability of parking spaces and maximizing the usage of parking space.
4. Management of public transportation: transport telematics may be used to improve public transportation systems by providing real-time monitoring of bus and rail timetables, increasing passenger experience, and optimizing route planning.

Benefits of telematics in transportation: there are several advantages of using telematics in transportation, including:

1. Transport telematics: this data may be used to monitor driver behavior, identify possible risks, and send real-time notifications to drivers, all of which contribute to a reduction in the likelihood that drivers would be involved in an accident.
2. An increase in productivity: the data obtained from transport telematics may be utilized to enhance vehicle performance, therefore lowering fuel consumption and optimizing the scheduling of maintenance tasks.
3. Cost reductions: the data provided by transport telematics may be utilized to optimize fleet management, which can lead to a reduction in operating expenses and an improvement in the bottom line.
4. Benefits for the environment: the data collected by transport telematics may be used to encourage the adoption of cleaner and more sustainable forms of transportation, therefore lowering the effect that transportation has on the environment.

Therefore, transport telematics is an essential technology for the future of transportation, since it will enable transportation systems that are safer, more efficient, and more environmentally friendly. The term "transport telematics" refers to an umbrella term that encompasses a wide range of subfields, including as telemetry systems, GIS, ITS, and V2V/V2I communication. Transport telematics may be used for a variety of applications, including traffic control, fleet management, intelligent parking systems, and administration of public transportation. Improved safety, better efficiency, cost savings, and environmental advantages are some of the upsides brought about by the use of transport telematics.

## 3.7 Telemetric Structure

The term "telemetric structures" refers to the many different pieces of hardware and software that come together to form a telematics system. This kind of system is used to gather and send data from distant places. The phrase "telemetric structure" is often used to refer to the physical and logical components that allow the collecting, processing, and transmission of data. These components might include sensors, communication devices, data storage systems, and software applications. In the next part, it will go into the many elements that make up telemetric structures in more depth [26, 27].

1. Sensors: sensors are physical devices that are used to gather data from the environment, such as temperature, humidity, pressure, or position. Examples of data that may be captured are temperature, humidity, pressure, and location. It is possible to link sensors to a variety of items, including automobiles, industrial, or infrastructure, in order to gather data and send it to a centralized server. Global positioning system (GPS) receivers, accelerometers, gyroscopes, pressure sensors, and temperature sensors are just a few examples of the many types of sensors that are often used in telematics systems.
2. Communication devices: communication devices make it possible for data to be sent from sensors to a central server or to other devices. These devices are able to make use of a variety of communication technologies, such as Wi-Fi, cellular networks, satellite communication, and even short-range communication technologies like Bluetooth and Zigbee. The needed distance and level of dependability for the telematics system should guide the selection of the appropriate communication technology.
3. Data storage systems: in order to save the information that was gathered by the sensors, data storage systems are used. These systems may make use of a variety of storage media, including hard disks, solid-state drives, cloud storage, or a combination of these. The quantity of data that is gathered, the processing needs, and the necessity for data security are three factors that should guide the selection of a data storage system.
4. Software applications: when it comes to processing and analyzing the data that was acquired by the sensors, software applications are employed. These applications may vary from simple data visualization tools to intricate analytical software that makes use of machine learning techniques to find patterns and outliers in the data. Examples of both types of applications are shown. The particular objectives of the telematics system and the amount of data analysis that is expected to be performed should guide the selection of the software application.
5. Telematics control unit: it is abbreviated as "TCU." A component of the telematics system that plays the role of the

system's nerve center is called the telematics control unit. The TCU is in charge of gathering data from the various sensors, storing that data, and then sending it on to the central server to be processed. In addition to this, the TCU has software programs that make it possible to handle and analyze data. The TCU has the capability of being included into a wide variety of equipment, such as automobiles, machines, or even infrastructure.

6. Central server: the central server is the primary component of the telematics system that is responsible for the processing and storage of data. The data is sent to the server from the TCU, where it is processed, and then stored for for analysis. Additionally, the central server has the capability of hosting software programs that provide real-time monitoring, data visualization, and data analysis.

The gathering, processing, and transmission of data from distant places is made possible by telematics systems, which rely on telemetric structures as their fundamental building block. Sensors, communication devices, data storage systems, software applications, telematics control units, and central servers are the components that make up telemetric structures. The particular objectives and specifications of the telematics system should guide the selection of the components. [28, 29]. Optimizing transportation networks, enhancing safety, boosting efficiency, and decreasing environmental impact are all important goals that may be accomplished with the use of telemetric structures.

## 3.8 ITS Taxonomy

Intelligent transportation systems (ITS) refer to a diverse collection of technologies and infrastructures that have been developed with the goal of enhancing the effectiveness, safety, and long-term viability of transportation networks. In order to get a deeper comprehension of the many ITS subsystems and applications, a taxonomy has been established to organize, categorize, and classify these various systems. In the next part, a comprehensive overview of ITS taxonomy will be presented.

1. Advanced traffic management systems (ATMS): an ATMS is a collection of technologies and systems that are intended to regulate the flow of traffic and improve the efficiency of

transportation networks. This covers the management of accidents and incidents, as well as traffic information and signal control systems.

2. Advanced traveler information systems (ATIS): ATIS is an abbreviation for "advanced traveler information systems," which refers to a collection of technologies and systems that are intended to offer passengers with real-time information on traffic conditions, travel times, and alternate routes. This includes things like changeable message signs, apps for mobile devices, and sophisticated navigation systems.
3. Advanced vehicle control systems (AVCS): AVCS stands for advanced vehicle control systems and refers to a collection of technologies and systems that are intended to enhance the efficiency and safety of motor vehicles. This includes devices that alert the driver if they are leaving their lane, technologies that prevent collisions, and adaptive cruise control.
4. Commercial vehicle operations (CVO): it is an acronym that stands for commercial vehicle operations, and it refers to a collection of technologies and systems that have been developed to enhance the effectiveness and safety of CVOs. Electronic logging devices, commercial vehicle inspections, and ITS for freight management are all included in this category.
5. Cooperative vehicle systems (CVS): CVS stands for "cooperative vehicle systems" and refers to a collection of technologies and systems that are meant to allow cars to interact with one another and with infrastructure. This comprises communication systems between vehicles, communication systems between vehicles and infrastructure, as well as platooning systems.
6. Enabling technologies (ET): ET stands for "enabling technologies" and refers to a collection of technologies and systems that are intended to make it possible to adopt ITS. This encompasses many kinds of communication technology, software platforms, and various types of data management systems.
7. Public transportation management (PTM): it refers to a collection of technologies and processes that have been developed with the intention of enhancing the efficacy and efficiency of public transportation networks. This includes passenger information that is updated in real time, automated vehicle locating

systems, and transportation services that adapt to meet changing demand.
8. Intelligent transportation systems (ITS): in rural areas, an ITS is a collection of technologies and systems that are intended to handle the specific problems posed by rural transportation networks. This comprises both weather information systems and emergency response systems as well as traveler information systems.
9. Security and emergency management (SEM): SEM is an acronym that stands for "security and emergency management." It refers to a collection of technologies and systems that are intended to enhance the safety and security of transportation networks. This comprises emergency alerting systems, surveillance systems, and management systems for incidents.
10. Transportation control centers (TMC): the transportation management centers (TMC) are a collection of technologies and systems that are intended to facilitate the centralized control and coordination of transportation networks. This comprises systems for managing traffic, responding to incidents, analyzing data, and dividing it up into sections.

The ITS taxonomy is a helpful framework that may be used to categorize and classify the many applications and components of ITS. The taxonomy is comprised of ten different categories that collectively cover a wide variety of ITS technologies and systems. These categories are as follows: traffic management, traveler information, vehicle control, commercial vehicle operations, cooperative vehicle systems, enabling technologies, public transportation management, rural ITS, security and emergency management, and transportation management centers [18–20]. It is possible for transportation agencies and organizations to effectively develop, deploy, and manage ITS systems in order to enhance the efficacy, safety, and sustainability of transportation networks provided they have a thorough grasp of the many components and uses of ITS.

## 3.9 ITS Application Areas and Uses

Intelligent transportation systems (ITS) include a vast array of application domains and applications, all of which are intended to enhance the effectiveness, safety, and longevity of transportation networks. In the next part, a comprehensive overview of the many application fields and applications of ITS will be presented.

1. Traffic management: it is one of the key application areas of intelligent transportation systems is traffic management. The flow of traffic is managed, congestion is reduced, and transportation networks are optimized with the help of technologies and systems provided by ITS. This covers the management of accidents and incidents, as well as traffic information and signal control systems.
2. Information: information for travelers another significant application area served by ITS is that of providing information for travelers. Real-time information regarding traffic conditions, travel times, and alternate routes is made available to passengers via the use of various ITS technologies and systems. This includes things like changeable message signs, apps for mobile devices, and sophisticated navigation systems.
3. Safety of motor vehicles: the safety of motor vehicles is an important area of application for ITS. The use of ITS technologies and systems has been shown to significantly cut down on the number of accidents that occur on the road. This includes devices that alert the driver if they are leaving their lane, technologies that prevent collisions, and adaptive cruise control.
4. Freight and commercial vehicle operations: another significant application area for ITS is freight and commercial vehicle operations. The operation of commercial vehicles may be made more effective while also being made more secure with the use of ITS technology and systems. Electronic logging devices, commercial vehicle inspections, and intelligent transportation systems for freight management are all included in this category.
5. Transit operations: the transit operations application area is an essential part of the ITS. The efficacy and efficiency of public transportation systems may be improved with the use of technologies and systems that are a part of ITS. This includes passenger information that is updated in real time, automated vehicle locating systems, and transportation services that adapt to meet changing demand.

6. Management of roads and infrastructure: ITS is also useful in the management of roads and infrastructure, which is an important application field. Roads, bridges, and other kinds of transportation infrastructure are kept in good shape via the employment of ITS technologies and systems for monitoring and maintenance. This consists of pavement management systems, road weather information systems, and asset management systems.
7. Emergency management: one of the most important areas of ITS use is emergency management. In the event of an emergency, technologies and systems that are part of an ITS are used to enhance the safety and security of transportation networks. This comprises emergency alerting systems, surveillance systems, and management systems for incidents.
8. Sustainability of the environment: the field of application of ITS that focuses on the preservation of the environment is expanding. The negative effects that transportation networks have on the surrounding environment may be mitigated with the use of ITS technology and systems. This encompasses alternative fuel car programs, traffic demand management systems, and electric vehicle charging infrastructure.
9. Mobility services: the field of application known as mobility services is one that is expanding at a fast rate within ITS. On-demand mobility services, such as ride-sharing and car-sharing programs, are provided by using various technologies and systems that are a part of ITS. Mobile apps, payment systems, and vehicle dispatching systems are all included in this category.

Intelligent transportation systems (ITS) include a vast array of application domains and applications, all of which are intended to enhance the effectiveness, safety, and longevity of transportation networks. These application areas and purposes include emergency management, environmental sustainability, mobility services, and traveler information. Freight and commercial vehicle operations, transit operations, road and infrastructure operations, freight and commercial vehicle operations, and freight and commercial vehicle operations all fall under this category. Transportation agencies and organizations may construct transportation networks that are more efficient, safer, and sustainable by utilizing the technology and systems described above [27–29]. These improvements will be of value to anybody who uses the systems.

## 3.10 Application in Urban Transportation

The development of ITS and telemetrics is making important contributions to the improvement of urban transportation networks. Transportation departments may improve safety, lessen the impact of congestion, and boost mobility in urban areas by making use of the technology available today. In the next part, a comprehensive description of how ITS and telemetrics might be used to urban transportation will be provided.

1. Traffic management: in the realm of urban transportation, traffic management is one of the most important application areas of ITS. Monitoring the flow of traffic and adjusting the timing of traffic signals using technologies and systems provided by ITS helps to increase mobility while also reducing congestion. In addition, ITS may be utilized for incident detection and response, allowing for faster resolution of any problems that may arise on the highways.
2. Information: information for travelers is another important application area of ITS in the context of urban transportation. Travelers may get information in real time regarding traffic conditions, trip times, and alternate routes thanks to ITS technologies and systems, which are employed to convey this information. This information may be provided to passengers in a variety of ways, including via mobile apps, variable message signs, and other modes of communication, all with the goal of assisting travelers in making well-informed choices on their trips.
3. Improvements to the efficacy and efficiency of transport operations: ITS is widely employed in public transport systems to make improvements to the efficacy and efficiency of transit operations. Automatic vehicle location (AVL) and real-time passenger information (RTPI) systems both make use of technologies and systems that fall under the ITS umbrella. Automatic vehicle location systems provide transit agencies the ability to monitor their cars in real time, while RTPI systems give passengers real-time information about the position of transit vehicles and the estimated time of their arrival. Public transportation will benefit from increased dependability and convenience as a result of this information, making it a more appealing choice for commuters.

4. Enhanced vehicle safety: ITS has a number of important applications, but one of the most important is enhanced vehicle safety. The use of ITS technologies and systems has been shown to significantly cut down on the number of accidents that occur on the road. This includes devices that alert the driver if they are leaving their lane, technologies that prevent collisions, and adaptive cruise control. Both motorists and pedestrians traveling on urban highways benefit from the increased safety afforded by these technological advancements.
5. Freight and commercial vehicle operations: ITS is used to improve the flow of freight and commercial vehicles in metropolitan areas in order to maximize efficiency. Electronic logging devices, commercial vehicle inspections, and intelligent transportation systems are all areas in which ITS technologies and systems are used for logistics and supply chain management. These technologies make the operations of freight and commercial vehicles more efficient, which in turn helps to minimize congestion and improves safety.
6. Road and infrastructure operations: the technologies and systems that are a part of ITS are used to monitor and maintain the status of roads, bridges, and other types of transportation infrastructure. This consists of pavement management systems, road weather information systems, and asset management systems. These solutions make the urban transportation infrastructure more reliable while also increasing its level of safety.
7. Mobility services: ITS technology and systems are used to deliver on-demand mobility services such as ride-sharing and car-sharing programs in order to fulfill the demand for these types of services. Mobile apps, payment systems, and vehicle dispatching systems are all included in this category. These systems provide modes of transportation other than the conventional usage of private vehicles, which in turn increases the amount of mobility available in metropolitan areas.

In summary, ITS and telemetrics are crucial technologies that have the potential to dramatically enhance urban transportation networks. These technologies have the potential to be used in a variety of application domains, including traffic management, traveler information, transit operations, vehicle safety, freight and commercial vehicle operations, road and infrastructure operations, and mobility service delivery. The use of these technologies allows transportation agencies and organizations to construct urban transportation networks that are more effective, safer, and sustainable, hence increasing the quality of life for all people of the metropolitan regions.

## References

1. Yu, H., Li, J., & Zhang, C. (2021). Vehicle platoon control with adaptive backstepping and V2V communication. *IEEE Trans. Intell. Transp. Syst.*, 22(8), 4801-4812.
2. Zhi, Y., Wang, X., Wang, J., & Han, S. (2021). An intelligent transportation system-based signalized intersection control approach. *J. Adv. Transp.*, 2021, 1-12.
3. Luo, X., Zhou, Y., Yang, D., & Liu, F. (2021). An improved vehicle detection method for intelligent transportation systems using faster R-CNN. *IEEE Access*, 9, 14775-14784.
4. Liu, X., Zhi, Y., Han, S., & Wang, X. (2021). Intelligent transportation system-based driving assistance for lane changing maneuver. *Transp. Res. Part C Emerging Technol.*, 129, 103135.
5. Shi, X., Zhou, Y., Yang, D., & Liu, F. (2021). A novel vehicle detection and tracking method for intelligent transportation systems based on faster R-CNN and Kalman filter. *IEEE Access*, 9, 16660-16670.
6. Zhang, Y., & Wang, X. (2021). An intelligent transportation system-based traffic signal control approach with traffic prediction. *IEEE Trans. Intell. Transp. Syst.*, 22(7), 4296-4307.
7. Shi, X., Zhou, Y., Yang, D., & Liu, F. (2021). A deep learning-based vehicle detection and tracking method for intelligent transportation systems. *IEEE Access*, 9, 33300-33309.
8. Jia, Y., Lin, X., Liu, Q., & Wang, J. (2021). Intelligent transportation system-based signal control considering heterogeneous traffic. *IEEE Trans. Intell. Transp. Syst.*, 22(6), 3777-3788.
9. Li, W., Zhao, Y., & Li, L. (2021). Real-time dynamic lane closure detection based on vehicle trajectory data in intelligent transportation systems. *Transp. Res. Part C Emerging Technol.*, 131, 103317.
10. Gong, C., Huang, X., & Wang, Q. (2021). Integrated vehicle routing and scheduling problem for urban solid waste collection: A heuristic algorithm based on intelligent transportation systems. *J. Cleaner Prod.*, 310, 127597.
11. Zhang, J., Chen, X., Lin, T., Liu, J., & Cheng, C. (2021). A distributed traffic signal control strategy for urban arterial roads based on the intelligent transportation system. *J. Adv. Transp.*, 2021, 1-16.

12. Zhang, Z., Gao, Y., Xu, G., Wang, H., & Lu, X. (2021). Robust traffic control based on connected and automated vehicles in intelligent transportation systems. *IEEE Trans. Intell. Transp. Syst.*, 22(5), 2845-2856.
13. Zhao, L., Zheng, Y., Zhang, Y., & Du, L. (2021). A multi-objective intelligent transportation system-based signal optimization model considering emissions reduction. *J. Cleaner Prod.*, 313, 127773.
14. Wang, X., Chen, J., & Wang, C. (2021). An intelligent transportation system-based simulation study of electric vehicle charging demand in public parking lots. *Energies*, 14(14), 4333.
15. Wang, X., Chen, J., & Wang, C. (2021). An intelligent transportation system-based simulation study of electric vehicle charging demand in public parking lots. *Energies*, 14(14), 4333.
16. Lin, X., Jia, Y., Li, S., & Cui, P. (2021). Optimization of a combined signal control strategy for intelligent transportation systems under heterogeneous traffic. *IEEE Trans. Intell. Transp. Syst.*, 22(5), 3047-3059.
17. Zhang, X., Xu, M., Li, Q., & Wang, H. (2021). A dynamic queue management strategy for congestion mitigation in intelligent transportation systems. *IEEE Trans. Intell. Transp. Syst.*, 22(4), 2288-2299.
18. Zhang, H., Liu, J., Zhou, X., & Luo, S. (2021). Smart driving strategy for automated vehicles in a connected intelligent transportation system. *Transp. Res. Part C Emerging Technol.*, 127, 103229.
19. Liu, H., & Chen, L. (2021). A distributed cooperative control approach for vehicle platooning in intelligent transportation systems. *IEEE Trans. Intell. Transp. Syst.*, 22(4), 2461-2471.
20. Tang, J., Han, S., Wang, X., & Liu, H. (2021). An intelligent transportation system-based real-time dynamic ride-sharing approach. *J. Adv. Transp.*, 2021, 1-14.
21. Li, H., Guo, J., Li, X., Li, G., & Li, S. (2021). A driver behavior prediction method for intelligent transportation systems based on dual path convolutional neural network. *IEEE Access*, 9, 13294-13303.
22. Song, L., Wang, X., & Han, S. (2021). Intelligent transportation system-based driving assistance for automatic emergency braking. *IEEE Trans. Intell. Transp. Syst.*, 22(3), 1615-1627.
23. Zhang, J., Huang, B., Xu, X., & Yao, X. (2021). A simulation-based optimization method for urban traffic management in intelligent transportation systems. *J. Adv. Transp.*, 2021, 1-13.
24. Zhang, H., Liu, J., Zhou, X., & Luo, S. (2021). Cooperative autonomous driving strategy for heterogeneous traffic in an intelligent transportation system. *Transp. Res. Part C Emerging Technol.*, 125, 103168.
25. Tang, J., Han, S., Wang, X., & Wang, H. (2021). Intelligent transportation system- based dynamic vehicle routing with shared mobility. *Transp. Res. Part C Emerging Technol.*, 127, 103231.

26. Chen, Y., Li, L., Liu, Y., & Li, Y. (2021). A novel cooperative control strategy for platoon formation in intelligent transportation systems. *J. Adv. Transp.*, 2021, 1-13.
27. Shen, Y., Zhu, Z., & Lv, Y. (2021). Multi-objective optimization of urban road network design based on the intelligent transportation system. *J. Adv. Transp.*, 2021, 1-13.
28. Yang, W., Liu, Y., & Yu, L. (2021). Design and simulation of an intelligent transportation system based on deep learning. *J. Adv. Transp.*, 2021, 1-12.
29. Wang, X., Zhou, Y., & Han, S. (2021). Intelligent transportation system-based dynamic vehicle routing with real-time traffic information. *Transp. Res. Part C Emerging Technol.*, 131, 103319.

# 4

# Use of ITS for Deployment of Sustainable Transportation Models


***Abstract***

In light of growing environmental concerns and the demand for effective mobility systems, the adoption of sustainable transport models has become essential. By incorporating cutting-edge technologies into transportation infrastructure and vehicles, intelligent transport systems (ITS) have emerged as a promising strategy to tackle these problems. In order to improve environmental performance, increase safety, and increase transportation efficiency, this chapter examines the use of ITS in the deployment of sustainable transportation models.

Implementing intelligent traffic management systems is one of the main ways that ITS supports sustainable transportation models. These systems make use of real-time traffic data to streamline travel times, lessen congestion, and improve traffic flow. Additionally, they make it easier to combine different forms of transportation like walking, cycling, and public transportation, fostering multimodal connectivity and promoting the use of eco-friendly transportation options [1, 2].

Additionally, ITS enables the installation of smart infrastructure elements like adaptive traffic control systems, variable message signs, and intelligent traffic signals. By giving drivers timely information, adjusting signal timing based on traffic conditions, and facilitating effective routing and navigation, these components improve safety. By enabling the implementation of intelligent parking systems, dynamic pricing mechanisms, and electric vehicle charging infrastructure, they also support the adoption of environmentally friendly vehicles.

The chapter also explores how ITS can encourage behavior change and promote eco-friendly travel options. Advanced traveler information systems (ATIS) give users access to real-time data on alternate modes of transportation, travel times, and carbon emissions related to various options. This data supports the transition to more environmentally friendly modes of transportation by assisting users in making knowledgeable decisions that are in line with sustainable transportation objectives.



Kundan Meshram. Urban Transportation Systems: Sustainability and Future Development, (63–98) © 2024 Scrivener Publishing LLC

The chapter also looks at how ITS can be combined with cutting-edge innovations like connected and autonomous vehicles (CAVs). Connected and autonomous vehicles can improve traffic efficiency by utilizing vehicle-to-vehicle and vehicle-to-infrastructure communication to optimize routes, cut fuel use, and decrease emissions. The development of mobility-as-a-service (MaaS) platforms, where users can easily access different modes of transport, furthering sustainability and lowering private vehicle ownership rates, is made possible by the integration of CAVs with ITS.

***Keywords:*** Intelligent transport systems, advanced traveler information systems, traffic congestion, advanced vehicle control system

## 4.1 Introduction

In the modern world, transport is essential for the movement of people and goods as well as for the advancement of the economy. However, traditional transportation methods that rely heavily on fossil fuels have significantly increased traffic congestion, air pollution, and carbon emissions. As a result, the idea of sustainable transportation has gained popularity as a way to deal with these issues and develop mobility systems that are more effective, environmentally friendly, and socially inclusive.

While simultaneously enhancing safety, accessibility, and overall transportation effectiveness, sustainable transportation models seek to reduce the detrimental effects of transportation on the environment. These models cover a wide range of tactics and programs, such as encouraging people to use public transportation, building bike lanes and walkways, implementing ITS, adopting electric vehicles, and encouraging people to make more environmentally friendly travel decisions.

Intelligent transport systems has become a potent tool for modernizing transport systems and assisting in the adoption of sustainable transport models in recent years. To enhance safety, effectiveness, and environmental performance, ITS entails the integration of cutting-edge technologies, data analytics, and communication networks into transportation infrastructure and vehicles.

The goal of this chapter is to provide a thorough investigation of the application of ITS in the introduction of sustainable transportation models. It explores the many ways that ITS can assist in achieving the objectives of sustainable transportation, such as lowering greenhouse gas emissions, promoting energy efficiency, reducing traffic congestion, boosting safety, and raising the general caliber of transportation services.

An extensive overview of sustainable transport models and the issues they aim to solve is given at the beginning of the chapter. It emphasizes the pressing need for sustainable mobility solutions in light of urbanization, climate change, and rising transportation needs. Sustainable transportation models seek to build a more livable and resilient future by reducing reliance on fossil fuels, promoting alternate modes of transportation, and improving system efficiency.

The chapter then looks at ITS' crucial contribution to the advancement of sustainable transportation goals. It covers how ITS makes use of innovations like real-time data collection, communication networks, and advanced analytics to enhance multimodal connectivity, optimize traffic flow, and enhance overall transportation operations. Transportation stakeholders can use data-driven insights to inform decisions and put forth practical plans for sustainable mobility by integrating ITS into traffic management systems, infrastructure elements, and traveler information systems [3, 4].

The chapter also looks at the specific uses of ITS in environmentally friendly transportation. It discusses how traffic congestion can be reduced, travel times can be optimized, and environmentally friendly transportation can be encouraged. It also looks at how ITS enables the deployment of smart infrastructure elements that improve safety, encourage the use of environmentally friendly vehicles, and make routing and navigation more effective.

The chapter explores the role of ITS in fostering behavior change and promoting sustainable travel options in addition to applications that are infrastructure-focused. It looks at how cutting-edge traveler information systems give people instant access to information about alternative modes of transportation, travel times, and environmental effects, enabling them to make well-informed decisions and select eco-friendly modes of transportation.

The chapter also emphasizes how ITS can be used in conjunction with cutting-edge technologies like connected and autonomous vehicles (CAVs). It looks at how CAVs can improve traffic efficiency by optimizing routes, consuming less fuel, and communicating with other vehicles and infrastructure. The combination of CAVs and ITS also creates opportunities for platforms that offer mobility-as-a-service (MaaS), allowing users to seamlessly switch between different forms of transportation while promoting sustainability and minimizing the need for private vehicle ownership.

In conclusion, implementing sustainable transportation models is essential for developing more environmentally friendly, secure, and effective transportation systems. By enhancing system performance, promoting behavior change, and incorporating emerging technologies, ITS offers a potent set of tools and technologies to support the realization of these models. Transportation planners and policymakers can create a sustainable future where transportation is both environmentally friendly and socially equitable for various scenarios by utilizing the potential of ITS.

## 4.2 How ITS can be Used to Maintain Sustainability

Intelligent transportation systems (ITS) provide a variety of capabilities and technologies that can be used to keep transportation systems sustainable. The ITS can make a significant contribution to lowering environmental impact, enhancing safety, increasing efficiency, and promoting sustainable travel options by integrating cutting-edge data collection, communication, and analysis tools. The effectiveness of ITS in maintaining transportation sustainability is examined in this section for different scenarios.

1. Real-time traffic management: real-time traffic management is one of the primary ITS applications for sustainability. Intelligent transportation systems can provide current information on traffic conditions, congestion levels, and travel times by gathering and analyzing data from various sources including sensors, cameras, and connected vehicles. With the help of this data, transportation agencies can take proactive steps to lessen traffic congestion, improve traffic flow, and cut emissions. By reducing idling time and fuel consumption, ITS can efficiently manage traffic and lessen environmental impact through adaptive ramp metering, dynamic traffic signal control, and congestion pricing.
2. Multimodal integration: in order to provide seamless and effective mobility options, sustainable transportation models emphasize the significance of integrating various modes of transportation. By enabling real-time information sharing, coordination, and synchronization across various

transportation modes, ITS plays a crucial role in facilitating multimodal integration. Examples of ITS applications that encourage easy transitions between modes, encourage the use of sustainable options like public transport, cycling, and walking, and lessen reliance on private vehicles include integrated fare payment systems, traveler information apps, and smart transit hubs.

3. Support for eco-friendly vehicles: one of the most important components of sustainable transportation is encouraging the use of eco-friendly vehicles, such as electric vehicles (EVs). By supporting EV charging infrastructure, ITS can help achieve this objective. Intelligent charging station deployment helps promote the use of EVs and lower greenhouse gas emissions from the transportation sector by managing peak demand, integrating renewable energy sources, and optimizing energy usage. Furthermore, ITS can make it easier to create priority zones and EV-only lanes, further encouraging the use of green transportation options.
4. Intelligent parking systems: as it affects travel patterns and traffic density, parking management is a crucial component of sustainable transportation. Intelligent parking systems that effectively direct drivers to available spaces can be created using ITS, cutting down on search time, traffic congestion, and unnecessary emissions. Intelligent transportation systems-enabled parking solutions maximize space utilization, encourage shared parking, and encourage the use of sustainable modes for the first and last mile of a journey by utilizing sensor technologies, real-time occupancy data, and smart payment systems [5, 6].
5. Advanced traveler information systems: information is essential for shaping travel habits and encouraging environmentally friendly decisions. Advanced traveler information systems (ATIS) that offer real-time data on alternate modes of transportation, travel times, emissions, and costs are made possible by ITS. These systems encourage the selection of sustainable modes and routes by arming users with accurate and thorough information, which reduces traffic, boosts air quality, and results in more effective transportation systems.
6. Data-driven decision making: for the planning of sustainable transportation, the gathering and analysis of transportation

data are essential. Traffic volumes, emissions levels, and travel patterns are just a few examples of the vast amounts of data that can be collected, integrated, and analyzed effectively thanks to ITS. Transportation planners and policymakers can pinpoint areas for improvement, put in place focused interventions, and assess the success of sustainable transportation initiatives by using data-driven decision-making processes. The iterative process made possible by ITS guarantees that sustainability objectives are continuously tracked and modified for maximum impact.

In summary, ITS offers a range of resources and innovations that can be used to keep transport systems sustainable. Real-time traffic management, multimodal integration, assistance for environmentally friendly vehicles, intelligent parking systems, advanced traveler information, and data-driven decision making are some of the ways that ITS helps to lessen traffic, improve air quality, increase safety, and encourage environmentally friendly modes of transportation. For the benefit of current and future generations, transportation stakeholders can build more effective, environmentally friendly, and socially equitable transportation systems by embracing the potential of ITS [7–9].

### 4.2.1 Services for Users' Components

User services are a critical component of ITS, which aim to offer people effective, secure, and environmentally friendly transportation options. These services include a wide variety of elements made to satisfy the various needs of users, such as real-time information delivery, customized trip planning, payment methods, and mobility management. The main ITS user service components are examined in this section along with their role in promoting sustainable transport and enhancing the overall user experience.

1. Real-time information: the availability of real-time information is essential for enabling travelers to choose their mode of transportation with knowledge. Intelligent transportation systems makes it easier to gather and disseminate real-time information on traffic patterns, public transportation schedules, traffic accidents, and alternate routes. Users can access this information via a variety of platforms, including mobile applications, websites, electronic message boards, and in-car

displays. Users can optimize their travel plans, choose the most effective modes of transportation, and avoid congestion with the help of accurate and up-to-date information, which reduces travel times, boosts reliability, and boosts overall system efficiency levels [10–12].

2. Personalized trip planning: a feature of user services, personalized trip planning takes into account the unique preferences, requirements, and constraints of every traveler. Intelligent transportation systems provides sophisticated algorithms and intelligent systems that create customized travel itineraries by taking into account variables like preferred travel mode, time constraints, financial constraints, and environmental criteria. In order to create seamless and effective travel experiences, these trip planning services give users options for combining various modes of transportation, such as public transportation, biking, walking, and car-sharing. Intelligent transportation systems encourages users to select environmentally friendly modes of transportation, optimize travel routes, and reduce environmental impact by enabling customized trip planning sets.
3. Integrated fare payment systems: in a transportation network, integrated fare payment systems are necessary for seamless travel between various modes of transportation. Intelligent transportation systems makes it possible to combine different fare payment methods, such as account-based systems, contactless smart cards, and mobile payment apps, onto a single platform. Users can easily pay for their travel, switch between different modes, and access discounted fares or bundled ticketing options thanks to this integration. Integrated fare payment systems promote the use of sustainable transport modes, lower entry barriers, and overall network efficiency by streamlining the payment process and providing incentives and rewards for different actions.
4. Mobility management: within user services, mobility management components work to maximize the use of transportation resources and encourage environmentally friendly travel practices. By offering tools for trip matching, scheduling, and coordination, ITS makes it easier to implement mobility management strategies like demand-responsive transportation, carpooling, and ride-sharing. By sharing rides, pooling resources, and planning travel routes, these

components help users reduce the number of vehicles on the road, ease traffic, and have a minimal negative impact on the environment. Mobility management services assist in integrating different mobility providers, advancing the idea of MaaS, and providing users with a smooth and effective journey across various modes of transportation.

5. User feedback and evaluation: to continuously improve the caliber and efficacy of transport systems, user feedback and evaluation components of user services are crucial. Intelligent transportation systems makes it possible to gather and analyze user feedback via surveys, ratings, and reviews. This feedback offers insightful information on user preferences, levels of satisfaction, and areas for development. Transportation stakeholders can effectively adapt services, spot bottlenecks, and respond to user concerns by incorporating user feedback into the planning and decision-making processes. A user-centric transport system that encourages sustainability, customer satisfaction, and scenarios for long-term engagement is made possible by ongoing evaluation and improvement of user services [12–14].

In summary, user services within ITS play a crucial role in improving the user experience, encouraging environmentally friendly transportation options, and maximizing transportation systems. The provision of real-time information, customized trip planning, integrated fare payment systems, mobility management, and user evaluation and feedback all help to increase efficiency, lessen congestion, promote the use of sustainable modes of transportation, and satisfy the various needs of transportation users. By utilizing these elements, ITS makes it possible for a user-centric strategy that promotes sustainability, improves mobility, and aids in the creation of effective and environmentally friendly transportation systems.

## 4.3 Advanced Traffic Management Systems

Traffic control is essential for streamlining traffic, reducing congestion, boosting safety, and advancing environmentally friendly transportation. In order to effectively monitor, regulate, and manage traffic on road networks, advanced traffic management systems (ATMS) rely on ITS technologies and data-driven approaches. This section examines the main ATMS features and functions, emphasizing their value in enhancing traffic operations and advancing sustainable transportation goals.

1. Traffic monitoring and data collection: to collect real-time data on traffic conditions, such as traffic volumes, speeds, and travel times, ATMS relies on advanced traffic monitoring technologies. This information is gathered from a variety of tools, including radar systems, loop detectors, traffic cameras, and vehicle probes. Advanced traffic management systems gives transportation agencies accurate and up-to-date information by continuously monitoring traffic parameters, empowering them to decide wisely and act quickly in the face of shifting traffic patterns. Real-time data collection is essential for efficient traffic management because it makes it easier to spot traffic jams, incidents, and unusual traffic patterns and enables prompt remedial action.
2. Traffic signal control: a crucial part of ATMS, traffic signal control optimizes signal timings based on current traffic conditions. Dynamic signal control strategies, which alter signal timings in response to shifting traffic demand, have taken the place of conventional fixed-time signal control. Advanced traffic management systems can maximize signal coordination, minimize stops, cut down on delays, and increase overall traffic flow efficiency by using traffic data and predictive algorithms. Adaptive traffic signal control systems can prioritize high-demand movements, grant transit vehicles green lanes, and encourage the use of sustainable transportation options like walking, cycling, and public transportation.
3. Incident management: by providing prompt detection, response, and resolution of incidents on road networks, ATMS plays a crucial role in incident management. Advanced traffic management systems can quickly locate incidents like accidents, breakdowns, or road hazards by fusing surveillance cameras, incident detection algorithms, and communication systems. Transportation authorities can quickly inform drivers, emergency personnel, and other key parties of alerts and advisories. Coordinated incident management and response shorten incident response times, ease traffic, increase safety, and boost the overall effectiveness of transportation systems.

4. Dynamic route guidance: a crucial component of ATMS, dynamic route guidance helps drivers choose the best routes based on current traffic conditions. Advanced traffic management systems can generate real-time route recommendations to avoid traffic, incidents, or road closures by using traffic data and predictive algorithms. This feature encourages efficient travel, shortens travel distances, consumes less fuel, and increases the dependability of transportation networks. In order to encourage travelers to select environmentally friendly options and lessen their reliance on private vehicles, dynamic route guidance systems can also incorporate information on alternate modes of transportation.
5. Traveler information systems: to give travelers up-to-date, accurate information, ATMS incorporates traveler information systems. Advanced traffic management systems distributes traffic updates, road conditions, travel advisories, and suggestions for alternate routes through a variety of platforms, including mobile applications, websites, dynamic message signs, and in-vehicle displays. By ensuring that travelers are informed, ATMS improves user experience, facilitates better travel planning, lowers uncertainty, and encourages the use of environmentally friendly transportation options. By distributing traffic among various routes and modes, traveler information systems also aid in the management of congestions [15–17].
6. Integration with emerging technologies: to further enhance traffic management and sustainability, ATMS has the potential to integrate with emerging technologies. For instance, real-time communication between vehicles and infrastructure is made possible by the integration of ATMS with CAVs, enhancing traffic coordination, optimizing routing, and enhancing safety. A proactive approach to managing traffic flow, incident detection, and signal control is made possible by the integration of ATMS with advanced analytics, machine learning, and artificial intelligence.

As a result, ATMS are a vital part of ITS, which support effective, secure, and environmentally friendly traffic operations. Advanced traffic management systems enhance traffic flow, lessens congestion, increases safety, and encourages the use of sustainable transportation modes through advanced data collection, dynamic signal control, incident management, dynamic route guidance, traveler information systems, and integration with emerging technologies. Transportation agencies can maximize the use of current infrastructure, improve user experience, and contribute to the creation of effective and sustainable transportation systems by utilizing ATMS capabilities.

## 4.4 Advanced Traveler Information Systems

Advanced traveler information systems (ATIS) are essential for improving the travel experience, encouraging environmentally friendly transportation options, and raising general transportation productivity. Advanced traveler information systems use ITS technologies from intelligent transportation systems to give users precise, real-time, and customized information about various aspects of their journeys. This section examines the essential elements and features of cutting-edge traveler information systems, emphasizing their importance in empowering users and advancing goals for sustainable transportations.

1. Real-time traffic information: real-time traffic information is one of the core elements of ATIS. Data on accidents, delays, road closures, and traffic congestion are also included. Advanced traveler information systems collect real-time traffic data from numerous sources by utilizing cutting-edge data collection techniques like sensors, GPS devices, and traffic cameras. Following that, this data is compiled, examined, and made available to users through a variety of channels, such as websites, mobile applications, dynamic message signs, and in-car displays. Travelers can avoid traffic, cut down on travel times, and generally improve travel efficiency by using real-time traffic information to plan their routes.
2. Alternative route suggestions: in addition to providing real-time traffic data, ATIS also suggest alternate routes based on the flow of traffic at the moment. The algorithms used by ATIS produce alternate routes that optimize travel times and reduce delays by taking into account traffic congestion, road closures, and incidents. By avoiding congested areas and choosing the best paths for their journeys, these alternate route recommendations help drivers choose more effective routes. Advanced traveler information systems improve

traffic flow, encourages the use of environmentally friendly modes of transportation, and encourages the use of less congested routes.

3. Multimodal trip planning: by incorporating data on various modes of transportation, such as public transportation, walking, cycling, and car-sharing services, ATIS supports multimodal trip planning. Travelers have access to thorough information about transit schedules, routes, fares, and connections through ATIS platforms. Advanced traveler information systems make it easier to plan door-to-door trips, promotes the use of sustainable modes, and lessens reliance on private vehicles by offering seamless integration of various modes of transportation. This multimodal trip planning feature improves the overall travel experience, encourages environmentally friendly transportation options, and helps to create integrated, effective transportation systems.
4. Personalized travel information: ATIS provide customized travel information based on each traveler's individual preferences and requirements. User profiles allow ATIS platforms to take into account things like preferred travel modes, accessibility needs, and time restraints. Advanced traveler information systems create tailored travel suggestions based on this data, including route suggestions, transit options, and real-time updates. Travelers are better able to make decisions that are in line with their unique preferences and priorities when they have access to personalized travel information, which also improves the user experience and lessens information overload.
5. Weather and environmental conditions: to provide complete travel information, ATIS incorporate data on weather and environmental conditions. Travelers can prepare for bad weather by using ATIS platforms, which integrate weather forecasts, air quality data, and other environmental information. This entails making suggestions for suitable modes of transportation, formulating alternate itineraries, and giving advice on potential delays or dangers. Advanced traveler information systems support sustainable travel options, increases safety, and lessens exposure to pollution by taking weather and environmental factors into account for different scenarios [17–20].

6. Integration with mobile and connected technologies: to improve accessibility and convenience for travelers, ATIS makes use of mobile and connected technologies. Access to real-time information, individualized trip planning, and navigation support are all made possible by mobile applications. When ATIS platforms are integrated with connected vehicles, there is seamless communication between the two, enabling real-time updates and suggestions to be sent right to the display or infotainment system of the vehicles and processes. With this integration, the user experience is improved, hands-free and distraction-free travel is encouraged, and adopting eco-friendly transportation options is made easier for different use cases [21].

Finally, by offering real-time, individualized, and comprehensive information about traffic conditions, alternate routes, multimodal options, weather, and environmental conditions, ATIS empower travelers. Advanced traveler information systems improve the travel experience, encourages environmentally friendly transportation options, and boosts overall transportation effectiveness by utilizing intelligent transportation systems technologies. Advanced traveler information systems promotes informed decision-making, route optimization, selection of environmentally friendly modes of transportation, and active involvement in the development of more environmentally friendly and effective transportation systems.

## 4.5 Advanced Vehicle Control System

The field of ITS, which focuses on improving vehicle performance, safety, and efficiency, has seen a significant advancement with the introduction of advanced vehicle control systems (AVCS). In order to monitor, regulate, and optimize different aspects of vehicle operation, such as acceleration, braking, steering, and stability, AVCS makes use of cutting-edge technologies and clever algorithms. This section examines the main AVCS functionalities and parts, emphasizing their value in enhancing vehicle control and advancing sustainable transportation goals.

1. Adaptive cruise control (ACC): an important aspect of AVCS that improves vehicle safety and effectiveness on highways and congested roads is adaptive cruise control (ACC). Adaptive cruise control uses radar or LiDAR-type sensors

to measure the distance and speed of the vehicle in front of it. Based on this data, ACC automatically adapts to changes in traffic conditions and modifies the vehicle's speed to maintain a safe following distance. Adaptive cruise control lessens the possibility of rear-end collisions, enhances traffic flow, and boosts fuel efficiency by maximizing vehicle spacing and minimizing unnecessary acceleration and braking.

2. Lane-keeping assist (LKA): it is an AVCS feature created to increase vehicle safety and lower the risk of lane departure accidents. Lane-keeping assist makes use of cameras or sensors to keep track of the position of the car within the lane. Lane-keeping assist provides corrective steering inputs to keep the vehicle within the lane boundaries if it starts to veer out of the lane without the driver signaling a lane change. Lane-keeping assist improves safety by preventing unintentional lane changes, which also lowers the risk of collisions and encourages more efficient traffic flow.
3. Electronic stability control (ESC): An essential AVCS feature that improves vehicle control and stability in hazardous driving situations or during emergency maneuvers. Electronic stability control makes use of sensors to track the motion of the car, keeping track of things like yaw rate, steering angle, and wheel speed. To assist the driver in regaining control and maintaining stability, the ESC selectively applies individual brakes and modifies engine torque when it detects a loss of control or departure from the intended path. Electronic stability control increases vehicle safety, lowers the chance of accidents, and improves all-around road safety by mitigating skidding and loss of control circumstances.
4. Autonomous emergency braking (AEB): AEB is a part of the AVCS that helps drivers reduce or avoid collisions in order to increase vehicle safety. Autonomous emergency braking makes use of sensors like radar or cameras to identify potential risks of collision with other cars, pedestrians, or obstacles. When a collision risk is identified, AEB automatically applies the brakes or offers braking assistance to lessen the force of the impact. Autonomous emergency braking increases reaction times, reduces collision risks, and encourages safer driving habits by aiding drivers in emergency situations.

5. Cooperative vehicle-infrastructure systems (CVIS): CVIS is an upcoming AVCS component that enables communication between vehicles and infrastructure, such as traffic lights, road signs, and roadside units. To transfer data between vehicles and infrastructure components, CVIS uses cellular networks or dedicated short-range communication (DSRC). Real-time traffic updates, information on the timing and phase of signals, and alerts about impending dangers or poor road conditions are all made possible by this communication. Cooperative vehicle-infrastructure systems promote efficient and sustainable driving practices by enhancing situational awareness, enhancing traffic flow, and providing vehicles with up-to-date information sets [22, 23].
6. Energy management and eco-driving assistance: to increase fuel efficiency and lessen the impact on the environment, AVCS can include energy management and eco-driving assistance features. In order to promote eco-friendly driving habits, these features offer drivers real-time feedback on their driving behavior, including acceleration, braking, and speed control. Energy management and eco-driving assistance help to reduce fuel consumption, lower emissions, and support sustainable transportation goals by enhancing vehicle performance and promoting fuel-efficient driving techniques.

In conclusion, advanced technologies and clever algorithms are used by AVCS to improve vehicle control, safety, and effectiveness. Advanced vehicle control systems improve vehicle performance, lowers the risk of accidents, improves traffic flow, and helps achieve sustainable transportation goals by integrating features like ACC, LKA, ESC, AEB, CVIS, and energy management/eco-driving assistance. The advancement of vehicle technology, improvement of the driving experience, and development of intelligent and sustainable transportation systems are all made possible by AVCS.

## 4.6 Commercial Vehicle Operational Management

A crucial component of the transportation sector is commercial vehicle operational management, which aims to maximize the effectiveness, efficiency, and safety of commercial vehicles like trucks, buses, and delivery vans. The timely and reliable transportation of both goods and people, as well as the reduction of operating expenses and emissions, all depend on the efficient management of commercial vehicles. The essential elements and tactics of commercial vehicle operational management are examined in this section for different use cases.

1. Fleet management: at the heart of commercial vehicle operational management is fleet management. It includes various tasks like purchasing, maintaining, scheduling, rerouting, and tracking vehicles. In order to track the locations of vehicles, gather information on their performance and fuel efficiency, and facilitate efficient decision-making, fleet management systems make use of technologies like Global Positioning System (GPS), telematics, and fleet management software. Fleet management maximizes fleet operations to increase productivity, decrease downtime, improve route planning, and reduce fuel usage and emissions.
2. Routine vehicle maintenance and diagnostics: these are essential for guaranteeing the dependability and safety of commercial vehicles. Preventive maintenance, regular inspections, and prompt repairs are all components of effective maintenance programs. Real-time data on vehicle health is provided by advanced diagnostic systems and telematics technologies, enabling preventative maintenance and reducing unforeseen breakdowns. In addition to decreasing vehicle downtime, proper maintenance procedures also improve fuel efficiency, increase vehicle longevity, and support the sustainability of the fleet as a whole.
3. Driver management and training: programs for optimizing driver performance, ensuring regulatory compliance, and encouraging safe driving practices are part of commercial vehicle operational management. Driver recruitment, training, performance evaluation, and recognition programs are all essential components of efficient driver management. Continuous driver training focuses on enhancing safe driving practices, observance of security procedures, and environmental awareness. Drivers who are motivated and well-trained increase customer satisfaction, decrease accidents, and improve fuel economy.
4. Route optimization: to increase effectiveness and lower operating costs, commercial vehicle routes must be optimized. Utilizing cutting-edge algorithms and software, route

optimization finds the most effective routes based on variables like distance, traffic, delivery/pickup schedules, and vehicle capacity. Route optimization lowers fuel consumption, improves delivery/pickup time accuracy, and lessens traffic congestion by minimizing mileage, idle time, and pointless stops. Optimized routes also help to lower emissions and increase the sustainability of transportation in general.

5. Load management and consolidation: effective load management and consolidation techniques reduce empty trips and maximize vehicle capacity utilization. To ensure the best payload per trip, this entails optimizing vehicle loading plans, coordinating multiple shipments, and consolidating loads. Reduced traffic volume, lower emissions, and improved overall transportation efficiency are all benefits of load management. For load management and consolidation efforts to be successful, cooperation and coordination between shippers, carriers, and receivers are essential.
6. Regulation compliance: the management of commercial vehicle operations calls for strict adherence to legal requirements and industry standards. This includes adhering to safety regulations, weight restrictions, driver hours-of-service laws, and emissions standards. Compliance guarantees that commercial vehicles operate within the law, enhancing road safety and reducing environmental effects. Effective commercial vehicle operational management also requires staying current with legislative amendments and industry best practices.
7. The incorporation of cutting-edge technologies: the management of operational processes for commercial vehicles greatly benefits from the integration of cutting-edge technologies. This covers the use of connected vehicle technologies, electronic logging devices (ELD), telematics systems, and ITS. Through the use of these technologies, fleet management systems, remote diagnostics, automated data collection, and real-time monitoring of vehicle performance are all made possible. Utilizing cutting-edge technologies improves operational effectiveness, supports sustainable transportation practices, and makes informed decisions easier.

In conclusion, commercial vehicle operational management includes a variety of tactics and elements designed to maximize the effectiveness, sustainability, and performance of commercial vehicles. Organizations can increase operational efficiency, lower costs, increase safety, and contribute to the creation of sustainable transportation systems by properly managing fleets, implementing maintenance programs, optimizing routes, training drivers, assuring regulatory compliance, and integrating cutting-edge technologies. In the scenarios of the transportation industry, commercial vehicle operational management is essential for achieving economic, environmental, and social goals.

## 4.7 Advanced Public Transportation Systems

The delivery of public transportation services is revolutionized by advanced public transportation systems (APTS), which increase their productivity, usability, and sustainability. Advanced public transportation systems raises the standard of public transportation, enhances the traveler experience, streamlines operations, and encourages the use of environmentally friendly modes of transportation by utilizing ITS technologies and cutting-edge methods. The main features and capabilities of modern public transport systems are examined in this section, which also emphasizes how important they are to the transformation of urban mobility.

1. Intelligent vehicle systems: to increase the effectiveness and dependability of public transport, APTS incorporate intelligent vehicle systems. These systems use cutting-edge passenger information technology, automatic vehicle location (AVL), and advanced vehicle tracking. Transit managers can better dispatch, schedule, and coordinate services by using AVL to track the location, speed, and performance of vehicles in real-time. Real-time passenger information systems improve overall passenger experience and cut down on waiting times by giving users accurate, current information about routes, schedules, delays, and service interruptions.
2. Fare collection and ticketing: the APTS depends on sophisticated fare collection and ticketing systems. To make fare payment simple and speed up transactions, these systems make use of contactless smart cards, mobile payment apps, or other electronic payment methods. They promote intermodal connectivity by enabling easy and seamless fare collection across various modes of transport. Advanced fare collection systems also make data collection and analysis

easier, allowing transit agencies to better plan services and allocate resources by gaining insights into customer behavior, demand trends, and revenue management.

3. Intelligent transit signal priority: an essential part of APTS that boosts the effectiveness and dependability of public transport services is intelligent transit signal priority (TSP). Transit vehicles and traffic signal controllers are connected by TSP using communication technologies. This decreases delays and enhances on-time performance by enabling traffic signals to give priority to incoming buses or trams. By providing more dependable and effective travel options, TSP improves service dependability, decreases travel times, and promotes the use of public transportation.
4. Demand-responsive transit: demand-responsive transit systems, which provide adaptable, on-demand transportation services tailored to passengers' needs, are incorporated into APTS. Based on passenger demand and in-the-moment requests, these systems optimize trip scheduling and vehicle assignments using cutting-edge algorithms and dynamic routing. Demand-responsive transit increases service accessibility, especially in sparsely populated areas or off-peak periods when fixed-route services might not be financially viable. Demand-responsive transit improves operational efficiency and contributes to sustainable transportation solutions by reducing empty trips and optimizing resource allocation.
5. Integrated mobility platforms: Modern public transport systems place a strong emphasis on combining different modes of travel into a seamless, interconnected network. Information from various transportation providers, such as buses, trains, taxis, ride-sharing services, and bike-sharing programs, is combined by integrated mobility platforms. These platforms offer a unified user experience by allowing travelers to plan, reserve, and pay for multimodal trips using a single interface. By enabling the use of public transportation in conjunction with other modes, reducing the need for private vehicles, and supporting a more effective and sustainable transportation ecosystem, integrated mobility platforms support sustainable transportation.
6. Data analytics and predictive modelling: to increase operational effectiveness, optimize service planning, and improve

decision-making, APTS uses data analytics and predictive modelling techniques. Transit agencies can learn more about travel patterns, customer preferences, and service effectiveness by analyzing data gathered from a variety of sources, such as vehicle sensors, fare collection systems, and passenger feedback. With the help of this data, optimization opportunities can be found, such as adding services during periods of high demand or modifying routes or schedules. With the aid of predictive modelling, future demand can be foreseen, enabling proactive service adjustments and resource allocation.

In conclusion, by combining intelligent vehicle systems, sophisticated fare collection, intelligent TSP, demand-responsive transit, integrated mobility platforms, and data analytics, APTS revolutionize public transportation. By including these elements, APTS improves passenger experience, optimizes operations, and encourages the use of sustainable transportation modes while also increasing the efficiency, accessibility, and sustainability of public transportation services. The development of more interconnected, effective, and environmentally friendly transport systems is made possible in large part by APTS.

## 4.8 Electronic Payment Systems

Electronic payment systems have completely changed, it conduct business by providing quick, safe, and practical payment options in a variety of sectors, including transportation. Electronic payment systems have become a crucial component of contemporary transport infrastructure, revolutionizing the way passengers pay for services and enabling increased efficiency and convenience. This section examines the different kinds, advantages, difficulties, and potential uses of electronic payment systems in the transportation industry.

Electronic payment system types include:

1. Contactless smart cards: in transport systems, contactless smart cards—such as transit cards or smart tickets—are frequently used. Passengers can tap or swipe these cards on reader terminals to pay for fares because the cards have embedded chips that store value or information. Transactions

made with contactless smart cards are quick and easy, eliminating the need for cash and paper tickets.

2. Mobile payment applications: To enable digital transactions, mobile payment applications make use of the capabilities of smartphones. To pay for transportation services, passengers can use mobile payment apps connected to their bank accounts, credit cards, or mobile wallets. These apps frequently make use of near field communication (NFC) technology, which enables customers to pay with QR codes or card readers by tapping their smartphones on them.
3. Near field communication (NFC): NFC technology enables contactless and secure communication between devices that are close to one another. Near field communication-enabled devices, such as smartphones or smartwatches, can be used to pay for transportation by simply tapping them on readers that are compatible. Due to its ability to speed up and simplify transactions, NFC is a popular option for electronic payment systems.
4. Digital wallets: also referred to as e-wallets, digital wallets securely store payment information on a mobile device or online platform. Users can access the information stored in the digital wallet to make payments by linking their credit cards, bank accounts, or prepaid cards. With digital wallets, users don't need to carry physical cards around because they can make purchases with just a few taps on their smartphones.
5. Biometric payment systems: for convenient and secure transactions, biometric payment systems use distinctive biological traits like fingerprints, iris scans, or facial recognition. Passengers using biometric identifiers can access and pay for services in the transportation industry with the help of biometric payment systems. These systems improve security, stop fraud, and do away with the need for tangible cards or gadgets.
6. Tokenization and virtual cards: tokenization entails the substitution of one-of-a-kind tokens for sensitive payment information, such as credit card numbers. For each transaction, a virtual card is generated, adding an extra layer of security and lowering the possibility of data breaches. Transport service providers have safe and practical payment options thanks to tokenization and virtual cards.

### 4.8.1 Electronic Payment Systems' Advantages

1. Convenience and ease of use: passengers can make payments easily and conveniently using electronic payment systems. Users can complete transactions quickly and easily without using cash or paper tickets with just a tap, swipe, or scan.
2. Seamless intermodal connectivity: by enabling users to use a single payment method on various modes of transportation, electronic payment systems enable seamless intermodal connectivity. This integration simplifies the payment process and promotes travel by multiple modes.
3. Shorter transaction times: using electronic payment methods speeds up transactions and cuts down on wait times at fare gates and ticket counters. The quick tapping or scanning of payment devices by passengers makes boarding quick and easy.
4. Enhanced security and fraud prevention: by reducing the reliance on cash transactions, which are vulnerable to theft and loss, electronic payment systems enhance security. Sensitive payment data is also protected by strong encryption and authentication protocols, reducing the risk of fraud and unauthorized access.
5. Better data collection and analytics: electronic payment systems produce useful data on travel trends, demand, and passenger behavior. Transit agencies can use this data to analyze trends, improve services, and better allocate resources for more effective operations.
6. Cost effectiveness and revenue management: Electronic payment systems help transport operators cut costs by eliminating the need for manual ticketing and cash handling procedures. Better fare pricing strategies and revenue optimization are also made possible by data analytics and revenue management tools.

While electronic payment systems have many advantages, there are a number of challenges and factors that must be taken into account for successful implementation.

1. Infrastructure compatibility: electronic payment systems need an infrastructure that is compatible, including network connectivity, card readers, and payment terminals. Coordinating

efforts among stakeholders and making sizable investments may be necessary to ensure seamless integration across transportation networks.

2. Interoperability and standardization: to give passengers a seamless payment experience across various transportation modes and service providers, interoperability between various electronic payment systems is crucial. Interoperability depends on standardizing protocols and interfaces.
3. Protecting passenger privacy and ensuring the security of payment: these data are of the utmost importance. To foster trust among passengers, strong encryption, authentication techniques, and adherence to privacy laws are required.
4. Adoption and acceptance: user acceptance and familiarity are key factors in the successful adoption of electronic payment systems. The key to widespread adoption is educating passengers about the advantages and simplicity of use, as well as addressing worries about security and privacy.
5. Accessibility and inclusion: all passengers, including those with disabilities or limited technology access, should be catered to by electronic payment systems. To provide equitable transport services, it is crucial to guarantee accessibility and inclusion for all users.
6. Regulatory and legal frameworks: to address issues like liability, consumer protection, and dispute resolution in electronic payment systems, it is essential to establish clear regulatory frameworks and guidelines.

Passengers now pay for transport services using electronic payment systems, which offer convenience, efficiency, and increased security. Numerous electronic payment methods, including contactless smart cards, mobile payment apps, and digital wallets, are now readily available, which has streamlined the payment process and improved the overall passenger experience. The adoption of electronic payment systems in transportation requires overcoming obstacles related to user acceptance, security, and infrastructure compatibility and interoperability levels. Electronic payment systems are poised to play a key role in defining the future of practical and sustainable transportation scenarios thanks to ongoing technological advancements and rising user acceptance levels.

## 4.9 Advanced Urban Transportation Models

Modern cities face a variety of complex problems, including traffic, pollution, and accessibility. Advanced urban transportation models are essential to solving these problems. In order to design transportation systems that are sustainable, effective, and centered on people, these models make use of cutting-edge technologies, creative strategies, and data-driven solutions. This section examines the idea of advanced urban transport models, as well as their elements, advantages, and contributions to the development of urban mobility in the future scenarios.

1. Integrated mobility solutions: modern urban transportation models place a strong emphasis on integrating different modes of transportation, such as walking, biking, and public transportation. In addition to offering passengers a variety of transport options and facilitating effective and sustainable travel decisions, integrated mobility solutions also promote seamless connectivity. These models promote the use of public transport and lessen reliance on private vehicles by providing integrated and multimodal journey planning, ticketing, and payment systems.
2. Smart infrastructure and intelligent transportation systems: to optimize traffic flow, increase safety, and boost operational effectiveness, advanced urban transportation models rely on smart infrastructure and ITS. These systems use sensors, cameras, and data analytics to manage signal timings, monitor traffic conditions, and give users real-time information. Cities can reduce traffic, shorten travel times, and improve the overall effectiveness of their transportation networks by integrating ITS with sophisticated traffic management systems, adaptive signal control, and congestion pricing.
3. Data-driven decision making: modern urban transport models depend heavily on data. By analyzing this data, transportation planners and operators can make well-informed decisions about service optimization, infrastructure planning, and resource allocation. These models harness data from various sources, including sensors, mobile applications, and social media. Data-driven models also allow for proactive management of the city's transport infrastructure. This enables real-time adjustments and adaptive planning to better meet locals' needs.

4. Mobility-as-a-service (MaaS): the idea of MaaS, which aims to provide a seamless and integrated transportation experience through a single platform, is frequently incorporated into advanced urban transportation models. Multimodal trip planning, booking, and payment are all possible with the help of MaaS platforms, which do away with the need for separate ticketing systems and offer a more seamless travel experience. Mobility-as-a-service models provide greater convenience, accessibility, and affordability for urban residents by incorporating public transit, shared mobility services, and other modes of transportation.
5. Low-emission and sustainable solutions: the reduction of carbon emissions and sustainability are given top priority in modern urban transportation models. In addition to encouraging the use of low-emission modes of transportation like walking and cycling, they also promote the use of electric buses and shared electric vehicles. To reduce environmental impact and enhance air quality in cities, these models also incorporate methods for effective route planning, optimized vehicle utilization, and the incorporation of renewable energy sources.

Advanced urban transport models are aware of the value of user-centric design and community engagement, which rank sixth. Residents, businesses, and community organizations are all included in the planning and decision-making process, as are other stakeholders. These models can be used to design transport systems that better serve the community, increase accessibility, and enhance overall quality of life by taking into account the distinct needs and preferences of various user groups.

A paradigm shift in how cities approach transportation planning and management can be seen in advanced urban transportation models. These models help cities build more effective, accessible, and sustainable transportation systems by integrating different modes of transportation, utilizing smart infrastructure and ITS, utilizing data-driven decision making, adopting MaaS concepts, prioritizing sustainability, and involving the community. Advanced urban transport models will be essential in determining the future of cities as urbanization grows and technology advances, ensuring mobility that is socially inclusive, environmentally friendly, and economically viable for real-world scenarios.

## 4.10 Security and Safety Systems

The development of sustainable transport networks depends heavily on ITS. It is essential to address the security and safety aspects of these systems as cities work to lessen traffic, increase efficiency, and reduce their environmental impact. This article examines the value of strong security and safety systems in ITS for environmentally friendly transportation, the difficulties encountered in this field, and the techniques and technologies used to reduce risks and guarantee the dependability and credibility of transportation infrastructure sets.

Security and safety in ITS are crucial: systems for sustainable transportation must include important components for security and safety. In the context of ITS, security entails defending communication, data, and transportation networks against online threats, unauthorized access, and malicious behavior. On the other hand, safety focuses on ensuring the wellbeing of drivers, pedestrians, and the entire community by putting in place procedures to avoid mishaps, deal with incidents, and reduce risks.

Security and safety issues for ITS:

1. Cybersecurity threats: as transport systems become more digital, cyber threats present a serious problem. The security and dependability of transport networks may be jeopardized by hackers who try to interfere with services, manipulate data, or gain unauthorized access to crucial systems.
2. Data privacy and protection: ITS produce a lot of data about user behavior, traffic flow, and vehicle data. To uphold public confidence and adhere to data protection laws, it is essential to ensure the privacy and protection of this data.
3. Infrastructure vulnerabilities: because of how ITS is interconnected, it is possible to physically attack transport systems by tampering with sensors or interfering with communication networks. Maintaining system integrity and ensuring safety depend on protecting infrastructure from physical threats.
4. User safety and awareness: increasing user safety and awareness is essential for avoiding mishaps and lowering risks. Essential ITS considerations include educating users about safe driving habits, ensuring appropriate signage, and putting protective measures in place for vulnerable road users.

### 4.10.1 Security and Safety Tactics for ITS

1. Access control and authentication: using strong access control mechanisms and authentication protocols, you can make sure that only people with permission can access important systems and data. Unauthorized access is prevented with the aid of user role management, secure login credentials, and multi-factor authentication.
2. Data security and encryption: using encryption methods to safeguard data both in transit and at rest helps thwart data breaches and unauthorized access. Sensitive information must be protected, and this requires secure data storage, reliable security updates, and powerful encryption algorithms.
3. Intrusion detection and prevention systems: the use of such systems enables the real-time detection and reaction to potential security breaches. Advanced monitoring systems are able to spot suspicious activity, notify administrators, and put precautions in place.
4. Incident response and recovery plans: creating thorough plans for incident response and recovery ensures prompt action in the case of safety or security incidents. These plans lay out processes for managing crises, organizing responses, and getting things back to normal.
5. Collaborative partnerships: strengthening security and safety initiatives is collaborating with public and private stakeholders, including law enforcement organizations, transportation companies, and technology providers. The overall security posture of ITS is improved by sharing information, participating in joint exercises, and establishing protocols for cooperation.
6. Training and education: providing users, maintenance staff, and system operators with training and education programs helps raise awareness of security and safety best practices. A culture of security and safety consciousness can be developed with the help of regular training sessions, workshops, and certification programs.

### 4.10.2 ITS Security and Safety Technologies

1. Artificial intelligence (AI) and machine learning: algorithms based on AI and machine learning can analyze massive amounts of data to spot anomalies, identify potential

security risks, and enable predictive maintenance for safer transportation systems.
2. Advanced sensor technologies: using advanced sensor technologies, such as radar, LiDAR, and video surveillance systems, improves real-time monitoring and makes it possible for quick incident detection and response.
3. Vehicle-to-everything (V2X) communication: V2X communication enables communication between vehicles and infrastructure, as well as between them and other vehicles and pedestrians. This technology makes it possible to avoid collisions, manage traffic, and improve overall road safety.
4. Remote sensing and surveillance systems: remote sensing technologies, like drones and satellite imaging, offer real-time monitoring of transportation infrastructure, enabling the early detection of physical vulnerabilities and potential security threats.
5. Blockchain technology: by providing decentralized and tamper-resistant data storage, blockchain technology improves the security and integrity of information exchange and transactions involving transportation. It can guarantee the immutability and privacy of data in ITS.

Systems for ensuring security and safety are essential parts of sustainable transportation networks. Intelligent transportation systems can provide dependable, resilient, and trustworthy transportation services by addressing cybersecurity threats, safeguarding data privacy, bolstering infrastructure, and promoting user safety. To ensure the security and safety of ITS, it is essential to implement strong security measures, make use of cutting-edge technologies, foster collaboration, and make investments in training and education levels. Prioritizing security and safety will make it possible to create effective, environmentally friendly, and people-centered transportation networks as cities continue to embrace sustainable transportation scenarios.

## 4.11 Urban Traffic Control

By maximizing traffic flow, reducing congestion, and raising the general effectiveness of urban transportation networks, urban traffic control is essential to achieving sustainable transportation goals. Intelligent transportation systems (ITS) have made traffic control systems more sophisticated by utilizing cutting-edge tools and data-driven strategies. This chapter examines the significance of urban traffic control in ITS for environmentally friendly transportation, the difficulties in managing urban traffic, and the methods and tools used to improve traffic flow and lessen congestion in urban areas.

Urban traffic control is crucial to ITS. For the development of sustainable transport systems, urban traffic control is crucial. Travel times, fuel consumption, emissions, and user satisfaction are all improved by efficient traffic flow. Urban traffic control helps create a more environmentally friendly and sustainable transportation environment by controlling traffic congestion and maximizing the capacity utilization of already-existing infrastructure.

### 4.11.1 Urban Traffic Control Challenges

1. Congestion: due to rising population, constrained road capacity, and ineffective traffic management, urban areas frequently experience high levels of congestion. Increased travel times, fuel waste, and environmental pollution are all effects of congestion.
2. Dynamic traffic conditions: events, accidents, construction, and other factors can cause traffic patterns in urban areas to change quickly. Real-time traffic management calls for flexible tactics and responsive infrastructure.
3. Multimodal integration: urban transportation networks combine a number of different modes, such as personal vehicles, public transportation, bicycles, and foot travel. Significant obstacles must be overcome in order to provide seamless intermodal connectivity and coordinate traffic control measures across these modes.
4. Safety considerations: it is important for urban traffic control to ensure the safety of all road users, including cyclists and pedestrians. Sustainable traffic management requires taking action to manage intersections, prioritize vulnerable road users, and prevent accidents.

### 4.11.2 Technologies and Strategies for ITS Urban Traffic Control

1. Adaptive traffic signal control: adaptive traffic signal control systems modify signal timings in response to traffic conditions using real-time data from sensors and cameras. These

programs enhance the efficiency of the entire network while streamlining traffic and minimizing delays.

2. Intelligent intersection management: to optimize intersection operations, intersection management systems use cutting-edge technologies like connected vehicle technology and vehicle detection sensors. Enhancing intersection efficiency and safety involves synchronizing traffic signals, putting in place dedicated turn lanes, and facilitating pedestrian-friendly crossings.
3. Management of traffic incidents: traffic incident disruptions and congestion are reduced by quick detection and action. Early detection, prompt action, and efficient incident management are made possible by the integration of incident detection systems, closed circuit television (CCTV) cameras, and communication networks.
4. Data-driven decision making: traffic control systems analyze traffic patterns, forecast congestion, and optimize traffic management strategies by using historical and real-time data. Traffic controllers can make well-informed decisions for effective traffic flow thanks to data analytics and machine learning algorithms.
5. Connected vehicle technology: using connected vehicle technology, traffic management systems, infrastructure, and vehicles can all communicate with one another. Improved traffic flow, lessened congestion, and increased safety are all benefits of sharing real-time information about things like traffic conditions, road hazards, and signal timing.
6. Integrated public transit systems: by integrating traffic control systems with public transit operations, public transit vehicles can be better coordinated and prioritized. Real-time passenger information systems, TSP lanes, and signal priority for buses all promote the use of public transport, which eases traffic from private vehicles.
7. Demand management techniques: using demand management techniques like congestion pricing, carpooling incentives, and flexible work schedules can help lower peak-hour traffic demand and spread out traffic more evenly throughout the day.
8. Collaborative partnerships: for effective urban traffic control, cooperation between transportation agencies, city planners, and technology providers is essential. The implementation

and efficacy of traffic management strategies are improved by public-private partnerships, data sharing, and collaborative projects and scenarios.

In ITS, urban traffic control is crucial to achieving the goals of sustainable transportation. Urban traffic control optimizes traffic flow, reduces congestion, and boosts overall network efficiency by utilizing adaptive strategies, utilizing cutting-edge technologies, and integrating multimodal transportation systems. Real-time optimization and responsive traffic control are made possible by the application of data-driven decision making, connected vehicle technology, and intelligent intersection management. By encouraging effective resource use and minimizing environmental impact, collaborative partnerships and demand management techniques also support sustainable transportation systems. Prioritizing efficient urban traffic control in ITS is essential for developing livable, sustainable, and interconnected urban environments and scenarios as cities continue to expand for different scenarios. The sustainable transportation scenarios is shown in Figure 4.1.

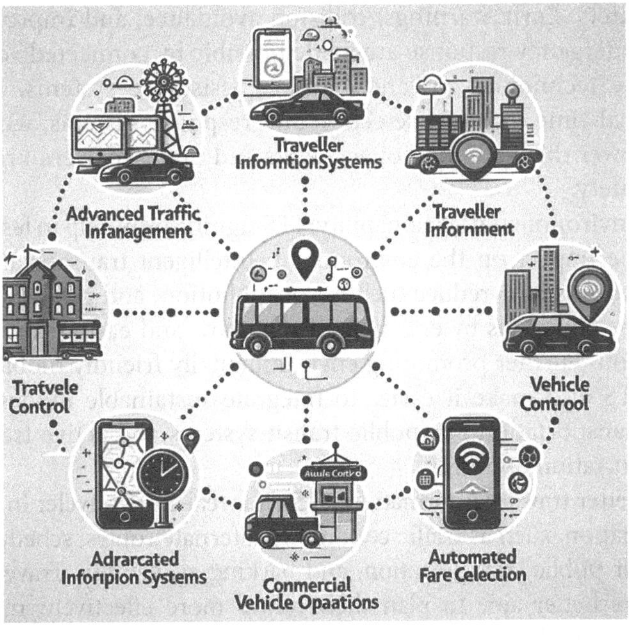


**Figure 4.1** Sustainable transportation scenarios.

## 4.12 Benefits and Limitations of ITS for Sustainable Transportation

The development of ITS has proven to be a game-changing means of achieving sustainable transportation objectives. Intelligent transportation systems seeks to improve the effectiveness, safety, and environmental sustainability of transportation systems through the integration of cutting-edge technologies, data-driven strategies, and creative solutions. This chapter examines the advantages and drawbacks of using ITS for environmentally friendly transportation, emphasizing both their potential benefits and implementation issues.

### 4.12.1 Benefits of ITS for Sustainable Transportation

1. Greater traffic efficiency: ITS allow for better traffic flow management, which cuts down on traffic jams and travel times. Dynamic traffic management is made possible by intelligent traffic signal control, adaptive traffic management systems, and real-time data analysis. This improves network capacity utilization and the efficiency of traffic operations.
2. Improved safety: ITS innovations help to increase road safety. Early warnings, collision avoidance, and improved emergency response are made possible by connected vehicle technology, advanced driver assistance systems, and real-time incident detection and response systems, which lower the likelihood of accidents and improve overall road safety.
3. Environmental sustainability: ITS significantly help to lessen the impact on the environment. Intelligent transportation systems help reduce fuel use, air pollution, and greenhouse gas emissions by enhancing traffic flow and easing congestion. Further promoting environmentally friendly mobility, ITS also make it easier to integrate sustainable modes of transportation like public transit systems and active transportation options.
4. Better traveler information: ITS offer real-time traveler information, such as traffic conditions, alternate routes, schedules for public transportation, and parking availability. Travelers are better able to plan their routes more effectively, make informed decisions, and select more environmentally friendly

modes of transportation when they have access to accurate and current information.

5. Effective logistics and freight: ITS increase the effectiveness of logistics and freight operations. Routes can be optimized, delivery times can be improved, and resource use can be improved with the help of intelligent freight management systems, smart logistics solutions, and real-time tracking and tracing. This lessens traffic, lowers transportation expenses, and has a minimal negative impact on the environment.
6. Data-driven decision making: ITS produce a ton of information about traffic patterns, traveler behavior, and transportation operations. Transportation planners and policymakers can gain useful insights, make knowledgeable decisions, and create evidence-based strategies for sustainable transportation planning and infrastructure development by utilizing this data through advanced analytics and data-driven decision making.

### 4.12.2 Constraints and Difficulties of ITS for Sustainable Transportation

1. Implementation costs: deploying ITS infrastructure and technologies can require a sizable upfront investment. For smaller cities and regions with tighter budgets, installing sensors, communication networks, ATMS, and maintaining these systems could be expensive.
2. Interoperability and standardization: integrating various ITS parts from various manufacturers and ensuring system interoperability can be difficult. Unified communication and coordination among various transport systems and stakeholders may be hampered by the lack of standardization across various regions and jurisdictions.
3. Privacy and data security issues: the gathering, storing, and analysis of enormous amounts of data in ITS raises privacy issues. Maintaining the public's trust and confidence in ITS systems depends on safeguarding personal information, preventing data breaches, and ensuring secure data management procedures.
4. Technical dependency and upkeep: ITS is heavily dependent on cutting-edge technologies, software programs, and communication networks. To avoid system failures and disruptions,

regular maintenance, software updates, and system reliability are crucial. Dependence on technology creates problems in places with poor infrastructure or limited connectivity.

5. User acceptance and adoption: user acceptance and adoption are essential for the successful implementation of ITS. To fully realize the advantages of ITS for sustainable transportation, behavior modification, user education about new technologies, and addressing potential user concerns and resistance are required.
6. Considerations for equity and accessibility: it is crucial to guarantee that all facets of the population have fair access to ITS benefits. For various scenarios, it is imperative to address inequities in digital connectivity, technology affordability, and accessibility barriers faced by vulnerable communities in order to prevent a worsening of already-existing transportation inequities.

Achieving goals for sustainable transport is significantly aided by ITS. Intelligent transportation systems can transform transportation systems by increasing traffic efficiency, boosting safety, promoting environmental sustainability, offering traveler information, optimizing freight operations, and assisting with data-driven decision making. To fully utilize the potential of ITS, however, issues with implementation costs, interoperability, privacy and security, technological dependencies, user acceptance, and equity must be resolved. To ensure the successful and inclusive deployment of ITS for sustainable transportations, collaborative efforts are needed from government agencies, technology providers, transportation stakeholders, and the general public scenarios.

## References

1. G. Rohith, K. B. Devika, P. P. Menon and S. C. Subramanian, "Sustainable Heavy Goods Vaehicle Electrification Strategies for Long-Haul Road Freight Transportation," in *IEEE Access*, vol. 11, pp. 26459-26470, 2023, doi: 10.1109/ACCESS.2023.3257431.
2. C. Chen, Y. Zhang, M. R. Khosravi, Q. Pei and S. Wan, "An Intelligent Platooning Algorithm for Sustainable Transportation Systems in Smart Cities," in *IEEE Sens. J.*, vol. 21, no. 14, pp. 15437-15447, 15 July15, 2021, doi: 10.1109/JSEN.2020.3019443.

3. M. Deveci, A. R. Mishra, I. Gokasar, P. Rani, D. Pamucar and E. Özcan, "A Decision Support System for Assessing and Prioritizing Sustainable Urban Transportation in Metaverse," in *IEEE Trans. Fuzzy Syst.*, vol. 31, no. 2, pp. 475-484, Feb. 2023, doi: 10.1109/TFUZZ.2022.3190613.
4. C. H. V. Cooper, "Exploring Potential Applications of Quantum Computing in Transportation Modelling," in *IEEE Trans. Intell. Transp. Syst.*, vol. 23, no. 9, pp. 14712-14720, Sept. 2022, doi: 10.1109/TITS.2021.3132161.
5. G. Xiu and Z. Zhao, "Sustainable Development of Port Economy Based on Intelligent System Dynamics," in *IEEE Access*, vol. 9, pp. 14070-14077, 2021, doi: 10.1109/ACCESS.2021.3051065.
6. H. Saboori, H. Mehrjerdi and S. Jadid, "Mobile Battery Storage Modeling and Normal- Emergency Operation in Coupled Distribution-Transportation Networks," in *IEEE Trans. Sustain. Energy*, vol. 13, no. 4, pp. 2226-2238, Oct. 2022, doi: 10.1109/TSTE.2022.3189838.
7. Y. Zhang, C. Wu, C. Qiao, A. Sadek and K. F. Hulme, "A Cognitive Computational Model of Driver Warning Response Performance in Connected Vehicle Systems," in *IEEE Trans. Intell. Transp. Syst.*, vol. 23, no. 9, pp. 14790-14805, Sept. 2022, doi: 10.1109/TITS.2021.3134058.
8. A. P. Cohen, S. A. Shaheen and E. M. Farrar, "Urban Air Mobility: History, Ecosystem, Market Potential, and Challenges," in *IEEE Trans. Intell. Transp. Syst.*, vol. 22, no. 9, pp. 6074-6087, Sept. 2021, doi: 10.1109/TITS.2021.3082767.
9. H. Hu, J. Du, X. Li, C. Shang and Q. Shen, "Risk Models for Hazardous Material Transportation Subject to Weight Variation Considerations," in *IEEE Trans. Fuzzy Syst.*, vol. 29, no. 8, pp. 2271-2282, Aug. 2021, doi: 10.1109/TFUZZ.2020.2997467.
10. H. Bi, W. -L. Shang, Y. Chen, K. Yu and W. Y. Ochieng, "An Incentive Based Road Traffic Control Mechanism for Covid-19 Pandemic Alike Emergency Preparedness and Response," in *IEEE Trans. Intell. Transp. Syst.*, vol. 23, no. 12, pp. 25092-25105, Dec. 2022, doi: 10.1109/TITS.2022.3191161.
11. M. Almasi, S. Khoshfetrat and M. Rahiminezhad Galankashi, "Sustainable Supplier Selection and Order Allocation Under Risk and Inflation Condition," in *IEEE Trans. Eng. Manage.*, vol. 68, no. 3, pp. 823-837, June 2021, doi: 10.1109/TEM.2019.2903176.
12. D. M. Nguyen, M. A. Kishk and M. -S. Alouini, "Toward Sustainable Transportation: Accelerating Vehicle Electrification With Dynamic Charging Deployment," in *IEEE Trans. Veh. Technol.*, vol. 71, no. 9, pp. 9283-9296, Sept. 2022, doi: 10.1109/TVT.2022.3180495.
13. H. Wang *et al.*, "Network-Wide Traffic Signal Control Using Bilinear System Modeling and Adaptive Optimization," in *IEEE Trans. Intell. Transp. Syst.*, vol. 24, no. 1, pp. 79-91, Jan. 2023, doi: 10.1109/TITS.2022.3215537.
14. Y. Song, P. Wu, Q. Li, Y. Liu and L. Karunaratne, "Hybrid Nonlinear and Machine Learning Methods for Analyzing Factors Influencing the Performance of Large-Scale Transport Infrastructure," in *IEEE Trans. Intell. Transp. Syst.*, vol. 23, no. 8, pp. 12287-12300, Aug. 2022, doi: 10.1109/TITS.2021.3112458.

15. I. Diab, A. Saffirio, G. R. C. Mouli, A. S. Tomar and P. Bauer, "A Complete DC Trolleybus Grid Model With Bilateral Connections, Feeder Cables, and Bus Auxiliaries," in *IEEE Trans. Intell. Transp. Syst.*, vol. 23, no. 10, pp. 19030-19041, Oct. 2022, doi: 10.1109/TITS.2022.3157080.
16. Y. Bichiou, M. Elouni, H. M. Abdelghaffar and H. A. Rakha, "Sliding Mode Network Perimeter Control," in *IEEE Trans. Intell. Transp. Syst.*, vol. 22, no. 5, pp. 2933-2942, May 2021, doi: 10.1109/TITS.2020.2978166.
17. J. Wang and Z. Qin, "Uncertain Programming Models for Sustainable Hub Covering Location Problem Over Incomplete Network," in *IEEE Access*, vol. 9, pp. 16092-16102, 2021, doi: 10.1109/ACCESS.2021.3051373.
18. Z. Bai, P. Hao, W. ShangGuan, B. Cai and M. J. Barth, "Hybrid Reinforcement Learning-Based Eco-Driving Strategy for Connected and Automated Vehicles at Signalized Intersections," in *IEEE Trans. Intell. Transp. Syst.*, vol. 23, no. 9, pp. 15850-15863, Sept. 2022, doi: 10.1109/TITS.2022.3145798.
19. P. Karczmarek, Ł. Gałka, A. Kiersztyn, M. Dolecki, K. Kiersztyn and W. Pedrycz, "Choquet Integral-Based Aggregation for the Analysis of Anomalies Occurrence in Sustainable Transportation Systems," in *IEEE Trans. Fuzzy Syst.*, vol. 31, no. 2, pp. 536-546, Feb. 2023, doi: 10.1109/TFUZZ.2022.3222025.
20. Z. Wang, Z. Sun, P. Liu, S. Wang and Z. Zhang, "Intelligent decision support platform of new energy vehicles," in *J. Syst. Eng. Electron.*, vol. 33, no. 4, pp. 785-791, August 2022, doi: 10.23919/JSEE.2022.000078.
21. X. Huang, Q. Liang, Z. Feng and S. Chai, "A TOD Planning Model Integrating Transport and Land Use in Urban Rail Transit Station Areas," in *IEEE Access*, vol. 9, pp. 1103-1115, 2021, doi: 10.1109/ACCESS.2020.3047207.
22. X. Huang, Q. Liang, Z. Feng and S. Chai, "A TOD Planning Model Integrating Transport and Land Use in Urban Rail Transit Station Areas," in *IEEE Access*, vol. 9, pp. 1103-1115, 2021, doi: 10.1109/ACCESS.2020.3047207.
23. Y. Zhou, H. -X. Li and S. -L. Xie, "Space-Decomposition-Based Spectral Modeling for Distributed Battery Thermal Dynamics," in *IEEE Trans. Transp. Electrif.*, vol. 8, no. 2, pp. 1634-1641, June 2022, doi: 10.1109/TTE.2021.3124894.

# 5

# Artificial Intelligence-Based Transportation System


***Abstract***

This book chapter provides an introduction to artificial intelligence (AI) and explores its role in optimizing various components of transportation systems. The chapter begins by discussing the basics of AI and its various applications in transportation systems. It then identifies the different components of transportation systems that require optimization, including congestion control, accident avoidance, and active alert systems. The chapter goes on to examine how AI can be used to optimize these components, specifically discussing the role of AI in congestion control, accident avoidance, and active alert system design. The chapter also explores the benefits and challenges associated with the use of AI in transportation systems. By highlighting the potential of AI in optimizing transportation systems, this chapter offers insights into how AI can be leveraged to improve the safety, efficiency, and effectiveness of transportation systems.



***Keywords:*** Artificial intelligence, geographics information system, public transit optimization, unmanned aerial vehicles, accident


## 5.1 Introduction to Artificial Intelligence

Computers, robots, and other devices may be taught to mimic human intellect via the study of artificial intelligence. The study of AI may also be referred to as "artificial intelligence." Artificial intelligence (AI) refers to the study of how machines can learn, reason, make choices, and carry out operations similar to those of the human brain in order to solve issues. In a nutshell, this research has led to the creation of advanced computer programs. The purpose of AI is to give computers capabilities that are comparable to human intellect. Learning, thinking, and problem-solving are all examples of these talents. The mind is immaterial and untouchable. Intelligence refers to a group of abilities related to the mind, such as the ability to reason, learn, solve problems, perceive, and communicate. These things make up the whole.


Kundan Meshram. Urban Transportation Systems: Sustainability and Future Development, (99–126)
 

The goals of AI research include, but are not limited to, the following: the ability to reason; the ability to represent knowledge; the ability to plan; the ability to learn; the ability to interpret natural language; the ability to realize; the ability to move and control objects. Long-term objectives have been set for the study of general intelligence. Methods include statistical analysis, computational intelligence, and the more conventional AI methodology of coding. We employ a wide variety of methods, from search and mathematical optimization to artificial neural networks (ANNs) to statistical, probabilistic, and economic approaches, in our investigations into the nature of AI. As computing has progressed, academics from many fields, such as physics, mathematics, psychology, linguistics, philosophy, and so on, have taken an interest in AI.

Fundamental to the field of artificial intelligence is the study and implementation of the rules that govern the functioning of the human brain (AI). Because of the progress achieved over the last 50 years, it is now a topic that transcends borders and has far-reaching consequences. Expert systems, knowledge base systems, intelligent database systems, and intelligent robot system implementations are just a few of the many areas where this technology is being put to use. The field of AI known as "expert systems" has been around the longest and has the most variety of uses. The knowledge management and decision-making technology of the 21st century is this. It is also the most exciting and profitable sector of the economy. Many uncertainties in civil engineering challenges, especially in engineering design, construction management, and program decision-making, need not only the use of mathematical, physics, and mechanics calculations, but also rely heavily on the expertise of engineers in the area. Various difficulties, for instance, were impacted by many uncertainties in engineering design, construction management, and program decision-making. Many challenges were influenced by varying degrees of uncertainty, for example, in the design of engineering projects, the management of construction projects, and the decision-making process for programs. The inadequacy and incorrectness of this information means that conventional methods of dealing with such situations are ineffective. Despite this, AI has several advantages that make it worthwhile to use. By modeling the skills of experts, we may tackle difficult issues with the same level of competency. In general, civil engineering offers a rich environment for the use of AI, both in general and in particular situations.

Multiple academics have been looking at the development of flexible structures for civil engineering, and only recently have they published their findings. The control framework of an active tensegrity structure combines self-diagnosis, multi-objective shape control, and reinforcement-learning processes. For modeling complicated systems using existing input- output data sets, adaptive neuro-fuzzy inference algorithms proved to be highly useful. This is due to the fact that artificial intelligence underpins many modern computer technologies, including adaptive neuro-fuzzy inference systems (AI). Systems of this sort are useful in civil engineering for simulating the nonlinear, complex, and ambiguous behavior of cement-based materials that have been exposed to single, dual, or multiple damage causes. This is true regardless of whether there is a single, double, or numerous damage variables at play. Researchers have created a neuro-fuzzy based prediction of the endurance of self-consolidating concrete when subjected to a variety of sodium sulfate exposure regimens. A study team has created an ANN that can predict the compressive strength of standard and high-strength self-compacting concrete (SCC) and high-performance concrete after 28 days (HPC). Both types of concrete make extensive use of fly ash in their aggregate mixes. Researchers determined the severity of the slope failure by using a method from the field of AI called back-propagation neural networks. The numerical evidence shows that ANNs may be helpful in assessing the potential for slope failure. A tunneling case study was used to illustrate the features of the integrated system after researchers presented a method for collecting data from experts to create the fuzzy expert system rules. This was accomplished with the help of a tunneling-themed case study. Cement-stabilized soil's maximum dry density (MDD) and unconfined compressive strength (UCS) may be predicted using one of two AI-based methods established by the research team. One is known as the MDD approach, while the other is known as the UCS method. The results of a research by civil engineering professors who use computer simulations to teach linear scheduling concepts and methods in their course "Construction Planning and Scheduling" have been published. Researchers took part in the investigation. To gauge the efficacy of using computer simulations in the instruction of linear scheduling, researchers compared students' evaluations with their responses to a questionnaire. Computing devices were employed to run the simulations. In their most recent paper, researchers describe a process for developing ontology-supported software applications. This method makes it feasible for the ontology to be updated while still being used by several programs at the same time. Application and ontology entered into a contract, which forms the foundation of the method. The terms of this Agreement are defined clearly and concisely in the form of integrity constraints. In the next section, we will take a high-level look at the parts of the transportation system that might stand to gain from the use of AI scenarios.

## 5.2 Components of Transportation System that Require Optimization

To ensure that transportation operations go without a hitch, several different components are required. These four factors are crucial in carefully crafted transportation situations, and they must be optimized for maximum efficiency. The term "modes" is used to refer to the wide variety of transportation options, the great majority of which are cars. Most of these modes of transportation are cars. While certain modes of transit are tailored to convey either passengers or cargo, others are able to accommodate both. The network of roads, canals, and other manmade waterways and land-based structures that connect and serve as departure and arrival points for various forms of public transit (such as ports or airports). Further, superstructures, which are often temporary assets, are considered to be part of infrastructures. Runways would be considered part of an airport's infrastructure, while the terminals and control equipment would make up the airport's superstructure. One kind of asset that fits the bill for infrastructure is a runway. The piers and waterways of a port are examples of the superstructure, while cranes and other yard machinery fall under the category of infrastructure. Locations are connected together to form a web that, when seen as a whole, represents the physical and administrative architecture of transportation systems. Discover which locations are linked and how those connections are maintained with the help of this system. The related sites are also listed. Some nodes in a network have more connections than others, which may indicate that they are more accessible (fewer connections). In the correct sequence, the flow of people, goods, and data across each network's backbone. Flows may start in one site, proceed to a different one, and terminate somewhere else entirely. It is frequently necessary to halt at a place that serves as an intermediate on the trip from one starting point to another. Passengers connecting flights between different airports, for example, may be required to do so at the hub airport.

The vast majority of individuals who are contemplating about going on a trip that will continue for a significant period of time are of the opinion that having to wait in traffic will be the single most challenging component of the trip. The process of generating judgements, which is used to manage route planning, has been substantially more effective as a direct result of advancements in technology, which have made the process considerably more efficient. Because of this, we are now in the position to plan our excursions in a way that is far more efficient than it was before. The user may request that AI-enabled route optimization algorithms produce the

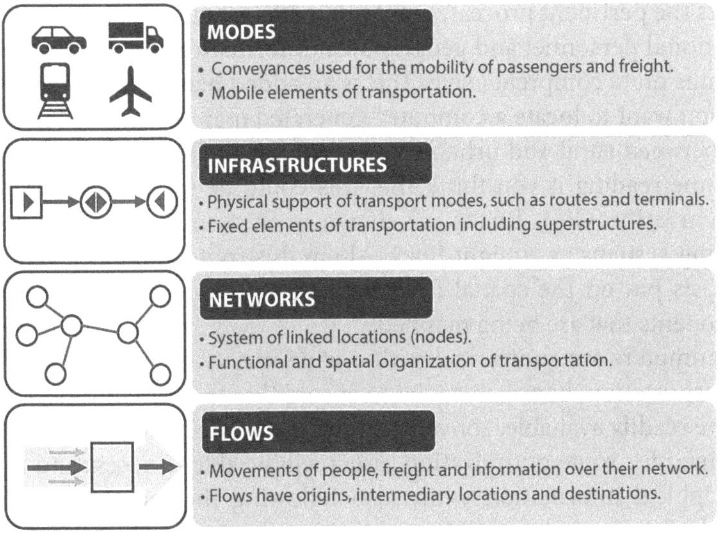


**Figure 5.1** Components in transportation that need optimizations.

shortest and least crowded routes that are available, and the algorithms will comply with the user's request if it is made. The components in transportation that need optimizations is shown in Figure 5.1.

"Route optimization system," which is also an acronym, is the name of a second system that was developed to assist passengers in arriving at their destinations within the allotted amount of time that was specified by their itineraries. This system was designed to help passengers arrive at their destinations on time. It is possible that the integration of both geographic information systems and AI into a single platform is what makes this system so effective as a trip planner. This combination links the map to a comprehensive database, allowing users to quickly and easily access the information they need along the most direct route.

In the context of this conversation, we are going to get acquainted with the ways in which AI and geographic information systems (GIS) work together to deliver the most time- and resource-effective itineraries. The software that is specifically developed to improve driving routes is going to be the focus of this section. After that, we will talk about the ways in which the development of these autonomous systems has been aided along its path toward its ultimate aim by the development of other things in the past.

A GIS, which is a kind of computer system, may be used to gather data on geographical areas, store that data, analyze that data, and display that analysis. The fact that the system is centralized gives it the ability to carry out all of these activities without any outside help. The system not only defines the pertinent processes and procedures, but it also takes use of the operational personnel and geographical information. In other words, the system is quite comprehensive. This is an added benefit to the situation.

If you want to locate a computer-generated map that can tell the difference between rural and urban regions, your best chance is to go via GIS. Continue reading if you think that this could be something that piques your curiosity (GIS). Roads are shown in these many different sorts of mapping systems as straight lines. Along this route, there is a significant emphasis put on the spatial link that exists between the many different components that are being mapped.

Commuters are confronted with a wide variety of challenges in terms of transportation; hence, it is vital to ensure that there are means of transit that are readily available, convenient, swift, and effective. If AI and GIS are combined for route optimization, users will be able to access and retrieve geographical information while also following the most efficient route possible. This is made possible by following the most efficient route possible. Geographic information systems are what are referred to as GIS. The improvement and simplification of travel itineraries are going to be necessary in order to make this a reality.

There is a great deal of uncertainty around the topic of minimizing the amount of time, travel, and financial resources required for the journey. The weather, the unpredictability of delays at crossings, the amount of traffic from a wide range of vehicles, and the demand for connections during both peak hours and off-peak hours are all potential contributors to the delay. Other potential contributors include the volume of traffic from a wide variety of vehicles.

The majority of nations' capital cities are often separated into three distinct highway portions, which are sometimes referred to as the "north," "west," and "southwest" areas of the city. Combining GIS and AI into route optimization systems allows for the calculation of journey times as well as the maintenance of arrival efficiencies, which are two of the system's core purposes. In order to convey the nature of the connections that exist between the three nodes, these systems make use of connection matrices. Because this system leverages wireless networking, cloud computing, and geographical databases, the user will always have instant access to the most up-to-date information, regardless of where they are or what time of day it is. This is made possible by the combination of these three technological advancements.

The employment of simulated multi-objective evolutionary algorithms, which are made feasible as a result of the AI that controls these systems, enables the creation of transportation networks that are extraordinarily effective. The clustering of ad hoc networks, which depend on a variety of various routing systems, would be able to help in the building of a system for optimizing routes over long distances. This would be possible because of the wide range of alternative routing approaches that ad hoc networks use. If this target could be reached, it would represent a significant achievement on the part of the individual.

It is possible that the transportation and logistics industries would benefit from the adoption of AI-powered trip planning tools due to the increased efficiency and accuracy that would result from the ability to consolidate information from multiple sources and use that information to form well-informed opinions about optimal routes. This ability would result from the ability to use the information to form well-informed opinions about optimal routes. This is because having access to this information allows for the creation of well-informed judgments about the most efficient routes. Companies in the travel and transportation industries that are now making investments in AI stand to earn major benefits as a result of the incredible availability of computing power and big data research. Because of this, it is possible for these companies to anticipate a rise in levels of client satisfaction as well as income. These types of benefits are within reach for organizations today because to the ease with which big data can be accessed.

It is likely that the sum of the transportation fees for the last leg of a trip will constitute a significant fraction of the total cost of transporting persons and commodities. It is conceivable that this alternative will turn out to be one that requires a significant financial investment. The process of reducing this cost component may benefit from optimizing the routes, and there is a probability that this will be the case. By making use of trip sheets and real-time data, AI systems are able to calculate an estimate of the amount of time that will be spent going along a certain route. The utilization of info that is up to date at the given moment in time makes this accomplishment conceivable.

Computers are continuously evaluating the data that is acquired from transportation networks in order to determine where and how much time may be saved along various routes. This analysis is being done in order to improve transportation efficiency. In order to determine how long a trip would take, a significant amount of information must be gathered regarding the topography, environment, and traffic along the route. If for any reason one of the access points is closed, the system may select on its own another point of admittance and path, and it may also provide this information to the user. There are several possible contexts in which this might take place.

When making deliveries, the navigation systems that are integrated into the dashboards of the trucks direct the drivers to the quickest and most efficient routes possible. These displays include an overwhelming amount of information. The information that is gathered by companies in the transportation industry may be put to use in the calculation of important performance indicators via the use of metrics such as the number of passengers transported and the typical driving speed of vehicles. When all of this data has been gathered, it may be used as the basis for end-to-end service planning benchmarks and the revisions that come after them. Mobility across all available channels is often required for travel agencies to have omnichannel capabilities if they are to satisfy the fundamental requirements for operations and realize overall success as a company. Because AI makes use of multi-objective adaptive algorithms, expanding into a multimodal network that integrates a number of various modes of transportation is not impossible but complicated. Although it is not impossible, it is hard. In order to effectively coordinate the utilization of a wide variety of motorized transportation modes and channels, conducting research that makes use of open geographic information and big data and that takes into consideration a number of different aspects of the natural environment might be helpful. Transportation planners may be able to enhance stock management and simplify a large number of route networks all at once with the assistance of AI by using analytical techniques that are now within their grasp. As a result of this, you may now choose between a large variety of different courses of action.

It is possible that the government will be able to start building the groundwork for the multi- modal network if it is able to make efficient use of the massive quantities of GIS information that it now has and is responsible for preserving. They are able to be of assistance in the invention and implementation of effective ways for controlling traffic in urban and rural areas and monitoring the maintenance of roads. Additionally, they are able to be of assistance in the monitoring of the condition of the roads. In addition to that, they are able to provide aid of this type. They are able to enable the coordinated working of several different kinds of public transportation by making use of this cutting-edge technology, which is something that is pretty helpful.

When it is necessary to do so, transportation businesses are able to exert control over their costs with the assistance of the insights that are offered by technology that utilizes AI. It is only because there are personnel, cars, and bus routes working together that this is even conceivable. It is feasible to apply capacity utilization algorithms to GIS data in a manner that is comparable to how these algorithms are used in the logistics sector. The application of these algorithms to GIS data may be done in a number of different ways. Because of this, there would be a reduction in the quantity of gas that would be consumed by automobiles that were sitting idle, which would be good financially.

Less-than-truckload shipments, often known as LTL shipments, provide logistics companies the option to save costs by shipping a lower number of items in a smaller number of vehicles over a route that generates less revenue. Because LTT reduces the number of stops that truckloads are required to make, a combination of GIS and AI might make it possible to distribute trucks in a manner that is economical with regard to the amount of gasoline that is used as well as the amount of space that is now available. This would be made possible by the fact that LTT reduces the number of stops that truckloads are required to make. As a consequence of recent developments in AI, transportation businesses may be able to gain an advantage through the use of simulations that carry out methods for optimizing costs. In addition, it is feasible to reduce the total number of insurance claims for property damage that are filed across the board for each and every mode of transportation. This is something that may be done. It is possible that if we get more skilled at preventing injuries, then conversations about the price of high-risk approaches will go with less friction.

One of the ways that machine learning systems add value throughout the course of time is via their ability to foresee constraints that are imposed on the allocation of resources within the transportation ecosystem. This is one of the ways that machine learning systems offer value. Planners have the power to implement preventative measures, which minimizes the likelihood that certain channels may get blocked as a consequence of the usage of such channels. When applying techniques of machine learning that have been adapted specifically for this function, the tasks are justifiable via the use of load pooling.

It is likely that planning transportation for all modes, including air, land, and waterways, will lead to a more effective use of the resources given by these systems. This might be the case if the planning includes all means of transportation. It has been figured out how to make the most out of everything that is accessible to a person, including the people, luggage-handling equipment, autos, and storage space that is available to them. This includes everything from the storage space to the staff to the vehicles. It's feasible that AI-powered engines that disperse resources might make transportation scheduling and maintenance operations more efficient when compared with existing scenarios.

## 5.3 Role of Artificial Intelligence in Optimization of these Components

In the contemporary world, where urbanization and technological advancements are reshaping the landscape of transportation, the integration of AI has emerged as a pivotal tool in optimizing various components of transportation systems. This chapter explores the multifaceted role of AI in enhancing the efficiency and effectiveness of transportation systems, particularly in the optimization of key components that demand constant refinement.

1. Artificial Intelligence-Powered Traffic Management
   Artificial intelligence plays a crucial role in optimizing traffic flow, a perennial challenge in urban areas. Machine learning algorithms analyze real-time data from various sources, such as sensors, cameras, and GPS devices, to predict traffic patterns, detect congestion, and optimize traffic signal timings. This not only reduces commute times but also decreases fuel consumption and emissions, contributing to environmental sustainability.
2. Route Planning and Navigation
   Artificial intelligence-driven route planning and navigation systems have revolutionized how people and goods move from one point to another. These systems take into account real-time traffic conditions, weather, road closures, and user preferences to offer optimized routes, thereby reducing travel time and fuel consumption. Moreover, they enhance safety by helping drivers avoid accidents and hazardous conditions.
3. Public Transit Optimization
   Artificial intelligence algorithms are employed to optimize public transit systems, including bus and train schedules. These systems adapt to changing demand patterns, ensuring that resources are allocated efficiently. Commuters benefit from reduced waiting times and more reliable services, leading to increased ridership and decreased congestion on roads.
4. Fleet Management
   In logistics and transportation, AI-driven fleet management solutions are invaluable. They optimize routes, monitor vehicle health, and provide predictive maintenance, resulting

in reduced operational costs and increased vehicle uptime. Furthermore, AI-powered telematics improve driver behavior and safety.

5. Demand Forecasting
   Artificial intelligence's predictive capabilities enable transportation planners to anticipate future demand accurately. This helps authorities make informed decisions about infrastructure investments, public transit expansion, and the deployment of shared mobility services. By aligning supply with demand, transportation systems become more efficient and cost-effective.
6. Environmental Sustainability
   Artificial intelligence-driven transportation systems contribute to environmental sustainability by reducing emissions and fuel consumption. They achieve this by optimizing traffic flow, promoting the use of electric and hybrid vehicles, and encouraging shared mobility options like carpooling and ride-sharing.
7. Enhanced Safety
   Artificial intelligence-based systems enhance safety by monitoring and analyzing real-time data to detect and predict accidents or potential hazards. Advanced driver-assistance systems (ADAS) use AI to provide features like lane-keeping, adaptive cruise control, and collision avoidance, reducing the risk of accidents. While AI offers tremendous benefits for transportation optimization, there are challenges to address. These include data privacy concerns, algorithmic bias, and the need for robust cybersecurity measures to protect critical transportation infrastructure from cyberattacks.

## 5.4 Congestion Control with Artificial Intelligence

It is possible for a country's economy, quality of life, and degree of safety to all benefit from having a transportation infrastructure that is well-managed and run efficiently. However, the demands that are placed on transportation continue to rise as a result of ongoing trends in population expansion, the development of new technologies, and the rising internationalization of the economy. These three factors, in particular, have continued to test the capabilities of the system to the point where it is on the verge of breaking down. The rate at which the number of vehicles on the road is rising is at times even larger than the rate at which the overall population is growing. This results in roadways that are more congested and unsafe because of the increased number of autos. Increasing the overall quantity of roads would not be adequate to tackle this problem anymore; we need a more comprehensive approach. It is not feasible to keep up with the rate at which automobiles are being created since the construction expenses are quite high and it takes a very long time for the product to be returned. One method for enhancing the management of feet is to see the road not as a highway for cars but rather as a highway for information, rather of viewing it as a highway for vehicles. This is one technique for improving the management of feet. The standard techniques of data analysis are struggling to keep up with the amount of data that is being consumed by the transportation system. In addition, the interaction between the numerous components of the system that are responsible for data creation has also contributed to this problem. On the other hand, machine learning is a data-driven solution that is able to satisfy the new system requirements. It is a subset of AI, which is a kind of computer intelligence. The goal of using machine learning is to automate the process of constructing analytical models by gaining a knowledge of the hidden patterns that are included within historical data. This is done in order to imitate the behavior of a system and to respond in the proper manner. When it comes to adapting to ever-increasing needs and requirements in transportation systems, the machine learning-based methodologies have had their relevance rethought as a result of the increased availability of more powerful computing resources and the collection of a vast amount of data. Machine learning solutions have already begun to make their promising imprints in the transportation industry, where it has been shown that they even offer a superior return on investment when compared to the more traditional solutions. Machine learning solutions have already begun to make their promising imprints in the transportation industry. Despite this, there is still a plethora of opportunity to employ and exploit machine learning technologies within the transportation difficulties, which begs for deeper inquiry into the matter. These solutions have as their primary goals the alleviation of traffic congestion, the enhancement of safety and the reduction of human error, the correction of inappropriate environmental impacts, the maximization of energy performance, and the enhancement of the productivity and efficiency of surface transportation. Additionally, these solutions have as their secondary goals the correction of inappropriate environmental impacts. In recent years, the implementation of strategies that are based on machine learning has emerged as a key component in the process of developing intelligent transportation. An improved deep learning model was used to study the complex linkages that exist within this framework between highways, transportation traffic, environmental variables, and traffic accidents. These relationships have been shown to be intertwined with one another. The proposed model incorporates both a supervised module for fine-tuning that is supervised and an unsupervised module for learning features that are not supervised. The supervised module is in charge of carrying out traffic accident prediction, whereas the unsupervised module is responsible for finding the functional connection that exists between the explanatory factors and the feature representations. Another research addresses the problem of "green wave bottleneck point" of the artery and signal interference by using a hierarchical traffic signal management approach that is based on reinforcement learning. This method is used to regulate the traffic signals. Both the prior study and this one were carried out by the same group of researchers. It is not feasible to provide useful guidance on travel if there is not a reliable way for calculating the amount of time required for trips. Travel time may be cut down in a number of different ways. Directly or indirectly. The monitoring of mobile phones, the employment of a probe vehicle, and the preservation of records at toll booths are some of the direct techniques. There are also a range of alternative technological options. When indirect approaches are used, the travel time is approximated by utilizing the observed traffic volume, speed, and occupancy in point sensors (such a loop detector or video camera) that are positioned along the vehicle track. These sensors are located beside the path that the vehicle takes. Data mining and machine learning have received an increased amount of public attention over the course of the last several years. As a result of the development of technologies for the acquisition of traffic information (such as data on global positioning systems (GPS) trajectories), we now have access to a significant amount of traffic data. This, in turn, paves the way for the development of a more accurate method of estimating travel time that is based on data mining. In contrast to traditional parametric models, data mining algorithms are able to study latent connections between variables. This is a significant advantage of the former. In light of this, the aforementioned special issue introduces an innovative way of data mining that makes use of random forests for the goal of computing travel durations. The approach was developed by the authors of the special issue. Random forests have been found to be capable of doing in-depth study on the influence of several parameters on journey durations. In the context of intelligent transport systems (ITS), it is essential to have an accurate prediction of future traffic conditions in order to relieve traffic congestion and respond effectively to traffic incidents. This is because ITS may help prevent and manage traffic problems. In this special issue, there is a presentation of a hybrid method for predicting traffic flow that makes use of the k-nearest neighbor (KNN) algorithm in conjunction with a long short-term memory network. This approach is based on the specialized problem (LSTM). On key roads and expressways in Minnesota, traffic flow prediction at two stations using detector data exemplifies the improved accuracy of traffic flow prediction. These stations are located on significant thoroughfares. Both of these stations may be found inside the greater metropolitan region of the Twin Cities. In a similar vein, unmanned aerial vehicles, also known as UAVs, offer a great deal of promise in terms of daily traffic management and control, in addition to the cost-effective collection of data on the current state of traffic. This is because unmanned aerial vehicles do not require a human pilot to operate them. Unmanned aerial vehicles (UAVs), on the other hand, present their very own distinct challenges, the most significant of which are connected to the size of the vehicles depicted in the photographs and the possibility of occlusion by shadows or other road objects, especially in conditions where there is a great deal of traffic congestion. In this framework, two learning-based frameworks are offered in order to detect automobiles and track the trajectories of vehicles in scenarios when there is a lot of traffic. Statistical machine learning approaches have also found their way into the picture in order to facilitate intelligent transportation. This is done in order to promote it. As part of one of the research studies that were described in the special issue, a random forest model that has been trained to predict with a high degree of accuracy the amount of time that a traffic incident last was included. The Bayesian optimization approach and the neighborhood components analysis were used during the training process for this model. In a different piece of research that focused on the application of statistical machine learning to the field of transportation, a naïve Bayesian classifier was the method of choice for data analysis. This classifier was used to determine the travel pattern of public transport commuters based on the smartcard data as well as the survey data that was acquired from commuters. Both sets of data were collected from riders of the public transit systems.

## 5.5 Accident Avoidance with Artificial Intelligence

Since there are now more cars on the road than ever before, there is also a greater likelihood of accidents happening on highways and in cities everywhere at any given time. One incident may disrupt transportation and logistics systems significantly [1]. Often shortened to "incidents," road traffic incidents (RTIs) are unplanned occurrences on the road network that pose a threat to drivers and passengers and create a halt in traffic flow. Accidents on the road are often referred to as RTIs. Rare but serious infections are occasionally called "events" in particular contexts. There are fundamentally three main types of disruptions to the flow of vehicular traffic: (1) a halt or block on the road; (2) constant or intermittent congestion; and (3) traffic offences. Vehicle traffic might be severely impeded by any one of these conditions. The United States Department of Transportation (DOT) reports that accidents, congestion, and reckless driving are the three most prevalent forms of road incidents [2] and that these three categories of events are closely related to each other [3]. These classes reflect the most likely varieties of vehicular mishaps that may occur [1, 2]. The vast majority of traffic incidents are car accidents, and the World Health Organization (WHO) ranked them as the tenth biggest cause of mortality worldwide in a report released in [4]. Furthermore, it was brought to everyone's attention that reckless driving, which is typically defined as a mental state in which the driver exhibits a wanton disregard for the rules of the road, is the leading cause of accidents; drivers make common driving mistakes like speeding, driving while impaired, and failing to wear safety equipment like helmets and seat belts. It was brought to everyone's notice that careless driving is the major cause of accidents, which is why this data was shared. It was also brought to everyone's attention that reckless driving, which is defined as a state of mind in which one has the attitude that one does not care about the safety of others on the road, is the primary cause of accidents. According to report 5, reckless driving is to blame for one-third of the over 13,000 annual fatalities caused by serious car accidents. Congestion, according to the Federal Highway Administration (FHWA), is another frustrating aspect of driving. When there are more vehicles than usual on a given stretch of roadway, we call that situation congestion. Therefore, travel times lengthen, often by a significant margin compared to normal or "free flow" rates. We may classify traffic jams into two categories: those that occur often due to a lack of road resources and those that occur seldom but are nonetheless substantial. A study [7] on urban mobility conducted by the Texas Transportation Institute (TTI) found that due to traffic congestion, 8.8 billion additional hours of travel time are lost annually, and 3.3 billion gallons of petroleum are wasted. In addition, the average vehicle commuter spends 54 hours more per year on the road than they would otherwise [8]. It is tried to determine whether there was any connection between the several traffic accidents that occurred at the same time and place, and they divided the accidents into two categories: primary and secondary. It has also been shown that the causes of 20%–50% of traffic accidents are the same, and that 50% of secondary accidents occur within 10 minutes after a major collision. To put it another way, if an event site clearance process is delayed by only 1 minute, it will take 2.8% longer to complete the task [10]. Furthermore, because to the interdependence of multiple RTIs, the boundary between distinct occurrences is murky. As a result, both the event type and the number of false alarms (FARs) rise (IT). It has also become harder to gauge the IS (incident severity) associated with RTIs, also known as the risk factor (RF). Implementing an IMS that can quickly and accurately detect the primary incident type (accident, congestion, and reckless driving) and its associated severity (low, medium, and high) and reporting the RTI with its details to the appropriate rescue departments to facilitate timely clearance can help reduce the risk of secondary incidents. This will allow the main event to be resolved more quickly. Existing incident management systems (IMS) for smart cities may be broken down into two groups: public relations (PR) incident management systems and post-incident management systems [11]. Managing before an issue occurs involves looking at potential causes and solutions. These devices are meant to alert drivers of approaching danger in the case of a collision. There are several assumptions that must be made before these systems may be certified as "zero occurrences" by the International Transport Forum (ITF) [12]. The absence of any accidents is the pinnacle of perfection, yet it is very difficult to attain due to the need for the whole of the system to function without flaw and in perfect synchrony at all times. In contrast, a traffic incident management system (TIMS) [13] has all the features necessary to handle traffic incidents after they have occurred. Core features include finding problems, reporting them, and fixing them. Specifically, it performs these actions. Automatic incident detection (AID) systems [14] employ parametric data acquired from on-road and in-vehicle sensing equipment to advance from nonautomatic incident detection (NID) systems (which depended on mobile phone calls from eyewitnesses) to identify incidents. These early, nonautomatic event detection systems were dependent on cell phone reports from witnesses. These technologies have progressed into what is now known as the automatic incident detection module. Next, algorithms for comparison, statistics, time series, and machine learning are used to the data to generate traffic patterns that are then used to classify events [15]. The reporting component alerts the relevant entities, such as medical facilities, insurance providers, police enforcement, the TCR, and nearby vehicles. The following are some examples of such groups: The incident sight must be cleared, which necessitates the assignment of tasks to the cleaning module [16, 17]. Ambulances, emergency teams, and updated signage are just few examples of what may be done (CMSs). This work is being done to restore the area to its pre-event conditions.

## 5.6 Active Alert System Design with Artificial Intelligence

Because of the extraordinary rate at which the world's population is growing, there will be an increased need for a diverse range of transportation choices. The number of people killed or injured in motor vehicle collisions has skyrocketed in recent years, which is very troubling and calls for immediate action. According to the findings of an exhaustive research that was carried out by the WHO, vehicle crashes come in at number seven on the list of the main causes of death around the globe. Millions of people lose their lives on a yearly basis as a result of these accidents. According to projections made by the Association for Safe International Road Travel (ASIRT), road accidents will eventually pass murder to become the fifth greatest cause of death in the world during the next few decades [1]. We were able to assemble this data as a result of the results of an inquiry [2]. The American Society for Injury Research and Prevention estimates that the expenditures connected with traffic crashes account for between 1% and 2% of national budgets globally. According to research that was carried out in 2019 by the Transport Research Wing (TRW) of India, there is an average of 1230 vehicle accidents that take place every single day in India. These accidents directly result in the deaths of 414 persons. According to the findings of this research, there were 51 accidents and 17 deaths for every hour of the year 2019 [2]. The roads of today are crowded with careless drivers who disobey traffic regulations in order to give in to their own whims and wishes, putting the lives of innocent bystanders in peril as a result. Accidents may be caused by a wide range of different things, including but not limited to driving at an unsafely fast speed, passing another vehicle at an unsafely close distance, using a mobile phone, driving in conditions with poor vision, and driving in conditions with adverse weather. Recent studies [2, 3] have revealed that speeding is the leading contributor to accidents that occur on the road. It is very necessary for the efficacy of the search and rescue operation [4, 5] to have accurate information about the location of the accident site. If the location of the event is inside a metropolitan area, even if there is heavy traffic, it will not take long for emergency personnel to reach at the scene. However, it might be difficult to give rapid emergency care in areas where there is minimal foot traffic or in close proximity to major thoroughfares. It has been seen that there may be significant delays in receiving medical assistance for catastrophic injuries, which can result in patients not making it when they should have. The amount of time it takes for an ambulance to arrive at the site of an accident, perform first aid, and transfer the patient to a hospital is strongly connected to the survival rates of the people who were injured in the accident. Both the business world and the academic world have seen an increase in the amount of interest they have in improving road safety as a direct consequence of the continually growing number of automobile collisions. It is essential to have a comprehensive system for identifying and warning of approaching collisions in order to cut down on the number of lives that are lost as a result of vehicular crashes. As soon as it is determined that there has been an accident, it will communicate this information to all of the emergency services that are located nearby. This includes hospitals, police stations, and fire departments. Over the course of the last several years, a wide variety of novel techniques to accident detection and warning have been put into practice. GPS is an acronym for "global positioning system," which describes the bulk of the systems that are now used to monitor vehicles. Some automobiles even come equipped with GPS systems that are factory-installed, allowing the driver to monitor their exact position and transmit that data to a distant server [6]. e-Notify, which is one of the most prevalent accident warning systems, mandates that an on-board unit, also known as an OBU, be placed in each and every vehicle [7, 8]. eCall is a mandatory safety feature that was developed by the European Commission and is required to be included in all newly built automobiles starting in the year 2015. When it is determined that an accident has occurred, the eCall system contacts the appropriate emergency services by calling the number 112 [9]. This number is comparable to 109 in the United States and 108 in India. Cloud computing [10] and big data [11–13] have made it possible to manage the massive data streams that are generated by sources such as satellites, radar, GPS [14–16] and cloud computing] [10, 14–16]. There has never been anything quite like this before [10–16]. The world we live in now is so technologically advanced that it is even possible, in principle, to monitor the whereabouts of each and every one of its constituent parts by using a GPS. In recent years, Internet of Things (IoT) and deep learning-based technologies have seen widespread use in the process of creating solutions for improving road safety in smart cities. Scholars have focused a lot of emphasis on the urgent problem of how to correctly identify and contact persons who have been involved in vehicle accidents because of the difficulties it presents. Over the last several decades, many accident detection and warning systems have been created. The methods used to locate and alert people to accidents may generally be divided into the following three categories: Hardware-based systems [22–26], an integrated accident detection system, and technology that employs cellphones as its nerve center [17–21].

Methodologies Developed Considering Mobile Devices 2.1 These programs largely depend on a smartphone and its assortment of built-in sensors to identify and record traffic events, which are subsequently sent to authorities. Global positioning systems and accelerometer data were used in the analysis outlined in Reference [17] to pinpoint the exact locations of the events under study.

This method does not make contact with emergency services; it is just effective for determining the existence of an accident. An accident detection system is given in the research in Reference [18]. This device uses a GPS module to locate an occurrence and sends text messages to surrounding hospitals alerting medical staff. The study's lone significant weakness was that it only used information from one sensor. Therefore, system operation will not be feasible if the sensor does not perform as anticipated. Another drawback is that it can result in more false alarms. The results of a study reported in Reference [19] suggest that mobile phones might be used to assist detect and report accidents. When deciding whether or not an automobile accident has occurred, G-force, noise level, and velocity of motion are all taken into consideration. The extraction of noise and gravitational force uses both the mobile sensor and the accelerometer (also known as a microphone). A call to the emergency number was made as soon as the situation was evaluated. Given that one of the most important factors in determining accidents is speed, a low-speed collision may increase the likelihood of false alerts from this device. An Android app that can identify problems and provide alerts when they happen was created by staff members in Reference [20]. Without a GPS module and a mobile sensor that monitored acceleration, they were unable to establish beyond a reasonable doubt that an accident had taken place. When an issue is detected, it promptly phones an emergency number and plays a personalized audio message (108). The study's main flaw is the use of only one sensor by the researchers. As a result, the whole system is worthless if the accelerometer sensor malfunctions, which is highly likely. An Android app was developed to identify when an accident would occur using just the smartphone's accelerometer and GPS based on the results presented in Reference [21]. The app will alert neighboring users and emergency services of the location of the issue in the case of an accident. This is the software's most important feature among its many beneficial ones. They only have one sensor to depend on as well. The smartphone-based system's dependency on smartphone technology is its biggest shortcoming. Each user will need a smartphone that is pre-installed with an accident detection app. The precision of the system depends only on the quality of the mobile sensor. At moderate speeds, the possibility of a false collision detection by the car rises. The automobile has more time to react, which is why this is the case. Techniques Relying on Physical Materials 2.2 Different accident detection systems employ a variety of sensor types to find occurrences and notify the proper persons. These sensors are often positioned on the outside of the car. A system for the early detection of automotive crashes is presented in Reference [22] and includes a vibration sensor and a microcontroller. This system cannot work without both components. The Atmel microprocessor AT89S52 is connected to the sensors, Global System for Mobile Communications (GSM), GPS, and any other modules that could be present. Most of the time, sensors will assess the data, which might include acceleration, force, vibration, and other similar occurrences, and then send the alert to all of the recorded emergency numbers until the disaster's location is identified. This process will continue until the specific site of the tragedy is discovered. At the moment, this technology cannot be used to support the SAR operations. The authors of the research published in Reference [23] developed a method to assist rescuers find and aid persons in need using the IoT and sensors like a GPS tracker and accelerometer to gather data that is then uploaded to the cloud in the case of an accident. The victim's emergency contacts will always be notified when a calamity strikes. In order for the ambulance to get to the accident site as quickly as possible, the message includes information on the accident's severity and location. This notice also contains the exact location of the occurrence. The study referenced under Reference [24] sought to develop an IoT-based system for monitoring and reacting to auto accidents. In the course of this work, a vibration sensor, a buzzer, and a GPS module are used to facilitate accident reporting and detection. A bell will ring one minute after the accident scene has been found. The buzzer will automatically send an online notice to the necessary emergency contacts after one minute. The IoT-based system for early accident detection and warning, similar to the one described in Reference [24], was demonstrated by researchers, according to Reference [25]. When an accident is discovered, customer service is informed of the details, and a representative then gets in touch with the closest hospital. This is the study's main qualification. Human participation is necessary for success. The notion of a completely automated system for accident detection and conducting rescue operations was proposed by the study findings given in Reference [26]. It would need a GPS element and a collision detection sensor to construct this system. The biggest disadvantage of an IoT-based system in comparison to a smartphone-based system is its high cost. The major issue with these sorts of systems is that. The accuracy of the system may be significantly increased by using a few different AI-based techniques.

Accident-Prevention and Detection Monitoring System in Its Entirety 2.3 To locate and rescue accidents, these specific ADRS employ a hybrid system that combines the IoT with deep learning. A unified system for tracking and reporting vehicle crashes would likely be advantageous, according to the evidence given in Reference [27]. An on-board unit, or OBU, would be used in this system to perform the aforementioned tasks. Several sensors gather and upload accident data, including the location, speed, kind of vehicle, and airbag deployment status, to a central database. Speed and the kind of vehicle may be noted, among other things. After that, it use machine learning and other cutting-edge techniques to accurately predict the accident. The main disadvantage of the method is that not all automobiles can run with an OBU. Deep learning was used in the study reported in Reference [26] to create a system for the quick identification of accidents and the rescue of victims. Now, every car has a built- in Wi-Fi web camera. This camera continuously monitors every vehicle on the road and routinely sends its observations to the cloud. A deep neural network is used to assess the video while it is being processed in the cloud. The findings are communicated to the control room next door if an accident is verified. Convolutional neural networks (CNN) are proposed as a method for accident detection in Reference [28], which can be found in the aforementioned literature review. In this paradigm, the accident group and the other group have distinct names. Convolutional neural networks may categorize live camera footage as "accident" or "non-accident." Real-time video feed from the camera is continuously sent to the server. They've chosen and categorized YouTube videos to create their own dataset for the model training. The authors claim that their accident detection approach has a 95% success rate. This device only has the ability to identify accidents; it is unable to take into consideration any attempts to save lives. A computer-vision-based accident detection and rescue system was created as a result of the study detailed in Reference [29]. It used a previously trained Yolo v3 model to identify events in real time. The Yolo model has been taught to recognize accidents under a range of circumstances, such as inclement weather, poor vision, etc. They expanded the dataset by experimenting with several photo enhancing methods. Real-time video feed from a camera positioned on the road is used to input data to the model. According to the study's author, the model has a 92.5% accuracy rate and a 7.5% false alarm rate. In the study published in Reference [30], IoT and machine learning were employed to detect accidents and activate the rescue module. They begin by gathering pertinent information about the event using the IoT toolset, which is then uploaded to the server for analysis. In this option, data processing takes place locally via fog computing, as opposed to being sent to the cloud or a server. However, the offered video data is probably too complex for the outcomes to be deemed satisfactory. This method is particularly good at detecting accidents because it uses machine learning. The development and publication of an ensemble deep learning approach based on the detection of accidents in dashboard camera data is found in Reference [31]. Once the data from the dashboard camera has been collected, this technique uses the gated recurrent unit (GRU) and the CNN to detect and characterize the accident. There are many applications that combine these two types of networks. The camera must be positioned on the dashboard, where it is more likely to be damaged in the case of an accident. This is the method's most evident disadvantage. This approach may produce more false positives due to its exclusive reliance on the camera and lack of IoT modules. This work, which was published in Reference [32], offers an automated approach for locating traffic accidents. Convolutional neural networks (CNN) and support vector machines are used in this technique to forecast the likelihood of accidents (SVM). The most important qualities from the whole pool of attributes were chosen for this research using a variety of feature selection strategies. According to the author, during the project's testing phase, they were able to achieve accuracy levels of up to 85%, which is too low for use in practical applications. The IoT component is also dissected into its component pieces. A. In order to determine the most significant factors contributing to traffic accidents and the recurrent shapes these factors took, a number of data mining approaches were used in the study detailed in Reference [33]. To complete the research, the author gathered data on auto accidents that occurred in each of the 15 wards of Rome Municipality between 2016 and 2019. The authors come to the conclusion that neural networks and decision trees are better suited for predictive analysis, while Kohonen networks and k-means algorithms are more suited for descriptive analysis. It have reached this conclusion as a consequence of the study. The research described in Reference [34] employed data mining methods on the FARS Fatal Accident dataset to identify the kind of relationship between the accident's cause and its environment. The a priori technique is used to create an association rule between the qualities and the aspects that are significant contributors to the accident. Drunk driving, bad weather, slippery roads, poor visibility, and a crowded roadway are all potential accident-causing variables. The following variables further aggravate this catastrophe: The clusters are then constructed using k-means clustering, and a classification model is built using a Naive

Bayes classifier. The concept being presented includes suggestions for safe driving. In Reference [35], a very analogous investigation of bus accidents took place. Between 2006 and 2019, they gathered data on every bus incident that took place in Victoria, Australia. A sophisticated set of association rules was established to identify the precise causes of bus accidents. Work in Reference [36] proposed a mechanism that could determine the frequency and velocity of accidents that a particular vehicle had experienced. This approach was invented for the software called "path analysis" is used to find this connection. Because the text clearly demonstrates the connection between the two circumstances, the author draws the conclusion that the car's high rate of speed was a contributing factor in the collision. Having fast access to a huge quantity of high-quality data for both training and testing is essential for creating effective deep learning models. The overwhelming majority of integrated accident detection systems make use of strategically placed cameras along highways. These cameras are positioned in key areas to capture occurrences as they happen. These cameras continuously monitor the flow of traffic and send their observations to the cloud [37]. Deep learning algorithms are used to examine the data to determine if an event was captured on camera. If the team wants to complete its study, they will need to get beyond the dataset's accident-related inaccessibility. The quality of the camera and a wide range of other environmental conditions, such as poor visibility, precipitation, fog, and so forth, have a significant impact on accuracy. Additionally, it could be quite challenging to keep roadside cameras operating properly. A deep neural network, more often referred to as CNN, was used to train the model, which accounts for the bulk of the progress gained thus far. Since there is not enough data to create a model from scratch, the accuracy of the accident detection procedure may be increased by using a pretrained model. This is because there is not currently enough information available for different scenarios.

## References

1. H. I. Ashqar, M. H. Almannaa, M. Elhenawy, H. A. Rakha and L. House, "Smartphone Transportation Mode Recognition Using a Hierarchical Machine Learning Classifier and Pooled Features From Time and Frequency Domains," in *IEEE Trans. Intell. Transp. Syst.*, vol. 20, no. 1, pp. 244-252, Jan. 2019, doi: 10.1109/TITS.2018.2817658.
2. H. Liu, Y. Tong, J. Han, P. Zhang, X. Lu and H. Xiong, "Incorporating Multi-Source Urban Data for Personalized and Context-Aware Multi-Modal

Transportation Recommendation," in *IEEE Trans. Knowl. Data Eng.*, vol. 34, no. 2, pp. 723-735, 1 Feb. 2022, doi: 10.1109/TKDE.2020.2985954.
3. D. Jackson, S. Belakaria, Y. Cao, J. R. Doppa and X. Lu, "Machine Learning Enabled Design Automation and Multi-Objective Optimization for Electric Transportation Power Systems," in *IEEE Trans. Transp. Electrif.*, vol. 8, no. 1, pp. 1467- 1481, March 2022, doi: 10.1109/TTE.2021.3113958.
4. D. Yang, L. Zhu, Y. Liu, D. Wu and B. Ran, "A Novel Car-Following Control Model Combining Machine Learning and Kinematics Models for Automated Vehicles," in *IEEE Trans. Intell. Transp. Syst.*, vol. 20, no. 6, pp. 1991-2000, June 2019, doi: 10.1109/TITS.2018.2854827.
5. L. Yang, M. Li, C. Shen, Q. Hu, J. Wen and S. Xu, "Discriminative Transfer Learning for Driving Pattern Recognition in Unlabeled Scenes," in *IEEE Trans. Cybern.*, vol. 52, no. 3, pp. 1429-1442, March 2022, doi: 10.1109/TCYB.2020.2987632.
6. M. Veres and M. Moussa, "Deep Learning for Intelligent Transportation Systems: A Survey of Emerging Trends," in *IEEE Trans. Intell. Transp. Syst.*, vol. 21, no. 8, pp. 3152-3168, Aug. 2020, doi: 10.1109/TITS.2019.2929020.
7. C. Chen, B. Liu, S. Wan, P. Qiao and Q. Pei, "An Edge Traffic Flow Detection Scheme Based on Deep Learning in an Intelligent Transportation System," in *IEEE Trans. Intell. Transp. Syst.*, vol. 22, no. 3, pp. 1840-1852, March 2021, doi: 10.1109/TITS.2020.3025687.
8. S. Shahriar, A. R. Al-Ali, A. H. Osman, S. Dhou and M. Nijim, "Prediction of EV Charging Behavior Using Machine Learning," in *IEEE Access*, vol. 9, pp. 111576-111586, 2021, doi: 10.1109/ACCESS.2021.3103119.
9. Y. Kan, Y. Wang, D. Wang, J. Sun, C. Shao and M. Papageorgiou, "A Novel Approach to Estimating Missing Pairs of On/Off Ramp Flows," in *IEEE Trans. Intell. Transp. Syst.*, vol. 22, no. 2, pp. 1287-1305, Feb. 2021, doi: 10.1109/TITS.2020.2989365.
10. M. Usman, M. A. Jan and A. Jolfaei, "SPEED: A Deep Learning Assisted Privacy- Preserved Framework for Intelligent Transportation Systems," in *IEEE Trans. Intell. Transp. Syst.*, vol. 22, no. 7, pp. 4376-4384, July 2021, doi: 10.1109/TITS.2020.3031721.
11. N. Kumar, S. S. Rahman and N. Dhakad, "Fuzzy Inference Enabled Deep Reinforcement Learning-Based Traffic Light Control for Intelligent Transportation System," in *IEEE Trans. Intell. Transp. Syst.*, vol. 22, no. 8, pp. 4919-4928, Aug. 2021, doi: 10.1109/TITS.2020.2984033.
12. N. S. Hadjidimitriou, M. Lippi, M. Dell'Amico and A. Skiera, "Machine Learning for Severity Classification of Accidents Involving Powered Two Wheelers," in *IEEE Trans. Intell. Transp. Syst.*, vol. 21, no. 10, pp. 4308-4317, Oct. 2020, doi: 10.1109/TITS.2019.2939624.
13. H. Yi and K. -H. N. Bui, "An Automated Hyperparameter Search-Based Deep Learning Model for Highway Traffic Prediction," in *IEEE Trans. Intell. Transp. Syst.*, vol. 22, no. 9, pp. 5486-5495, Sept. 2021, doi: 10.1109/TITS.2020.2987614.

14. F. Jaffar, T. Farid, M. Sajid, Y. Ayaz and M. J. Khan, "Prediction of Drag Force on Vehicles in a Platoon Configuration Using Machine Learning," in *IEEE Access*, vol. 8, pp. 201823-201834, 2020, doi: 10.1109/ACCESS.2020.3035318.
15. J. Li, X. Pei, X. Wang, D. Yao, Y. Zhang and Y. Yue, "Transportation mode identification with GPS trajectory data and GIS information," in *Tsinghua Sci. Technol.*, vol. 26, no. 4, pp. 403-416, Aug. 2021, doi: 10.26599/TST.2020.9010014.
16. Y. Wu *et al.*, "Person Reidentification by Multiscale Feature Representation Learning With Random Batch Feature Mask," in *IEEE Trans. Cognit. Dev. Syst.*, vol. 13, no. 4, pp. 865-874, Dec. 2021, doi: 10.1109/TCDS.2020.3003674.
17. B. Sliwa, R. Falkenberg, T. Liebig, N. Piatkowski and C. Wietfeld, "Boosting Vehicle- to-Cloud Communication by Machine Learning-Enabled Context Prediction," in *IEEE Trans. Intell. Transp. Syst.*, vol. 21, no. 8, pp. 3497-3512, Aug. 2020, doi: 10.1109/TITS.2019.2930109.
18. T. Kim, L. F. Vecchietti, K. Choi, S. Lee and D. Har, "Machine Learning for Advanced Wireless Sensor Networks: A Review," in *IEEE Sens. J.*, vol. 21, no. 11, pp. 12379- 12397, 1 June1, 2021, doi: 10.1109/JSEN.2020.3035846.
19. Y. Qin, H. Luo, F. Zhao, C. Wang, J. Wang and Y. Zhang, "Toward Transportation Mode Recognition Using Deep Convolutional and Long Short-Term Memory Recurrent Neural Networks," in *IEEE Access*, vol. 7, pp. 142353-142367, 2019, doi: 10.1109/ACCESS.2019.2944686.
20. X. Kong, Y. Zhang, W. L. Eisele and X. Xiao, "Using an Interpretable Machine Learning Framework to Understand the Relationship of Mobility and Reliability Indices on Truck Drivers' Route Choices," in *IEEE Trans. Intell. Transp. Syst.*, vol. 23, no. 8, pp. 13419-13428, Aug. 2022, doi: 10.1109/TITS.2021.3124221.
21. N. O. Alsrehin, A. F. Klaib and A. Magableh, "Intelligent Transportation and Control Systems Using Data Mining and Machine Learning Techniques: A Comprehensive Study," in *IEEE Access*, vol. 7, pp. 49830-49857, 2019, doi: 10.1109/ACCESS.2019.2909114.
22. L. Zhu, F. R. Yu, Y. Wang, B. Ning and T. Tang, "Big Data Analytics in Intelligent Transportation Systems: A Survey," in *IEEE Trans. Intell. Transp. Syst.*, vol. 20, no. 1, pp. 383-398, Jan. 2019, doi: 10.1109/TITS.2018.2815678.
23. J. Casavílca Silva *et al.*, "Light-Field Imaging Reconstruction Using Deep Learning Enabling Intelligent Autonomous Transportation System," in *IEEE Trans. Intell. Transp. Syst.*, vol. 23, no. 2, pp. 1587-1595, Feb. 2022, doi: 10.1109/TITS.2021.3079644.
24. D. Panovski and T. Zaharia, "Long and Short-Term Bus Arrival Time Prediction With Traffic Density Matrix," in *IEEE Access*, vol. 8, pp. 226267-226284, 2020, doi: 10.1109/ACCESS.2020.3044173.
25. A. Haydari and Y. Yılmaz, "Deep Reinforcement Learning for Intelligent Transportation Systems: A Survey," in *IEEE Trans. Intell. Transp. Syst.*, vol. 23, no. 1, pp. 11-32, Jan. 2022, doi: 10.1109/TITS.2020.3008612.

26. T. Mao, A. -S. Mihăită, F. Chen and H. L. Vu, "Boosted Genetic Algorithm Using Machine Learning for Traffic Control Optimization," in *IEEE Trans. Intell. Transp. Syst.*, vol. 23, no. 7, pp. 7112-7141, July 2022, doi: 10.1109/TITS.2021.3066958.
27. J. Zhao, L. Zhang, J. Ye and C. Xu, "MDLF: A Multi-View-Based Deep Learning Framework for Individual Trip Destination Prediction in Public Transportation Systems," in *IEEE Trans. Intell. Transp. Syst.*, vol. 23, no. 8, pp. 13316- 13329, Aug. 2022, doi: 10.1109/TITS.2021.3123342.
28. X. Ma, H. Zhong, Y. Li, J. Ma, Z. Cui and Y. Wang, "Forecasting Transportation Network Speed Using Deep Capsule Networks With Nested LSTM Models," in *IEEE Trans. Intell. Transp. Syst.*, vol. 22, no. 8, pp. 4813-4824, Aug. 2021, doi: 10.1109/TITS.2020.2984813.
29. Y. Alghorani, A. S. Chekkouri, D. A. Chekired and S. Pierre, "Improved S-AF and S- DF Relaying Schemes Using Machine Learning Based Power Allocation Over Cascaded Rayleigh Fading Channels," in *IEEE Trans. Intell. Transp. Syst.*, vol. 22, no. 12, pp. 7508-7520, Dec. 2021, doi: 10.1109/TITS.2020.3003820.
30. A. Aksjonov, P. Nedoma, V. Vodovozov, E. Petlenkov and M. Herrmann, "Detection and Evaluation of Driver Distraction Using Machine Learning and Fuzzy Logic," in *IEEE Trans. Intell. Transp. Syst.*, vol. 20, no. 6, pp. 2048-2059, June 2019, doi: 10.1109/TITS.2018.2857222.
31. N. Marković, P. Sekuła, Z. Vander Laan, G. Andrienko and N. Andrienko, "Applications of Trajectory Data From the Perspective of a Road Transportation Agency: Literature Review and Maryland Case Study," in *IEEE Trans. Intell. Transp. Syst.*, vol. 20, no. 5, pp. 1858-1869, May 2019, doi: 10.1109/TITS.2018.2843298.
32. S. Jain, M. Singhal and S. Shukla, "Comments on "Traffic Sign Recognition Using Kernel Extreme Learning Machines With Deep Perceptual Features"," in *IEEE Trans. Intell. Transp. Syst.*, vol. 20, no. 10, pp. 3759-3761, Oct. 2019, doi: 10.1109/TITS.2018.2850057.
33. S. Aradi, "Survey of Deep Reinforcement Learning for Motion Planning of Autonomous Vehicles," in *IEEE Trans. Intell. Transp. Syst.*, vol. 23, no. 2, pp. 740-759, Feb. 2022, doi: 10.1109/TITS.2020.3024655.
34. R. Zhang, A. Ishikawa, W. Wang, B. Striner and O. K. Tonguz, "Using Reinforcement Learning With Partial Vehicle Detection for Intelligent Traffic Signal Control," in *IEEE Trans. Intell. Transp. Syst.*, vol. 22, no. 1, pp. 404-415, Jan. 2021, doi: 10.1109/TITS.2019.2958859.
35. P. He, A. Wu, X. Huang, J. Scott, A. Rangarajan and S. Ranka, "Truck and Trailer Classification With Deep Learning Based Geometric Features," in *IEEE Trans. Intell. Transp. Syst.*, vol. 22, no. 12, pp. 7782-7791, Dec. 2021, doi: 10.1109/TITS.2020.3009254.

36. A. Qayyum, M. Usama, J. Qadir and A. Al-Fuqaha, "Securing Connected & Autonomous Vehicles: Challenges Posed by Adversarial Machine Learning and the Way Forward," in *IEEE Commun. Surv. Tutorials*, vol. 22, no. 2, pp. 998-1026, Secondquarter 2020, doi: 10.1109/COMST.2020.2975048.
37. Z. Wang, R. Chu, M. Zhang, X. Wang and S. Luan, "An Improved Selective Ensemble Learning Method for Highway Traffic Flow State Identification", in *IEEE Access*, vol. 8, pp. 212623-212634, 2020, doi: 10.1109/ACCESS.2020.3038801.

# 6

# Introduction of Signal Processing for Sustainable Transport


***Abstract***

This chapter aims to provide a thorough examination of signal processing methods and uses in relation to environmentally friendly transportation systems. Due to growing concerns about environmental degradation, energy consumption, and traffic congestion, sustainable transport has become a crucial area of research and development. The effectiveness, safety, and overall sustainability of transportation systems are greatly improved by signal processing, a potent tool for signal analysis and manipulation.

This chapter starts off with an overview of sustainable transport, highlighting the major difficulties and goals in this area. It talks about how these problems require data-driven, intelligent strategies to be solved. The creation of intelligent transportation systems is made possible by signal processing techniques, which offer a way to extract useful information from a variety of sources, including sensors, cameras, and communication networks.

Later in the chapter, the basics of signal processing are covered, including key ideas like signal representation, filtering, feature extraction, and classification. It examines various signal processing techniques, such as Fourier transform, wavelet analysis, and time-frequency analysis, which are frequently used in transportation applications. The discussion focuses on their applicability and efficiency in deciphering and analysing signals related to transportation, such as data on traffic flow, vehicle sensor data, and environmental sensor datasets and samples.

The chapter also examines particular signal processing applications in environmentally friendly transportation systems. In order to improve traffic flow and use less energy, it looks at how signal processing techniques can be used for traffic management, congestion detection, and prediction. The chapter also looks at how signal processing is used in intelligent transportation systems to improve coordination and safety through vehicle-to-vehicle and vehicle-to- infrastructure communication.

The chapter also discusses the difficulties and constraints of signal processing in sustainable transportation. It discusses concerns about privacy, real-time processing, and data quality and offers potential solutions and future research directions to address these issues.



Kundan Meshram. Urban Transportation Systems: Sustainability and Future Development, (127–142) © 2024 Scrivener Publishing LLC

In conclusion, this chapter offers a thorough overview of signal processing methods and how they are used in environmentally friendly transportation systems. It draws attention to the role signal processing plays in addressing the difficulties associated with sustainable transportation while enhancing effectiveness, safety, and environmental sustainability levels. This chapter lays the groundwork for future investigation and study in the fascinating area of signal processing for environmentally friendly transportations [1–3].

***Keywords:*** Signal processing, linear time-invariant, image processing, traffic flow

## 6.1 Introduction

Due to the urgent need for effective, environmentally friendly, and secure transport systems, the field of sustainable transport has attracted a lot of attention recently. Signal processing is one discipline that must be integrated in order to achieve sustainability in transportation, and it is one of those disciplines that is essential to this process. The development of intelligent transportation systems and the ability to extract useful information from transportation data are both made possible by signal processing, which includes both image and signal analysis techniques.

With a focus on its uses in urban transportation systems, this chapter offers a thorough introduction to signal processing for sustainable transportation. It starts off by giving a general overview of signal processing and how it relates to sustainable transportation. The use of signal processing techniques enables the real-time monitoring, intelligent control, and optimisation of transportation systems by processing and analysing data from a variety of sources, including sensors, cameras, and communication networks.

The basics of image processing, a crucial component of signal processing, are then covered in the chapter because they are crucial for analysing the visual data obtained by cameras or other imaging devices in urban transportation systems. Important ideas like image representation, enhancement, filtering, feature extraction, and classification are covered in the discussion. It is essential to comprehend these fundamentals in order to effectively use image processing techniques to address transportation-related issues.

The chapter also explores one-dimensional (1-D), two-dimensional (2-D), and three-dimensional (3-D) fundamental signal characteristics. We can understand the complexities and intricacies involved in signal processing for sustainable transport by looking at various types of signals encountered in transportation systems, such as traffic flow data, vehicle sensor data, and environmental sensor datasets and samples.

System classification is yet another important topic covered in this chapter. In order to analyse and model the behaviour of transportation systems, it is essential to have a basic understanding of how systems are classified, especially linear time-invariant (LTI) and linear shift-invariant (LSI) systems. This information forms the basis for developing signal processing techniques and algorithms that are suitable for handling the dynamics and peculiarities of urban transport systems.

The use of image processing methods in urban transport systems is then highlighted in the chapter. Traffic monitoring, vehicle detection, licence plate recognition, traffic sign recognition, and surveillance systems all rely heavily on image processing. Traffic management authorities can make informed decisions in real-time and optimise traffic flow for increased efficiency and safety by utilising image processing algorithms to extract pertinent information from images and video streams.

With an emphasis on urban transportation systems, this chapter concludes by offering a thorough introduction to signal processing for sustainable transportation. The chapter lays the groundwork for understanding the subsequent application of signal processing techniques in the context of sustainable transport by looking at the fundamentals of signal processing, image processing, and the characteristics of various signals encountered in transportation. This chapter is a helpful resource for academics, professionals, and decision-makers who want to use signal processing to overcome the limitations of urban transportation systems and find sustainable transportation solutions.

## 6.2 Signal Processing Overview

The analysis, manipulation, and interpretation of signals are the focus of the fundamental engineering and mathematical discipline of signal processing. Any type of information that varies with time, space, or any other independent variable is referred to in this context as a signal. Techniques for signal processing make it possible to extract useful data, patterns, and features from signals, enabling decision-making in a variety of areas, including communications, image processing, audio processing, and, most significantly, transportation systems.

Signal processing is essential for addressing the complex issues related to transportation efficiency, safety, and environmental sustainability in the context of sustainable transportation. Signal processing techniques enable the development of intelligent transportation systems that can enhance traffic management, optimise energy consumption, and improve overall transportation performance by processing and analysing signals obtained from various sources such as sensors, cameras, and communication networks [4, 5].

In order to understand and interpret underlying phenomena or to make predictions, signal processing's main goal is to extract meaningful information from signals. Signal processing techniques use a number of crucial steps, such as signal acquisition, representation, filtering, feature extraction, and classification, to accomplish this.

Signal acquisition is the procedure of obtaining signals through the use of sensors or other measurement tools. Signals can be gathered from a variety of sources in transportation systems, including traffic flow sensors, vehicle detectors, environmental sensors, and video cameras. These signals could consist of information on vehicle speed and occupancy, pollution levels, and traffic flow. The forecasting and maintenance operations is given in Figure 6.1.

**Figure 6.1** Forecasting and maintenance operations.

Signal representation entails converting the acquired signals into a format that is appropriate for processing and analysis. The representation of a signal can be 1-D, 2-D, or 3-D, depending on the type of signal. While video streams from traffic cameras are typically represented as 2-D images or 3-D video sequences, traffic flow data that is collected at a specific location can be represented as a 1-D time series. In order to enhance or extract certain signal components while reducing unwanted noise or interference, filtering is a crucial step in the signal processing process. Techniques for filtering data can be used to reduce noise in sensor measurements, smooth out fluctuations in data on traffic flow, or pick out important details in pictures or videos taken by transportation systems. The process of feature extraction entails locating and separating pertinent data or patterns from signals. This can be done in transportation systems by identifying patterns of traffic congestion, detecting vehicle movements, or extracting particular traits from visual data, such as object detection or licence plate recognition. By effectively representing and categorising signals, feature extraction paves the way for additional analysis and decision-making.

Assigning signals or signal segments to predetermined categories or classes is a fundamental component of signal processing. Classification algorithms can be used to automatically recognise and categorise various traffic conditions, pinpoint unusual occurrences, or differentiate between various vehicle types. Classification is essential to traffic management because it enables intelligent systems to make decisions in real time based on the signals that have been analysed.

Finally, signal processing offers the methods and tools required for deciphering and extracting useful data from signals acquired in transportation systems. Sustainable transport systems can be made more effective, safe, and environmentally sustainable by utilising signal processing techniques like signal acquisition, representation, filtering, feature extraction, and classification scenarios. The use of signal processing in transportation helps to optimise traffic flow, lessen congestion, use less energy, and improve all aspects of transportation performance levels.

## 6.3 Fundamentals of Image Processing

Specifically focusing on the analysis, manipulation, and interpretation of visual data captured by cameras or other imaging devices, image processing is an essential part of signal processing.

Image processing techniques are essential for extracting useful information from visual data in the context of sustainable transportation in order to improve the effectiveness, safety, and sustainability of urban transportation systems.

The foundations of image processing cover a variety of ideas and methods that make it possible to extract useful information from images like traffic scenes, security footage, or aerial imagery. To effectively use image processing techniques in the context of transportation systems, it is crucial to comprehend these fundamentals [6].

Image representation: image representation entails the conversion of visual data into a form that can be processed and analysed. Images are represented as arrays of pixels in digital image processing, where each pixel corresponds to a distinct point in the image and contains data about its colour or intensity. The number of pixels in an image determines its resolution, with higher resolutions offering more accurate representations.

Image enhancement techniques work to enhance an image's visual appeal or extract certain features to make it easier to understand. By adjusting the image's contrast, brightness, or colour balance, enhancement techniques can improve the image's visual appeal or draw attention to certain details. Image enhancement in transportation systems can help increase object visibility, facilitate better traffic flow analysis, or improve surveillance capabilities.

Image filtering: to achieve desired effects, image filtering involves applying mathematical operations to an image. Spatial filters, which alter each pixel in an image based on its surrounding pixels, and frequency domain filters, which alter the frequency content of an image using methods like Fourier analysis, are examples of common filtering techniques. In transportation applications, filtering can be used for noise reduction, edge detection, or image smoothing.

Image feature extraction is the process of locating and extracting pertinent visual cues or patterns from an image. Objects, textures, shapes, or other visual characteristics that convey significant information can be represented by these features. Techniques for feature extraction can be applied to camera feeds to analyse traffic flow patterns, identify and track moving objects, and detect road signs or traffic signals.

Image segmentation is the process of breaking an image down into useful areas or objects. To distinguish between various regions of interest, it attempts to group pixels with similar characteristics. A traffic scene's vehicles can be separated from the background using segmentation techniques, as can the boundaries of roads and the foreground and background of objects in surveillance footage. The analysis and classification tasks that follow require accurate image segmentation.

Image classification entails grouping images or portions of images into predefined categories or classes in accordance with their visual content. It is possible to train classification algorithms, like machine learning or pattern recognition methods, to identify and classify particular objects, traffic situations, or events captured in images. Applications like vehicle recognition, traffic sign detection, or anomaly detection in transportation systems all benefit from image classification.

The foundational concepts of image processing offer a basis for visual data analysis and interpretation in the context of sustainable transportation. Researchers and practitioners can effectively use image processing techniques to extract useful information from images captured in transportation systems by understanding the concepts of image representation, enhancement, filtering, feature extraction, segmentation, and classification. By putting these fundamental ideas into practise, it is possible to create intelligent transportation systems with improved capabilities for traffic analysis, object detection, surveillance, and raising the overall performance levels of transportation systems.

## 6.4 Fundamental Signals (1-D, 2-D, and 3-D)

Signals are crucial elements of signal processing, and depending on the type of data being analysed, they can exist in different dimensions. Understanding the fundamentals of 1-D, 2-D, and 3-D signals is essential in the context of sustainable transportation and the application of signal processing techniques. The type of data being processed and the number of independent variables involved are reflected in these various signal dimensions and scenarios [7–9].

One-dimensional signals (also referred to as 1-D signals) are representations of data that vary along a single independent variable, usually time. one-dimensional signals can be used to represent different types of sensor data in the context of sustainable transportation, including traffic flow rates, vehicle speed measurements, and environmental sensor readings. A time series that depicts traffic flow over a certain period, for instance, can be regarded as a 1-D signal.

Signals in two dimensions: these signals typically represent spatial information and involve data that varies across two independent variables. Two-dimensional signals in transport systems frequently take the form of pictures or other visual data that has been recorded by cameras or other imaging devices. Road maps, aerial photographs, and traffic surveillance cameras are typical sources of 2-D signals in the context of sustainable transportation. These signals offer important details regarding the spatial distribution of objects, the state of the roads, or traffic patterns.

A video is a temporal sequence of 2-D signals made up of images obtained from cameras positioned at various locations. In transportation systems, video data is frequently used for surveillance, object detection, and traffic monitoring. Two-dimensional signals must be analysed and processed using methods that are specifically designed for image data, such as spatial filtering, feature extraction, and image processing algorithms.

Signals in three dimensions: 3-D signals contain data that changes in three independent variables. Three-dimensional signals frequently represent spatiotemporal information in the context of sustainable transportation, combining the advantages of both 1-D temporal data and 2-D spatial datasets and samples. For instance, multiple cameras recording a scene from various angles over time can produce 3-D signals. These signals offer in-depth knowledge of the dynamics and motion of objects in the transportation environment.

Applications like traffic flow analysis, object tracking, or behaviour modelling benefit greatly from 3-D signals. These signals allow for a more thorough understanding of transportation phenomena and enable more precise predictions and decision-making by taking into account both the spatial and temporal dimensions.

Processing 1-D, 2-D, and 3-D signals: each signal dimension requires a specific set of techniques and algorithms for processing 1-D, 2-D, and 3-D signals in the context of sustainable transportation. It is necessary to modify signal processing methods like filtering, feature extraction, and classification to take into account the unique properties and structures of the various signal dimensions.

When processing 1-D signals, for instance, time-series analysis techniques like Fourier analysis or wavelet transforms are frequently used. These techniques make it possible to find periodic patterns, frequency components, or anomalies in traffic flow data or sensor measurements.

Contrarily, 2-D signal processing methods, such as image filtering, segmentation, or object recognition, concentrate on sifting through images taken by cameras or other imaging equipment used in transportation systems to extract pertinent datasets and samples. To effectively analyse and interpret visual data for various scenarios, these techniques make use of spatial properties like pixel intensities, texture features, or shape characteristics.

When processing 3-D signals, both the spatial and temporal dimensions must be taken into account. The dynamic behaviour and movement patterns of objects in transportation systems are frequently studied using methods like spatiotemporal filtering, trajectory analysis, or motion estimation sets.

In conclusion, effective signal processing techniques in the context of sustainable transport require an understanding of the fundamentals of 1-D, 2-D, and 3-D signals. Each signal dimension offers particular information and poses particular difficulties that call for specialised processing methods. Researchers and practitioners can gain useful insights from transportation data by utilising the appropriate algorithms and methodologies for each signal dimension, enabling improved traffic management, improved safety measures, and more sustainable transportation systems.

## 6.5 Classification of Systems

The classification of systems is a fundamental idea in the field of signal processing that aids in comprehending and analysing the behaviour of various system types found in various domains, such as sustainable transportation. A process or a piece of equipment that converts an input signal into an output signal is known as a system. By categorising and characterising systems, it can create algorithms and signal processing techniques that are suitable for handling particular system types.

Systems with linear time invariance (LTI): systems can be divided into categories that are fundamentally based on their linearity and time-invariance characteristics. The response to a sum of input signals is equal to the sum of individual responses to each input signal, which is a characteristic of linear systems. The output of a time-invariant system is constant regardless of the timing of the input signal. Applications for sustainable transportation and signal processing frequently use LTI systems.

Linear time-invariant systems have a number of distinctive qualities that make them particularly well-suited for processing and analysis. These qualities consist of stability, time-invariance, and linearity. Mathematical analysis and the use of linear algebra techniques are made easier by the linearity, which guarantees that the system response is proportional to the input signal. The system's behaviour must be time-invariant in order for time-domain and frequency-domain analysis techniques to be applied. Stability guarantees that for bounded input signals, the system's output will remain bounded, preventing undesirable oscillations or unstable behaviour.

Systems that are linear shift-invariant (LSI): another significant system classification that is encountered in signal processing is LSI systems. Like LTI systems, LSI systems have linearity and time invariance, but they also have shift invariance. A time shift in the input signal has no impact on the system's response if it is shift-invariant. The behaviour of transportation systems, such as traffic flow dynamics, vehicle response traits, or sensor measurements, is frequently modelled and analysed using LSI systems.

System classification applications in sustainable transportation The design and application of signal processing techniques for sustainable transportation heavily depends on the classification of systems. Researchers and practitioners can create appropriate algorithms and methodologies to tackle particular challenges by understanding the traits and behaviours of systems used in transportation. Here are a few illustrations:

1. Analysis of traffic flow: an LTI system can be used to model traffic flow, with the input signal denoting traffic demand or control measures and the output signal denoting the resulting traffic flow. Traffic engineers can learn more about congestion patterns, traffic bottlenecks, or the efficacy of control strategies by examining the system's response.
2. Modelling the way that vehicles react to environmental or control inputs requires an understanding of system dynamics and system classification. It is possible to forecast vehicle behaviour and improve transportation systems for increased safety, fuel efficiency, and comfort by thinking of vehicles as LTI or LSI systems.
3. Sensor data processing: signals produced by sensors used in transportation systems, such as environmental sensors and traffic detectors, must be processed and interpreted. The behaviour of these sensors can be categorised as LTI or LSI systems, which enables the creation of signal processing methods to remove noise, calibrate measurements, or find anomalies.

In conclusion, a key idea for comprehending and analysing the behaviour of the various types of systems encountered in sustainable transportation is the classification of systems in signal processing. Researchers and practitioners can create efficient signal processing techniques and algorithms to handle particular challenges in transport systems by classifying systems as LTI or LSI and taking into account their linearity, time-invariance, and stability properties. The classification of systems makes it easier to analyse traffic patterns, model vehicle responses, and process sensor data, all of which help create smart, sustainable transportation systems [10–12].

## 6.6 Characteristics of LTI/LSI Systems

The fundamental classes of LTI and LSI systems in signal processing each have unique properties. Understanding these features is crucial for signal processing and analysis across a range of applications, including sustainable transportation. Let's explore the main traits of LTI and LSI systems:

Linearity: linearity is one of the main traits of LTI and LSI systems. According to the definition of linearity, the response of the system to a linear combination of input signals must be the same as the linear combination of the responses of the system to each individual input signal. In other words, the output of the system is a direct function of the input signal. The mathematical property of superposition holds for a linear system, making it simple to analyse using methods from linear algebra.

Time-invariance: LTI and LSI systems have this important trait in common as well. The output response from time-invariant systems is constant regardless of when the input signal is applied. To put it another way, if the input signal is advanced or delayed in time, the output response will also be advanced or delayed, maintaining the same shape and characteristics. The use of time-domain and frequency-domain techniques is made possible by this property, which makes it easier to analyse and process signals without the use of laborious time-dependent transformations.

Shift-invariance (LSI systems): one distinctive quality of LSI systems is shift invariance. It denotes that even when the input signal is time-shifted, the system's response is unaffected. The input signal's time shift will be reflected in the output signal as well. Shift-invariance is especially important in temporal or spatial domains. For instance, a shift-invariant system in image processing would yield the same results regardless of where the image's main subject was located.

Linear time-invariant and LSI systems are causal, which means that their output response is dependent only on the input values of the past and present, not on the values of the future. Causality states that the input signal up to that point alone determines the system's output at any given time. Real-time applications require this property to ensure that the system's response is both physically meaningful and consistent with the temporal sequence of events.

Stability: For LTI and LSI systems, stability is an essential quality. When the output response stays within the bounds of the input signals, the system is said to be stable. Stability makes sure that there are no undesirable oscillations or uncontrollable reactions in the system's behaviour. In order to ensure the dependable and predictable operation of control systems, traffic management algorithms, or signal processing techniques, stability analysis is crucial in applications for sustainable transportation.

Impulse response and convolution: the output response of an LTI or LSI system when an impulse function (such as the Dirac delta function) is used as the input signal is known as the impulse response. The system's behaviour is captured in the impulse response, which also offers important details about its characteristics like frequency response or system dynamics.

When any input signal is applied, the output response of an LTI or LSI system is calculated using the convolution operation. Convolution, a basic signal processing operation, is crucial to the analysis and processing of signals in LTI and LSI systems.

In conclusion, it is critical to comprehend the properties of LSI and LTI systems for signal processing and analysis across a variety of fields, including sustainable transportation. The development of efficient signal processing techniques, control systems, and algorithms for traffic analysis, vehicle response modelling, and sensor data processing is made possible by a number of key factors, including linearity, time-invariance, shift-invariance (for LSI systems), causality, stability, impulse response, and convolution properties. Researchers and practitioners can design and implement intelligent and sustainable transport systems with increased performance and efficiency by taking advantage of the properties of LTI and LSI systems.

## 6.7 Application of Image Processing in Urban Transportation Systems

In order to ensure effective traffic flow, increased safety, and enhanced mobility, urban transportation systems must be monitored, analysed, and managed effectively. Techniques for image processing have become useful resources for gathering and analysing visual data in urban transportation systems. Transportation authorities and researchers can gain useful information from images taken by surveillance cameras, drones, or other imaging devices by utilising image processing algorithms and methodologies. Here, it look at an instance of how image processing is applied to urban transport systems: automatic traffic analysis and monitoring.

Automatic monitoring and analysis of traffic conditions is a prominent use of image processing in urban transport systems. Traffic authorities can gather pertinent data about traffic flow, congestion, and incidents by placing cameras in strategic locations. This data can then be processed and analysed. The use of image processing methods in this situation has the following advantages:

1. Traffic flow monitoring: from the captured images, image processing algorithms can quickly identify and track moving objects. Traffic flow parameters like speed, density, and flow rate can be estimated by looking at the movement patterns and vehicle density. This data offers important insights into bottlenecks, congestion, and general traffic conditions. Authorities can implement congestion management strategies, improve signal timings, and make wise decisions for traffic management thanks to traffic flow monitoring.
2. Incident detection: image processing techniques can be used to find and categorise on-road incidents like collisions, mechanical failures, or roadside debris. Algorithms can identify unusual patterns, modifications in vehicle behaviour, or the presence of obstacles by continuously analysing the image datasets and samples. When incidents are discovered early, authorities can react quickly, send out emergency services, and reduce the effects on traffic and safety.
3. Automated vehicle classification and counting: image processing algorithms can classify and count vehicles based on their size, shape, and other characteristics. For the design of infrastructure, capacity analysis, and transportation planning, this information is essential. Authorities can decide on expanding the road network, allocating lanes, or putting in place specific vehicle restrictions by accurately estimating the number and types of vehicles.
4. Automatic licence plate recognition: using image processing techniques, licence plates can be read and recognised automatically from images. Identification of vehicles, tracking of their movements, and enforcement of traffic laws are all made possible by licence plate recognition. Applications like toll collection, parking control, or the identification of vehicles used in criminal activity benefit the most from its use.
5. Pedestrian and cyclist detection: the use of vehicles is not the only mode of urban transportation. The detection and tracking of cyclists and pedestrians can both be done using image processing. Transportation authorities can design bicycle-friendly infrastructure, increase pedestrian safety measures, and improve non-motorized mode of transportation mobility in cities by recognising their presence and examining their behaviour.

These are only a few uses for image processing in urban transport systems. Machine learning, deep learning, and computer vision algorithm improvements have greatly improved image processing's capacity to extract useful information from visual datasets and samples [10–12].

To sum up, image processing is essential for urban transport systems because it makes automatic traffic monitoring and analysis possible for different use cases. Transportation authorities can gather real-time data about traffic flow, incidents, vehicle counts, and pedestrian/cyclist movements by examining images taken from security cameras or other imaging devices and scenarios [13–15]. Decision-makers are now better equipped to put into practice efficient traffic management techniques, improve infrastructure, and increase overall mobility and safety in urban areas. The use of image processing methods in urban transportation systems demonstrates how computer vision technologies have the potential to revolutionize how it track, evaluate, and manage urban mobility levels.

## References

1. M. Deveci, A. R. Mishra, I. Gokasar, P. Rani, D. Pamucar and E. Özcan, "A Decision Support System for Assessing and Prioritizing Sustainable Urban Transportation in Metaverse," in *IEEE Trans. Fuzzy Syst.*, vol. 31, no. 2, pp. 475-484, Feb. 2023, doi: 10.1109/TFUZZ.2022.3190613.
2. N. B. Manjong, A. S. Oyewo and C. Breyer, "Setting the Pace for a Sustainable Energy Transition in Central Africa: The Case of Cameroon," in *IEEE Access*, vol. 9, pp. 145435- 145458, 2021, doi: 10.1109/ACCESS.2021.3121000.
3. Y. Xue, S. Ji, G. Zhu and P. Zhao, "Solving the Sustainable Automobile Production- Distribution Joint Optimization in the Physical Internet-Enabled Hyperconnected Order- to-Delivery System by I-NSGAIII," in *IEEE Access*, vol. 11, pp. 7471-7494, 2023, doi: 10.1109/ACCESS.2023.3237735.
4. R. Satymov, D. Bogdanov and C. Breyer, "The Value of Fast Transitioning to a Fully Sustainable Energy System: The Case of Turkmenistan," in *IEEE Access*, vol. 9, pp. 13590- 13611, 2021, doi: 10.1109/ACCESS.2021.3050817.
5. Y. Song, P. Wu, Q. Li, Y. Liu and L. Karunaratne, "Hybrid Nonlinear and Machine Learning Methods for Analyzing Factors Influencing the Performance of Large-Scale Transport Infrastructure," in *IEEE Trans. Intell. Transp. Syst.*, vol. 23, no. 8, pp. 12287-12300, Aug. 2022, doi: 10.1109/TITS.2021.3112458.
6. D. G. Broo and J. Schooling, "A Framework for Using Data as an Engineering Tool for Sustainable Cyber-Physical Systems," in *IEEE Access*, vol. 9, pp. 22876-22882, 2021, doi: 10.1109/ACCESS.2021.3055652.

7. X. Huang, Q. Liang, Z. Feng and S. Chai, "A TOD Planning Model Integrating Transport and Land Use in Urban Rail Transit Station Areas," in *IEEE Access*, vol. 9, pp. 1103-1115, 2021, doi: 10.1109/ACCESS.2020.3047207.
8. M. Khemir, M. Rojas, R. Popova, T. Feizi, J. F. Heinekamp and K. Strunz, "Real-World Application of Sustainable Mobility in Urban Microgrids," in *IEEE Trans. Ind. Appl.*, vol. 58, no. 2, pp. 1396-1405, March-April 2022, doi: 10.1109/TIA.2021.3132847.
9. J. Jiang, W. Tu, H. Kong, W. Zeng, R. Zhang and M. Konecny, "Large-Scale Urban Multiple-Modal Transport Evacuation Model for Mass Gathering Events Considering Pedestrian and Public Transit System," in *IEEE Trans. Intell. Transp. Syst.*, vol. 23, no. 12, pp. 23059-23069, Dec. 2022, doi: 10.1109/TITS.2022.3198178.
10. P. Wu, P. Wang, H. -L. Chi, Y. Zhong and Y. Song, "Exploring Factors Affecting Transport Infrastructure Performance: Data-Driven Versus Knowledge-Driven Approaches," in *IEEE Trans. Intell. Transp. Syst.*, vol. 23, no. 12, pp. 24714-24726, Dec. 2022, doi: 10.1109/TITS.2022.3192895.
11. I. Zubin, N. Van Oort, A. Van Binsbergen and B. Van Arem, "Deployment Scenarios for First/Last-Mile Operations With Driverless Shuttles Based on Literature Review and Stakeholder Survey," in *IEEE Open J. Intell. Transp. Syst.*, vol. 2, pp. 322-337, 2021, doi: 10.1109/OJITS.2021.3106164.
12. A. Karam and K. H. Reinau, "A Real-Time Decision Support Approach for Managing Disruptions in Line-Haul Freight Transport Networks," in *IEEE Trans. Intell. Transp. Syst.*, vol. 23, no. 12, pp. 24765-24777, Dec. 2022, doi: 10.1109/TITS.2022.3193956.
13. T. Cristóbal, A. Quesada-Arencibia, G. S. De Blasio, G. Padrón, F. Alayón and C. R. García, "Using Data Mining to Estimate Patterns of Contagion-Risk Interactions in an Intercity Public Road Transport System," in *IEEE Access*, vol. 10, pp. 99150-99167, 2022, doi: 10.1109/ACCESS.2022.3206838.
14. A. Saha, V. Simic, T. Senapati, S. Dabic-Miletic and A. Ala, "A Dual Hesitant Fuzzy Sets-Based Methodology for Advantage Prioritization of Zero-Emission Last-Mile Delivery Solutions for Sustainable City Logistics," in *IEEE Trans. Fuzzy Syst.*, vol. 31, no. 2, pp. 407-420, Feb. 2023, doi: 10.1109/TFUZZ.2022.3164053.
15. R. Mendes Borges and E. Quaglietta, "Assessing Hyperloop Transport Capacity Under Moving-Block and Virtual Coupling Operations," in *IEEE Trans. Intell. Transp. Syst.*, vol. 23, no. 8, pp. 12612-12621, Aug. 2022, doi: 10.1109/TITS.2021.3115700.

# 7

# Geographic Information System-Based Transportation System

***Abstract***

In this chapter, delve into the world of transportation systems that are based on geographic information systems (GIS), and investigate the essential role that GIS plays in analyzing, visualizing, and managing spatial data related to transportation infrastructure. In light of the rapid changes taking place throughout the world at the present time, the first section of this chapter provides an overview of the significance of effective transport systems. It draws attention to the difficulties associated with managing complex transportation networks and lays the groundwork for comprehending how GIS can be used to address these difficulties in an efficient manners and processes [1, 2].

After that, the chapter gives an in-depth explanation of GIS, placing particular emphasis on the system's capacity to combine spatial and attribute data, thereby enabling the creation of visual representations and facilitating decision-making procedures. It investigates the various sources of GIS data, including government agencies, private organizations, research institutions, and platforms that rely on crowdsourcing for their information and scenarios. For the purpose of utilizing GIS in the analysis and planning of transportation, the availability and accessibility of these data sources are absolutely necessary for different use cases.

The chapter then delves into the pivotal function that GIS plays in the transportation sector. This demonstrates how GIS has revolutionized transportation planning and management by giving professionals the ability to evaluate and optimize existing infrastructure, plan new routes, analyze traffic patterns, and forecast future demands. It is possible for policymakers, urban planners, and transportation authorities to make well-informed decisions with the help of GIS. These decisions can improve efficiency, cut down on congestion, raise safety standards, and encourage the development of sustainable transportation systems.

The evaluation of highways and railroads by means of GIS constitutes a significant portion of this chapter's content. It emphasizes the capability of GIS to conduct in-depth analyses of transportation infrastructure, such as road conditions, traffic flow, accident hotspots, and maintenance requirements. Geographic information systems make it easier to identify problem areas by integrating both real-time data and historical records. This opens the door for preventative maintenance

Kundan Meshram. Urban Transportation Systems: Sustainability and Future Development, (143–160) © 2024 Scrivener Publishing LLC

and improvements that are specifically targeted. In addition, GIS plays an essential part in determining the effect that proposed road and railway developments will have on the entire transportation network.

A case study on the application of GIS in smart cities is included in this chapter so that readers can gain some insight into its practical application. It investigates the possibility of integrating GIS with other smart city components, such as Internet of things devices, real-time data feeds, and intelligent transportation systems. The case study demonstrates how GIS contributes to the development of smart cities by providing the ability for transportation planning, traffic management, and sustainable urban mobility. This is accomplished through an analysis of real-world examples.

In summing up, the focus of this chapter has been on the significant part that GIS plays in transportation systems. It places an emphasis on the significance of spatial data, investigates the various sources of GIS data, and demonstrates the numerous ways in which GIS can be utilized in transportation analysis and planning. Transportation professionals can improve the efficacy, safety, and sustainability of transportation networks by making use of GIS. This chapter serves as a foundation for further exploration and implementation of GIS-based transport systems. It paves the way for smarter, more connected cities as well as improved transport experiences for individuals as well as communities alike for a variety of different scenarios.

***Keywords:*** Geographic information systems, traffic management, smart city

## 7.1 Introduction to Geographic Information System

The geographic information system (GIS) has emerged as a powerful tool for analyzing, managing, and visually representing spatial data in today's world, which is both technologically advanced and highly interconnected. A comprehensive comprehension of our complicated world can be attained through the use of GIS, which combines geographical information with attribute data. The GIS has found applications in a wide variety of fields, ranging from mapping and urban planning to environmental analysis and disaster management sets.

### 7.1.1 The Development of GIS and Its Definition

The idea of a GIS can be traced back to the fields of cartography and computer science. Before the invention of computers, maps were traditionally drawn out on paper. However, with the advent of computers came a quantum leap in the ability to store, manipulate, and analyze spatial data. The evolution of GIS into their current form can be credited to developments in computer technology, database management systems, and remote sensing techniques.

The term "geographic information system" (GIS) refers to a system that records, stores, manages, examines, and presents both spatial and non-spatial data that is connected to specific geographic locations. To enable spatial analysis and decision making, it requires the coordinated effort of a number of different components, including hardware, software, data, and trained personnel.

### 7.1.2 Constituent Parts of a GIS

A GIS is made up of several key components that, when combined, provide an all-encompassing understanding of spatial relationships.

#### *7.1.2.1 Data*

A GIS requires many different kinds of data, including geographic data (like maps, satellite imagery, and aerial photographs), attribute data (like population statistics and land use classification), and spatial data (like coordinates and topology).

#### *7.1.2.2 Software*

The GIS software gives users the ability to input, manipulate, analyze, and view spatial data. There are a wide variety of GIS software options available, both commercial and open-source, with each one providing a different set of functionalities and tools.

#### *7.1.2.3 Hardware*

A GIS can be deployed on a wide variety of hardware platforms, ranging from standard desktop computers to high-performance servers. The complexity and scope of the GIS application will determine the hardware configuration that is necessary for different use cases [2–4].

#### *7.1.2.4 Methods and Techniques*

Geographic information systems make use of a wide variety of methods and techniques, including spatial analysis, network analysis, geocoding, and geostatistics, among others. Users are given the ability to perform spatial queries, generate maps, conduct proximity analyses, and model geographic phenomena thanks to these methods.

### 7.1.3 The Importance of GIS and Its Uses

Understanding the spatial relationships, patterns, and trends of geographic features is one of the primary functions of GIS. Users are given the ability to visualize and analyze data in the context of a spatial environment, which helps them gain valuable insights and make more informed decisions. Numerous and varied applications can be found for GIS, including the following.

#### *7.1.3.1 Urban Planning Geographic Information Systems (GIS)*

It is helpful in urban planning because they allow for the analysis of land use, transportation networks, the distribution of populations, and infrastructure. It contributes to the planning of effective cities, the enhancement of resource utilization, and the administration of urbanization.

#### *7.1.3.2 Management of the Environment*

The GIS is helpful in evaluating and monitoring environmental factors such as biodiversity, changes in land cover, and the distribution of natural resources. It contributes to the planning of sustainable development, disaster management, and conservation efforts.

#### *7.1.3.3 Public Health*

The GIS plays a significant role in public health by analyzing disease patterns, locating high-risk areas, and optimizing the distribution of healthcare resources. It is helpful in epidemiological studies, as well as the planning and preparation of emergency responses and disease surveillance.

#### *7.1.3.4 Management of Natural Resources*

Geographic information systems (GIS) are useful for the management and analysis of natural resources like forests, water bodies, and mineral deposits. Planning, monitoring, and evaluating the impact on the environment are all made easier as a result.

### 7.1.4 Obstacles to Overcome and Possible Future Paths

Despite the fact that GIS has fundamentally changed how we analyze and manage spatial data, there are still issues that need to be resolved. The availability and quality of data are two of these challenges, as well as interoperability between the various GIS systems, concerns over data privacy, and the requirement of skilled GIS professionals. However, exciting new developments in the field are on the horizon thanks to recent advances in technology such as mobile GIS, cloud computing, and remote sensing scenarios.

Emerging technologies such as artificial intelligence, machine learning, and big data analytics are expected to enhance the capabilities of GIS in the future, which indicates that the future of GIS holds enormous potential. The use of GIS will continue to be indispensable in resolving difficult spatial issues, facilitating decision-making based on evidence, and laying the groundwork for sustainable development scenarios.

In conclusion, GIS has fundamentally altered the ways in which we engage with and make sense of the world that surrounds us. Its capability of integrating spatial and attribute data provides valuable insights that can be applied to a wide variety of different applications. Global information systems will continue to be an essential tool for addressing spatial challenges, fostering the making of informed decisions, and developing a better understanding of our dynamic and interconnected planet for real-time scenarios as we move forward during research use cases [5, 6].

## 7.2 Sources of GIS

The GIS draws on a wide variety of data sources in order to deliver information that is both accurate and comprehensive regarding the surrounding area. The creation and upkeep of GIS databases, which provide the basis for spatial analysis, visualization, and decision-making processes, are made possible by the contributions of these sources. This report examines the various sources of GIS data, focuses on the significance of those sources, and discusses the types of data that are provided by those sources.

1. Government agencies: significant amounts of data for GIS come from a variety of governmental agencies operating at different levels, including national, regional, and local authorities. These organizations are responsible for the collection, management, and dissemination of a vast array of spatial information, including that which is pertinent to infrastructure, land use,

demographics, transportation networks, and environmental factors. Topographic maps, aerial photographs, and satellite imagery are all examples of the types of geospatial data that can be obtained from national mapping agencies. In addition, departments that are responsible for transportation, urban planning, the environment, and natural resources are tasked with the collection and maintenance of GIS data that is specific to each of their respective domains.

2. Private organizations: additionally, a significant contribution to the production of GIS data is made by non-government organizations. For the purposes of meeting their operational requirements, businesses operating in industries such as telecommunications, utilities, transportation, and real estate collect spatial information. For example, companies that provide telecommunication services keep databases that contain information on network coverage, cell towers, and signal strength. Data on power distribution networks, water supply systems, and sewage infrastructure can be obtained from utility companies through the use of GIS. Data on routes, vehicle tracking, and traffic flow are provided by various private companies involved in the transportation, logistics, and delivery services industries. When these datasets are made available to others or shared, they develop into valuable sources of GIS information sets [7, 8].
3. Educational and scientific establishments dedicated to research: scientific studies, field surveys, and academic research are all ways that research institutions and universities contribute to the data that makes up GIS. These organizations frequently run data collection campaigns in order to collect information on a variety of subjects or to carry out spatial analysis for the purposes of research. Studies on biodiversity, ecological systems, climate change, and socio-economic indicators are some examples of topics that have been researched and used to generate GIS data. These datasets, which are typically put through rigorous scientific methods and other quality assurance processes, contribute to the expansion of the knowledge base and enhance the accuracy of the information that is obtained from GIS.
4. Open data initiatives: the amount of GIS data that is readily available to the general public has significantly increased as a result of open data initiatives, which have been spearheaded by governments and organizations that are not-for-profit.

Open data portals give users access to a wide variety of spatial datasets, enabling them to download the data and apply it in a variety of contexts. The topics that are covered by these datasets include things like transportation networks, administrative boundaries, land cover, population demographics, and environmental data. Open data initiatives promote transparency, encourage citizen engagement, and foster innovation by enabling users to analyze and visualize GIS data for their specific needs in a manner that is tailored to their individual requirements [9–12].

5. Data collected from the public: crowd-sourced data has emerged as a valuable source of information for GIS as a result of the widespread availability of smartphones and internet connectivity. Volunteers are able to contribute to mapping efforts through the use of platforms such as OpenStreetMap (OSM), which enables them to update and enrich spatial data with specifics on roads, buildings, points of interest, and other geographical features. This approach to collaborative work not only improves the accuracy and coverage of GIS data, but it also makes it possible for communities to actively participate in the process of creating and maintaining spatial information that is pertinent to their specific locations.
6. Commercial data providers: commercial data providers provide GIS datasets and services that are specialized and tailored to the needs of particular businesses and applications. These service providers collect datasets, aggregate them, and keep them up to date using a combination of remote sensing, surveying, and data acquisition techniques. Examples of this type of data layer include high-resolution satellite imagery, data collected using LiDAR technology, aerial photographs, and specialized data such as demographic information, business listings, or environmental indices. Commercial data providers offer datasets that are dependable and up to date, as well as datasets that cater to specific needs. These datasets are frequently utilized by industries including environmental consulting, marketing, real estate, and urban planning.

For the successful application of GIS in a variety of settings, it is essential to have a wide variety of data sources available, as well as easy access to those sources. The rich tapestry of GIS information is woven together with contributions from a variety of sources, including commercial data providers, open data initiatives, crowd-sourced data, and government agencies and private organizations and research institutions. These sources provide users with the ability to analyze, visualize, and make informed decisions based on comprehensive spatial information by offering a wide range of data types. These data types include spatial data, attribute data, and remotely sensed data. As the field of GIS continues to advance, collaboration and the integration of data from numerous sources will become increasingly important in order to ensure that GIS datasets are accurate, up to date, and relevant for a variety of applications.

## 7.3 Role of GIS in Transportation

Transportation networks connect people, goods, and services across geographic areas and are therefore essential to the functioning of modern societies. Planning, management, and decision-making regarding the city's transport infrastructure must be done effectively in order to ensure economic growth,

**Figure 7.1** GIS application for ITS.

contribute to effective urban planning, and guarantee that cities continue to function normally. The technology known as GIS is an essential component in the solution to the complex problems that transportation professionals are currently facing. This report explores the various applications of GIS in the transportation sectors as well as the benefits that it provides. GIS application for intellegent transport system (ITS) is shown in Figure 7.1.

1. Data integration and visualization: the GIS serves as a foundation for integrating a wide range of spatial and non-spatial data associated with transportation networks. It gives professionals in the transportation industry the ability to combine different types of data sources, including maps, satellite imagery, aerial photographs, road networks, traffic data, and information about land use. Global information systems enable a comprehensive understanding of transportation networks, including road conditions, traffic congestion, accident hotspots, and the requirements for infrastructure, by providing a visual representation of this integrated data in the form of maps, charts, and graphs.
2. Analyzing the network and establishing routes: GIS makes network analysis easier, which helps improve efficiency and make the most of available transportation options. Professionals in the transportation industry can use GIS to determine the routes that are the quickest or shortest between two locations, taking into consideration factors such as traffic patterns, road conditions, and speed limits. Routing algorithms that are based on GIS can help determine the most efficient routes for vehicles, which can cut down on travel time, fuel consumption, and emissions. This capability is especially useful for emergency response management, as well as the planning and management of public transport and logistics and delivery services.
3. Strategies for managing traffic and analyzing congestion: the use of GIS makes it possible to monitor, analyze, and manage traffic in real time. Global information systems allow for the identification of congested areas, the analysis of traffic patterns, and the projection of future demand thanks to the integration of data from sensors, traffic cameras, and other monitoring systems. The use of GIS in traffic management centers allows for the monitoring of traffic conditions, the detection of incidents, and the dissemination of information

to drivers, which in turn enables drivers to make informed route choices. Models and simulations that are based on GIS can be of assistance in determining how various traffic management strategies, road closures, or the construction of new infrastructure will affect the flow of overall traffic sets [12–14].

4. Management of assets and preventative maintenance: the management of assets and the upkeep of transportation infrastructure are both significantly aided by the use of GIS. Global information systems enable transportation authorities to prioritize maintenance activities, identify infrastructure deficiencies, and plan for rehabilitation or replacement projects. This is accomplished by integrating data on road conditions, pavement quality, signage, bridges, and other assets. Asset management systems that are based on GIS can help maximize the use of available resources, extend the useful life of assets, and improve the safety and dependability of transportation networks.
5. Demand forecasting and planning for the transport system: through the analysis of spatial data on population demographics, land use, and transportation networks, GIS contributes to the process of transportation planning. Planners can use GIS to assess the impact of new developments, evaluate accessibility to services, and locate areas that lack adequate transportation infrastructure. Models and simulations built on top of GIS provide assistance in forecasting the demand for transport and planning for the requirements of future infrastructure. Because of this, policymakers are able to make well-informed decisions regarding road expansions, investments in public transit, and initiatives to promote sustainable transportation.
6. A risk assessment and procedures for dealing with emergencies: when conducting a safety analysis, GIS can be used to pinpoint areas prone to accidents, investigate the factors that may have contributed to those accidents, and put in place targeted safety measures. Global information systems assist transportation authorities in identifying high-risk areas, such as intersections that have a history of accidents or hazardous road segments, which enables the transportation authorities to implement safety improvements in those areas. Global information systems provide real-time situational

awareness during emergencies. This enables authorities to respond effectively by rerouting traffic, coordinating evacuation routes, and providing up-to-date information to the general public.

The GIS plays an essential and multi-faceted part in the transportation industry. The GIS allows transportation professionals to make informed decisions, optimize routes, manage traffic, maintain infrastructure, and plan for future needs by integrating and analyzing spatial data. The GIS also enables them to plan for future needs. The use of GIS improves the effectiveness, safety, and sustainability of transportation systems, all of which contribute to increased mobility, economic growth, and an overall improvement in quality of life. The GIS will play an increasingly crucial role in shaping the future of transportation, fostering smarter cities, and supporting the ever-growing transportation demands of our interconnected world for real-time scenarios as technology continues to evolve, and these are all goals that GIS will help achieve for real-time scenarios.

## 7.4 Assessment of Roads, and Railways Using GIS

Connectivity is underpinned by transport infrastructure, such as roads and railways, which also plays an important part in both the expansion of the economy and the movement of people within society. It is absolutely necessary to conduct routine inspections of the state of these infrastructure systems and to monitor both their performance and their condition in order to guarantee their safety, effectiveness, and long-term viability. The technology of GIS provides powerful tools and methods for evaluating and analyzing roadways and rail lines. This piece examines the various ways in which GIS can be applied to the analysis of roadways and railroads, focusing on the GIS's positive effects and important contributions to transportation management use cases.

1. Data integration and visualization: GIS makes it possible to integrate a wide variety of data sources, such as spatial data (such as maps and satellite imagery), attribute data (such as traffic volume and road conditions), and infrastructure data (such as signage and bridges). A comprehensive view of the road and railway networks can be obtained using GIS thanks to the combination and visualization of the respective datasets. Transportation professionals are able to identify spatial

relationships, conduct pattern analyses, and visualize the condition and performance of the infrastructure as a result of using this technology.

2. Condition assessment: GIS are useful for evaluating the state of transportation infrastructure such as roads and railways because they integrate data on aspects such as pavement quality, surface distresses, and structural integrity. Geographic information systems help identify areas that are in need of repair or maintenance by performing spatial analysis on the data. It enables the authorities in charge of transport to effectively prioritize their needs and distribute their resources by identifying critical sections or segments that require immediate attention. An assessment of the network's condition based on GIS data helps to improve overall network performance, as well as to optimize and extend the lifespan of infrastructure and maintenance activities [15, 16].
3. Performance analysis: GIS makes it easier to conduct analyses of performance metrics associated with road and rail networks. Geographic information systems make it possible to evaluate the effectiveness and dependability of a system by integrating data on factors such as the volume of traffic, travel times, and levels of congestion. It assists in locating bottlenecks, conducting traffic pattern analyses, and measuring the reliability of travel times. For the purpose of transportation planning, GIS-based performance analysis provides valuable insights, highlighting areas that can benefit from capacity enhancements or improvements.
4. Evaluation of safety: GIS plays an important part in determining whether or not roads and railways are safe to use. Geographic information systems help identify high-risk areas and analyze the factors that contribute to those areas by integrating data on the locations of accidents, traffic violations, and the characteristics of the roads. It gives transportation authorities the ability to prioritize safety improvements and put in place targeted measures such as improvements to signage, changes to road geometry, or traffic calming strategies. The use of GIS in safety evaluation contributes to the reduction of accidents, the improvement of safety for road users, and the creation of safer transportation systems.

5. Asset management: GIS facilitates asset management for transportation networks like roads and railways by integrating data on infrastructure components like bridges, culverts, and signage. It helps with inventory management, condition evaluation, and planning maintenance. The use of GIS enables transportation authorities to more effectively track and manage assets, prioritize maintenance activities, and allocate resources. It helps maximize the useful life of infrastructure components, reduce the likelihood of disruptions, and guarantee the overall safety and functionality of transportation networks.
6. Analysis of the impact on the environment: GIS is helpful in analyzing the effect that roads and railways have on the surrounding environment. The integration of data on land use, ecological features, and sensitive areas is made possible through the use of GIS, which makes it possible for transportation authorities to evaluate the potential environmental impacts and find ways to mitigate those impacts. An analysis based on GIS can help identify environmentally sensitive areas, evaluate alternative routes, and reduce the amount of ecological disturbance caused. It does so by taking into account environmental factors and encouraging the development of environmentally friendly infrastructure. This helps to make sustainable transport planning possible.
7. Planning and decision making: GIS plays an essential part in the planning and decision making processes pertaining to transportation, specifically roads and railways. Geographic information systems enable more informed decision-making regarding route planning, infrastructure investments, and transportation policies by integrating data on population demographics, land use, and transportation networks. Models and simulations that are based on GIS can be of assistance in determining what the future demands will be, determining the impact of proposed projects, and optimizing transportation systems to meet both current and future requirements.

The use of GIS for the evaluation of roadways and railways provides useful insights and methods that are driven by data for the management of transportation systems. Geographic information systems make it possible to integrate, visualize, and analyze a wide variety of datasets, which in turn helps with condition assessment, performance analysis, safety evaluation, asset management, environmental impact assessment, as well as planning and decision making. Transportation professionals are able to make informed decisions, optimize resources, and ensure the safety, efficiency, and sustainability of road and railway infrastructure systems when they utilize GIS technology sets.

## 7.5 Case Study of Smart City GIS

In recent years, the idea of "smart cities" has gained a lot of traction as urban areas have become increasingly focused on utilizing technology and data in order to enhance the standard of living of their citizens. The technology known as GIS is an essential component in both the conceptualization and execution of smart city initiatives. This case study investigates the use of GIS in the context of smart cities, highlighting the benefits of GIS as well as its contributions to urban planning, service delivery, and citizen engagement scenarios.

1. This case study focuses on a made-up city called Smartville, which underwent a transformation to become a smart city in order to improve its livability, sustainability, and efficiency. Congestion in the streets, ineffective management of the city's resources, and low levels of citizen participation were just some of the problems the city was facing. As a solution to these problems, Smartville came to the conclusion that it needed to implement GIS technology as an essential part of its smart city infrastructure.
2. Data integration: Smartville's implementation of a comprehensive GIS platform included the integration of a variety of data sources that were pertinent to urban planning and the provision of services. This included geographical data such as maps and satellite imagery, administrative data such as land use and zoning, infrastructure data such as transportation networks and utility networks, and socioeconomic data such as population demographics and employment statistics. By integrating and visualizing these datasets, Smartville was able to gain a comprehensive view of the spatial and non-spatial information pertaining to its city. This made it easier for the city to make informed decisions and plans.

3. Urban planning and development: GIS was an integral part of Smartville's various urban planning endeavors. The city utilized GIS to conduct an analysis of population demographics, patterns of land use, and transportation networks in order to locate areas in need of infrastructure improvements or redevelopment. The optimization of land use planning, the identification of suitable locations for new developments, and the determination of the impact that proposed projects will have on the city's overall structure were all helped by spatial analysis based on GIS. Urban planners made use of GIS to run simulations of potential future growth, which helped them ensure sustainable development and efficient use of resources.
4. Transportation management: Smartville's transportation management system was improved thanks to the utilization of GIS. The city could monitor traffic flow, identify hotspots for congestion, and provide motorists with dynamic routing information if it integrated real-time data from sensors, traffic cameras, and mobile applications. Algorithms based on GIS were used to calculate the most efficient routes, taking into account the current state of traffic in addition to recommending alternative modes of transportation, such as taking the bus or riding a bicycle. Because of this, there was less congestion, improved travel times, and an overall increase in the efficiency of the transportation system.
5. Service delivery and resource management: the use of GIS made it possible for Smartville to improve both their service delivery and their resource management. The city's utility infrastructure was integrated with GIS, which enabled real-time monitoring of water, electricity, and waste management systems. Geographic information systems were able to assist in the identification of regions with high resource usage by analyzing patterns of consumption. This allowed for more targeted conservation efforts and improved resource allocation. Maintaining and managing the city's infrastructure, as well as minimizing downtime and ensuring optimal performance, were all made easier with the assistance of asset management systems that were based on GIS.
6. Enhanced citizen engagement and participation in decision-making processes: through the use of GIS, Smartville was able to improve the level of citizen engagement and

participation in decision-making processes. Residents of the city were given the ability to access spatial information, participate in surveys, and report issues thanks to the implementation of a user- friendly GIS portal by the city. Through crowd-sourcing initiatives, the general public is given the opportunity to visualize planned developments, provide feedback, and contribute data. Residents were given the ability to actively participate in the development of their city through the use of tools based on GIS, which helped to foster a sense of ownership and promoted transparency in governance.

7. Efforts towards sustainability and environmental management: the GIS was an essential component in Smartville's efforts towards sustainability and environmental management. The city might be able to identify areas for environmental preservation and sustainable development if it were to integrate environmental data such as the quality of the air, the amount of green space, and the ecological zones. The use of GIS-based analysis helped to identify energy-efficient routes, optimal locations for installing renewable energy systems, and the implementation of green infrastructure initiatives for different scenarios. As a consequence, this led to a smaller negative impact on the environment, better air quality, and increased resistance to the effects of climate change levels [17, 18].

The demonstration of the significant benefits and contributions of GIS technology to urban planning, service delivery, citizen engagement, and sustainable practices was made possible by the implementation of GIS in Smartville's smart city initiatives. Integrating and analyzing various datasets provided decision-makers with actionable insights, which enabled them to make decisions based on evidence and improved resource allocation. By utilizing GIS, Smartville was able to transform into a city that is more livable, efficient, and sustainable, serving as an example for other urban areas that are just beginning their journeys towards becoming smart cities.

## References

1. P. Wu, P. Wang, H. -L. Chi, Y. Zhong and Y. Song, "Exploring Factors Affecting Transport Infrastructure Performance: Data-Driven Versus

Knowledge-Driven Approaches," in *IEEE Trans. Intell. Transp. Syst.*, vol. 23, no. 12, pp. 24714-24726, Dec. 2022, doi: 10.1109/TITS.2022.3192895.
2. Y. Sun, Y. Hua, L. Mou and X. X. Zhu, "CG-Net: Conditional GIS-Aware Network for Individual Building Segmentation in VHR SAR Images," in *IEEE Trans. Geosci. Remote Sens.*, vol. 60, pp. 1-15, 2022, Art no. 5201215, doi: 10.1109/TGRS.2020.3043089.
3. J. Luo, D. Zhou, Z. Han, G. Xiao and Y. Tan, "Predicting Travel Demand of a Docked Bikesharing System Based on LSGC-LSTM Networks," in *IEEE Access*, vol. 9, pp. 92189- 92203, 2021, doi: 10.1109/ACCESS.2021.3062778.
4. Y. Sun *et al.*, "Long-Term Matching Optimization With Federated Neural Temporal Difference Learning in Mobility-on-Demand Systems," in *IEEE Internet Things J.*, vol. 10, no. 2, pp. 1426-1445, 15 Jan.15, 2023, doi: 10.1109/JIOT.2022.3206491.
5. Q. Ouyang, T. Sun, Y. Xue and Z. Liu, "Long Short-Term Memory and Graph Convolution Network for Forecasting the Crude Oil Traffic Flow," in *IEEE Access*, vol. 10, pp. 18922- 18932, 2022, doi: 10.1109/ACCESS.2022.3150852.
6. B. Hernández, A. Alkayas, E. Azar and A. T. Mayyas, "Mathematical Model for the Placement of Hydrogen Refueling Stations to Support Future Fuel Cell Trucks," in *IEEE Access*, vol. 9, pp. 148118-148131, 2021, doi: 10.1109/ACCESS.2021.3123901.
7. Y. Zou, J. He and H. Chen, "Directional-Layered Space-Time Densities: A New Spatiotemporal Trajectory Aggregation and Geographic Visualization Approach," in *IEEE Access*, vol. 9, pp. 9026-9047, 2021, doi: 10.1109/ACCESS.2020.3045805.
8. S. Villarroya, J. R. R. Viqueira, J. M. Cotos and J. A. Taboada, "Enabling Efficient Distributed Spatial Join on Large Scale Vector-Raster Data Lakes," in *IEEE Access*, vol. 10, pp. 29406-29418, 2022, doi: 10.1109/ACCESS.2022.3157405.
9. D. Xiao, X. Xu, C. Ma and N. Lyu, "Addressing Driving Actions of At-Fault Older Drivers: Bayesian Bivariate Ordered Probit Analysis," in *IEEE Access*, vol. 9, pp. 45803-45811, 2021, doi: 10.1109/ACCESS.2021.3067011.
10. H. Jang, K. Yu and S. Park, "Managing 3D GIS Data for Indoor Environment Using Property Graph Database," in *IEEE Access*, vol. 11, pp. 37216-37228, 2023, doi: 10.1109/ACCESS.2023.3266519.
11. M. Luckner, I. Krzemińska, P. Wawrzyniak and J. Legierski, "Estimating Population Density Without Contravening Citizen's Privacy: Warsaw Use Case," in *IEEE Trans. Syst. Man Cybern.: Syst.*, vol. 52, no. 7, pp. 4494-4506, July 2022, doi: 10.1109/TSMC.2021.3096754.
12. K. Han, S. M. Yu, S. -L. Kim and S. -W. Ko, "Exploiting User Mobility for WiFi RTT Positioning: A Geometric Approach," in *IEEE Internet Things J.*, vol. 8, no. 19, pp. 14589-14606, 1 Oct.1, 2021, doi: 10.1109/JIOT.2021.3070367.
13. L. Zong, H. Wang and G. Luo, "Transmission Control Over Satellite Network for Marine Environmental Monitoring System," in *IEEE Trans. Intell. Transp. Syst.*, vol. 23, no. 10, pp. 19668-19675, Oct. 2022, doi: 10.1109/TITS.2022.3145881.

14. D. Jato-Espino, A. Roldan-Valcarce, F. Collazos-Arias and J. Rodriguez-Hernandez, "Spatial Statistical Modeling of Rockfall Hazard in a Mountainous Road in Cantabria (Spain)," in *IEEE J. Sel. Top. Appl. Earth Obs. Remote Sens.*, vol. 14, pp. 6729-6742, 2021, doi: 10.1109/JSTARS.2021.3075112.
15. Z. Liu, Z. Gong, J. Li and K. Wu, "mT-Share: A Mobility-Aware Dynamic Taxi Ridesharing System," in *IEEE Internet Things J.*, vol. 9, no. 1, pp. 182-198, 1 Jan.1, 2022, doi: 10.1109/JIOT.2021.3102638.
16. S. Basu, D. Bera and S. Karmakar, "Detection and Intelligent Control of Cloud Data Location Using Hyperledger Framework," in *IEEE Trans. Consum. Electron.*, vol. 69, no. 1, pp. 76-86, Feb. 2023, doi: 10.1109/TCE.2022.3201932.
17. D. Lu, Y. Li and F. Guo, "Evaluating Spatial and Temporal Characteristics of Population Density Using Cellular Data," in *IEEE Trans. Intell. Transp. Syst.*, vol. 23, no. 6, pp. 5717-5726, June 2022, doi: 10.1109/TITS.2021.3056434.
18. V. Singh, V. Singh and E. Estrada, "Construction and stochastic scale-free modelling of empirical, global, index-case SARS-CoV-2 transmission network," in *J. Complex Netw.*, vol. 10, no. 1, pp. 1-15, Jan. 2022, doi: 10.1093/comnet/cnab047.

# 8

# Deployment of Sustainability for Non-Motorized Transportation Systems

***Abstract***
This chapter presents an in-depth investigation of the many different facets connected to the incorporation of sustainability in non-motorized transportation (NMT) systems. NMT is an abbreviation that stands for non-motorized transportation modes. Some examples of NMT include walking, cycling, and the use of various types of micro-mobility devices. This chapter examines key components of next-generation transportation (NMT), categories of NMT, planning strategies for smart cities that facilitate NMT, the effect that NMT planning has on healthcare outcomes, and the utilization of artificial intelligence (AI) and machine learning (ML) in integrating sustainability in NMT systems. This is done in recognition of the significance of NMT in the promotion of sustainable urban mobility sets. Specifically, this chapter focuses on the integration of NMT with sustainability levels [1–3].

***Keywords:*** Non-motorized transportation system, artificial intelligence, micro-mobility

## 8.1 Introduction

Non-motorized transportation systems underline the relevance of infrastructure, such as pedestrian-friendly walkways, designated bicycle lanes, and secure parking facilities. Additionally, it emphasizes the value of supporting active transportation via public awareness campaigns and community involvement. Following that, the classification of NMT is investigated. This offers a framework within which to comprehend the several modes and user profiles that are involved. This chapter gives insights into the specific qualities and concerns connected with each of the categories by classifying NMT into numerous kinds, such as walking, cycling, and micro-mobility. These types include walking, cycling, and micro-mobility. In addition to that, the chapter goes into the development of smart cities in order to make NMT more accessible. This chapter

Kundan Meshram. Urban Transportation Systems: Sustainability and Future Development, (161–180) © 2024 Scrivener Publishing LLC

investigates the incorporation of NMT infrastructure and services into urban planning procedures. It places an emphasis on the need for comprehensive urban design strategies that provide accessibility, connection, and safety to NMT the highest priority. In the context of providing assistance for NMT planning and operations, the role of technology is another topic that is examined. Some examples of this function include intelligent transportation systems and real-time data analysis. In addition to this, the chapter explores the influence that NMT planning has on the outcomes of healthcare. It investigates the ways in which the promotion of NMT might result in greater physical activity, less air pollution, and enhanced public health. In addition, the potential for NMT to make a contribution to the prevention and treatment of chronic illnesses, as well as its favorable impacts on mental well-being, are investigated for a variety of diverse situations. Lastly, the chapter investigates how approaches from artificial intelligence (AI) and machine learning (ML) might be used to integrate sustainability into NMT systems. This chapter addresses the ways in which these technologies may assist the optimization of NMT infrastructure, improve route planning, increase safety measures, and allow data-driven decision-making processes. For real-time use cases, the potential of AI and ML to support the creation of smart NMT systems that adapt to user preferences and dynamically react to changing urban environments is also being examined. These systems would adapt to user preferences and would dynamically respond to changing urban circumstances. In a nutshell, the purpose of this chapter is to provide an all-encompassing review of the implementation of sustainability for NMT systems. It analyzes the use of AI and ML for integrating sustainability into NMTs, focuses on the components and categories of NMT, investigates the design of smart cities to assist NMT, examines the influence that NMT planning has on healthcare, and emphasizes the components and categories of NMT. In the context of smart cities, the insights that are offered in this chapter are intended to make a contribution to the development of sustainable urban mobility policies by advocating NMT as a transportation alternative that is both practical and kind to the environment and scenarios.

## 8.2 Components of NMT

Systems of NMT are very important to the development of environmentally responsible mobility in urban settings. Walking, cycling, and the use of micro- mobility devices such as scooters and skateboards are all examples of non-motorized modes of transportation that fall under the umbrella term of non-motorized modes of transportation (NMT). It is vital to have a good understanding of the major components that contribute to the functioning, safety, and convenience of NMT systems in order to properly build and execute these systems. This section offers a comprehensive review of the components that combine to make up NMT systems.

1. Infrastructure: the physical foundation for mobility that is both secure and efficient is provided by infrastructure, which serves as the linchpin of NMT systems. The following are important parts of the infrastructure:
   a. Pedestrian facilities: pathways, sidewalks, and crosswalks specifically designated for pedestrian use that are kept in good repair, have adequate lighting, and are easily accessible to all users, including those with physical impairments. It is important that these facilities be smoothly connected with other forms of transportation in order to provide pedestrians with the ability to navigate the urban environment in a comfortable and secure manner.
   b. Bicycle lanes and tracks: dedicated lanes for cyclists that are separated from lanes used by motor vehicle traffic in order to offer a safe and uninterrupted route for riding. Bicycle lanes may be physically segregated from highways or incorporated into the urban fabric, such as via cycle-friendly streets and greenways. Both of these options have their advantages and disadvantages for different scenarios.
   c. Bicycle parking: safe and convenient parking facilities for bicycles, including racks, stands, and stations for bike-sharing programs. In order to encourage more people to adopt cycling as a viable method of transportation, these facilities should be strategically situated near important destinations, transit hubs, and public places.
2. Precautionary measures: when it comes to the effective implementation of NMT systems, one of the most important factors is ensuring the users' safety. Among the preventative safeguards are:
   a. Traffic calming: the process of implementing measures to slow the speeds at which vehicles travel, such as speed bumps, roundabouts, and streets that are smaller, in order to make the environment safer for walkers and bicycles.

   b. Signalization and signage: the installation of traffic signals, pedestrian crossings, and signage that is particularly tailored for NMT users in order to improve the visibility of NMT users and assure interactions that are seamless and predictable with other motorists and pedestrians.
   c. Separation from motorized traffic: the creation of physical barriers, such as curbs or bollards, to divide NMT infrastructure from lanes for motorized vehicles, therefore lowering the probability of crashes and other types of conflicts.
3. Accessibility and connection: in order to promote their use by a wide variety of user groups, NMT systems need to put accessibility and connection at the forefront of their design. Among the most important factors to take into account are the following:
   a. Universal design: making sure that the NMT infrastructure is accessible to persons who have physical impairments by installing things like ramps, tactile pavement, and audio signals at pedestrian crossings.
   b. Seamless intermodal integration: this refers to the facilitation of seamless transfers between NMT modes and other modes of public transportation, such as buses, trains, and trams, by providing integration points, dedicated parking places, and clear signs.
   c. Final-mile connection: an initiative that focuses on enhancing the connection between key transit hubs and destinations through NMT infrastructure, with the goal of making it simpler for commuters to finish the first and final portions of their trips utilizing NMT modes of transportations [4–6].
4. Education and awareness: it is essential to raise public awareness and educate the general public about the advantages of NMT as well as the best ways to implement it in order to encourage its adoption. Components that fall under this category include:
   a. Public awareness campaigns: running campaigns to educate the public on the merits of non-motorized modes of transportation, safety guidelines, and the environmental

and health benefits connected with active modes of transportation.

b. Training programs: offering training programs for cyclists and pedestrians to boost their confidence and understanding while utilizing NMT modes, such as bicycle handling skills, traffic regulations, and pedestrian safety levels.

c. Community engagement: involving local communities, neighborhood organizations, and advocacy groups in the process of developing and implementing NMT infrastructure by soliciting their comments, addressing any issues they may have, and involving them in the process.

Cities are able to establish transportation networks that are sustainable, inclusive, and centered on people by taking into consideration and implementing the aforementioned components into the design and development of NMT systems. Non-motorized transportation systems are successful because of the effective infrastructure, safety measures, accessibility, connection, and education that all contribute to their success. These factors make NMT systems appealing, safe, and convenient alternatives for urban transportation situations.

## 8.3 Categories of NMT

The term "non-motorized transportation" (NMT) refers to a broad variety of means of travel that are propelled by the force of the human body. These modes provide environmentally friendly and physically beneficial alternatives to the use of motorized vehicles. For the purpose of designing and creating transportation systems that cater to the varied requirements and preferences of users, having a solid understanding of the different classifications of NMT is necessary. This section offers a comprehensive summary of the many NMT subcategories for readers to peruse.

1. Walking: walking, also known as natural movement therapy (NMT), is the most simple and accessible kind of NMT. It involves the mobility of persons on foot. It is a convenient means of conveyance that may be incorporated into a variety of contexts, including day-to-day activities, leisure pursuits, and travel over shorter distances. Walking may be beneficial for a variety of reasons, including getting to and from work

or school, doing errands, getting exercise, and participating in leisure activities. It has a low threshold for required infrastructure and is often used as the basis for additional NMT modes.

2. Cycling: cycling, or just riding a bicycle for transportation reasons, is referred to as "cycling." It provides a means of non-motorized transport that is both quicker and more effective in comparison to walking, allowing users to cover more distances with the same amount of effort. Cycling is an excellent mode of transportation, as well as a fun and healthy form of pleasure and exercise. In order to guarantee both convenience and security, specialized infrastructure, such as bicycle lanes, walkways, and parking facilities, is required. Different users have varying tastes and requirements, and manufacturers have responded to those differences by developing a wide variety of bicycles, including regular bicycles, cargo bikes, and electric bicycles.
3. Micro-mobility: micro-mobility refers to small and lightweight means of NMT that are often used for travel over shorter distances inside metropolitan settings. Instruments such as electric scooters, skateboards, hoverboards, and kick scooters are included in such categories and scenarios. Micro-mobility solutions have seen a surge in popularity over the last few years as a result of their ease of use, adaptability, and little impact on the environment. They are especially useful for first- and last-mile connection, enhancing other forms of transportation, and alleviating congestion in densely populated metropolitan areas, all of which are possible because to their adaptability levels.
4. Users of wheelchairs and other mobility aids: users of wheelchairs and other mobility aids constitute a significant segment within NMT. Wheelchairs, mobility scooters, and several other types of assistive equipment are necessities for navigating the environment for those who are disabled or have restricted mobility. To ensure accessibility and encourage the participation of individuals with disabilities in NMT systems, it is essential to provide barrier-free infrastructure, which may include ramps, elevators, and paved areas with raised textures.
5. Skating and rollerblading: skating and rollerblading both require using inline skates or rollerblades as a mode of

transportation and for recreational purposes, respectively. Individuals who are adept in skating and like it as a form of transportation might choose one of these modes of transportation as an alternative, despite the fact that it is not as common as walking or cycling. Skaters may be integrated into NMT networks safely via the use of dedicated skate tracks or shared-use routes.

6. Pushcarts and handcarts: pushcarts and handcarts are wheeled devices that are controlled manually and are used to convey products and freight. They are often used in retail sales on the street, in delivery services, and in other types of low-scale companies. Users of pushcarts play an important part in the urban economy and need the existence of suitable infrastructure, such as dedicated loading and unloading spaces, in order to have their operations facilitated.

It is essential to keep in mind that the aforementioned classifications of NMT are not incompatible with one another, and people may make use of more than one mode, dependent on the demands of their particular situations. For instance, a person may walk to a local bus stop, ride a bicycle for longer distances, and use a kick scooter for shorter journeys within the same set of neighborhoods and potential situations. As a result, the development of sustainable and multimodal urban transportation systems requires the construction of a comprehensive and integrated NMT network that can support a variety of different types.

City planners and policymakers can develop NMT systems that cater to a diverse range of users, promote active mobility, and contribute to the creation of vibrant and livable cities by taking into consideration the unique characteristics, infrastructure requirements, and user profiles associated with each of the categories and scenarios [7–9].

## 8.4 Planning Smart Cities to Facilitate NMT

The idea of smart cities seeks to improve the quality of life for citizens by integrating cutting-edge technology, data-driven decision-making, and environmentally responsible urban planning and development. Facilitating NMT modes such as walking, cycling, and other micro-mobility devices is an essential component of smart city design. This component is also known as "NMT." Cities have the ability to construct transportation networks that are sustainable, efficient, and focused on the needs of the people if they give NMT a higher priority in their urban planning scenarios. This section offers a comprehensive review of the planning solutions that smart cities may use to make room for NMT sets [10–12].

1. Integrated urban design: smart city planning for NMT should place a priority on integrated urban design that encourages connectedness, accessibility, and safety for NMT users. Among the most important concerns are:
   a. NMT infrastructure: the process of designing and putting into action a complete network of dedicated pedestrian walkways, cycling lanes, and other forms of micro-mobility infrastructure such as scooter-sharing stations. This infrastructure need to include well planned connections to significant attractions, nodes of public transit, and residential areas.
   b. Mixed-use development: the encouragement of mixed-use development, which includes the combination of residential, commercial, and recreational spaces in close proximity to one another. Because of this, there will be less of a need to drive over great distances, and people will be more likely to choose walking or cycling as feasible forms of transportation for their day-to-day activities.
   c. Land use planning: incorporating land use planning concepts that encourage the creation of compact, walkable communities that are equipped with amenities, green areas, and access to public transit. This encourages walking and biking as modes of transportation while simultaneously reducing dependency on motorized cars.
2. Intelligent transportation systems (ITS): utilizing ITS may improve the effectiveness and safety of NMT in smart cities. Key factors include:
   a. Real-time information: providing real-time information on NMT infrastructure, such as the availability of bicycle-sharing systems and parking spaces, as well as updates on the conditions of the routes. Users are given the ability to make educated judgments and more successfully plan their travels as a result of this knowledge.
   b. Smart traffic management: the implementation of smart traffic management systems that provide priority to NMT modes at junctions, alter the timing of traffic signals depending on demand, and promote interactions

that are both safe and seamless between NMT users and drivers of motorized vehicles.
   c. Data analytics: the process of using data analytics to obtain insights into NMT use trends, detect bottlenecks or safety problems, and improve infrastructure development and resource allocation.
3. Connectivity: connectivity in the final mile smart cities should place an emphasis on enhancing connectivity in the final mile, which will help bridge the gap between public transportation and NMT modes. The following are important strategies:
   a. Intermodal integration: the creation of seamless links between NMT infrastructure and key transportation hubs, such as bus stops, railway stations, and metro stations. This involves the provision of safe parking for bicycles, hubs for micro-mobility, and dedicated walkways for pedestrians.
   b. Shared mobility solutions: the integration of ride-sharing, e-scooter sharing, bike sharing, and other forms of shared mobility with public transit networks. This promotes commuters to integrate non-motorized modes of transportation (NMT) with public transit in order to travel in an efficient and environmentally friendly manner.
4. Safety and security: it is essential, for the general adoption of NMTs, to place a high priority on the safety and security of its users. The following are important measures:
   a. Illumination and surveillance: ensuring that the NMT infrastructure, including walkways, crossroads, and parking lots, are adequately illuminated. Increasing the level of security and deterring any possible threats by installing surveillance systems and emergency call boxes.
   b. Separation from motorized traffic: the process of separating NMT infrastructure from motorized vehicle traffic by constructing physical barriers, dedicated lanes, or traffic-calming techniques. This lowers the likelihood of accidents occurring and helps to provide a risk-free setting for those who use NMT.
   c. Safety education: conducting educational campaigns and awareness initiatives to encourage safe NMT practices among users, motorists, and pedestrians in the

community. This involves teaching users about the laws of the road, the correct manner to use equipment, and the importance of mutual respect on shared roads.

5. Engagement with the community and participation in activities: it is very essential for effective NMT integration that the community be included in the planning and decision-making process, and that they be engaged in the process. The following are important strategies:
   a. Participatory planning: encouraging public participation in planning processes via workshops, surveys, and consultations in order to collect insights, feedback, and ideas from residents, community groups, and NMT advocates.
   b. Collaboration with local NMT organizations, advocacy groups, and NGOs: collaborating with local NMT organizations, advocacy groups, and NGOs to harness their knowledge, encourage community ownership, and ensure that NMT planning matches with the needs and ambitions of the communities.
   c. Continuous evaluation and input: the implementation of mechanisms to routinely analyze the efficiency of the NMT infrastructure, obtain input from users, and make required modifications and enhancements based on increasing demands and changing urban dynamics. By implementing these design concepts, smart cities have the potential to efficiently support NMT, hence generating environmentally friendly and livable urban settings that place a priority on active mobility, decrease congestion, improve air quality, and boost the overall quality of life for people. Smart city planning for the incorporation of NMT encourages a transition toward environmentally friendly modes of transportation while simultaneously fostering cities that are healthier, greener, and more vibrant for real-time scenarios [13–15].

## 8.5 Effect of NMT Planning on Healthcare

The design of NMT, which focuses on encouraging and supporting modes such as walking, cycling, and micro-mobility devices, has the potential to have a substantial influence on both public health and the results of healthcare. Cities have the ability to create settings that encourage active lives, lower the number of sedentary behaviors, and enhance the general health of the people by giving NMT more priority in urban design and transportation infrastructure. This section presents a comprehensive review of the ways in which NMT planning has affected healthcare provisions.

1. Regular physical exercise and fitness: NMT planning both promotes and makes it easier for people of all ages and fitness levels to participate in regular physical exercise. The following are important benefits:
   a. An increase in the amount of time spent engaged in physical exercise: the use of active means of transportation, such as walking or cycling, provides chances for the incorporation of regular periods of physical exercise into daily routines. people may enhance their cardiovascular health, boost their muscle strength, better control their weight, and minimize their chance of developing chronic illnesses such as heart disease, diabetes, and obesity by incorporating physical exercise into their daily activities, including their commutes. This can help people fulfill the required activity recommendations.
   b. Physical fitness and endurance: participating in NMT on a consistent basis may increase muscular strength and endurance, as well as enhance aerobic fitness. Walking and cycling, in particular, contribute to lower body strength, balance, and coordination, which all contribute to promoting healthy aging and lowering the risk of falls among older persons.
2. Positive implications for mental health and overall well-being: the NMT planning process has beneficial effects on both mental health and overall well-being. The following are important benefits:
   a. Stress reduction: active transportation modes give a chance for stress reduction by introducing physical exercise into daily routines and providing a respite from the stresses of commuting in motorized cars.
   b. Health benefits: active transportation modes offer a variety of health benefits, including cardiovascular health, weight loss, and improved mental health. Walking and cycling in areas that are pedestrian-friendly and have plenty of green space may have a relaxing impact, enhance one's mood, and lower one's stress levels.

c. Mental health: regular physical exercise, which is a component of NMT, has been related to better mental health outcomes, such as decreased feelings of depression and anxiety as well as enhanced cognitive performance. The increased possibility of social engagement that comes with walking or cycling may also help strengthen social relationships and support networks, both of which contribute to an overall improvement in mental health.

3. Impact on air quality and the environment: the implementation of NMT planning results in cleaner and healthier settings, which in turn has a beneficial effect on the general population's health. The following are important benefits:
   a. Decreased emissions of pollutants: for decreased emissions of pollutants such as particulate matter, nitrogen oxides, and carbon dioxide, active transportation modes over motorized vehicles is encouraged. This leads to decreased air pollution. A significant reduction in the prevalence of respiratory illnesses, including asthma, allergies, and cardiovascular disease, has been linked to improvements in air quality.
   b. Contribution to the mitigation of climate change: NMT planning makes a contribution to the mitigation of climate change by decreasing dependency on motorized transportation. Reduced emissions of greenhouse gases and energy usage have a beneficial effect on public health because they lower the likelihood of climate-related health problems. These problems include heat-related illnesses, infectious diseases, and food insecurity.
4. Public health and prevent injuries: measures to improve public health and prevent injuries are given priority in effective NMT planning road safety measures are given priority in effective NMT planning, which contributes to the reduction of injuries and an improvement in public health. among the most important concerns are:
   a. Separation of non-motorized transportation infrastructure: dedicated bicycle lanes, pedestrian walkways, and separated NMT infrastructure lower the danger of accidents and conflicts with motorized vehicles, which improves the safety of NMT users.
   b. Traffic calming measures: the implementation of traffic calming measures, such as lowered speed limits, traffic

signal optimization, and improvements to road design, contribute to safer NMT settings and lower injury rates.

   c. Inclusive design: the planning for NMT should give priority to inclusive design principles, which will ensure that people with disabilities, older persons, and vulnerable groups can utilize the infrastructure. This encourages equal access to active transportation alternatives, enhances public safety, and lowers the obstacles that prevent people from participating.

5. Cost of healthcare NMT planning: a decrease in the cost of healthcare: NMT planning may assist in the reduction of healthcare expenses by assisting in the prevention of chronic illnesses, the improvement of overall health outcomes, and the reduction of the load on healthcare systems. The following are important benefits [16–18]:
   a. Prevention of chronic illnesses: the regular physical activity that is linked with NMT has been shown to lessen the chance of developing chronic illnesses such as obesity, diabetes, cardiovascular disorders, and certain forms of cancer. Non-motorized transportation planning makes a contribution to the long-term enhancement of health and the avoidance of chronic illnesses by encouraging more physically active lives.
   b. Decreased healthcare use: NMT planning may lead to decreased healthcare use by lowering the prevalence of chronic illnesses and increasing general well-being. This reduction in healthcare consumption can include fewer hospitalizations, trips to emergency rooms, and the related medical expenses for treating chronic ailments.
   c. Higher productivity: active transportation options, such as walking, biking, and public transit, have been shown to lead to higher productivity and decreased absenteeism in the office. It has been shown that engaging in regular physical exercise may increase cognitive function, focus, and general job performance, all of which can have a favorable effect on economic production sets.

In conclusion, NMT planning has far-reaching implications on healthcare outcomes, including the promotion of physical activity, the improvement of mental health, the enhancement of air quality, the assurance of road safety, and the reduction of expenditures associated with healthcare as shown in

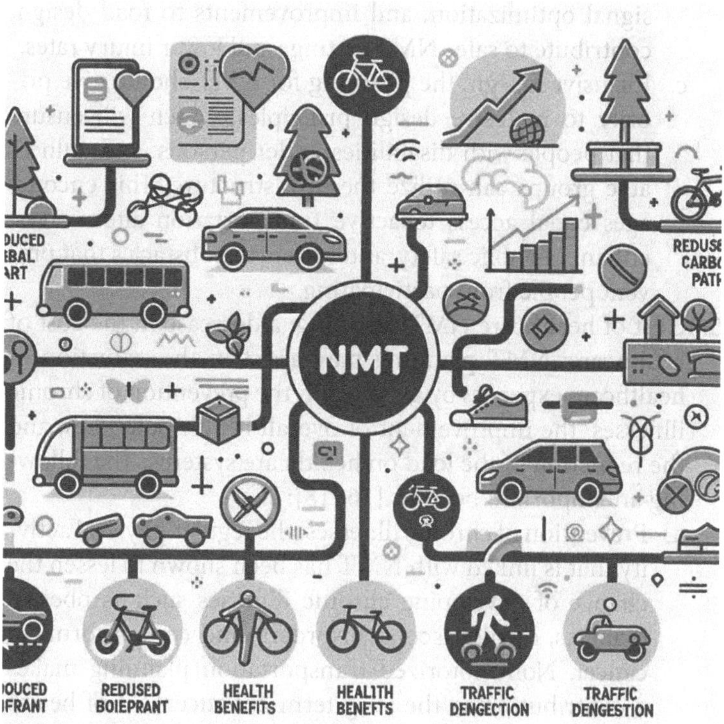


**Figure 8.1** NMT systems.

Figure 8.1. Cities have the ability to create surroundings that are more livable, healthier, and have a beneficial influence on the citizens' well-being and quality of life when they give NMT a higher priority and incorporate it into their urban planning and transportation infrastructure. It is possible that the combination of NMT planning with healthcare strategies and policies will result in considerable long-term health advantages, producing communities that are both sustainable and resilient for real-time scenarios.

## 8.6 Use of Artificial Intelligence and Machine Learning for Integrating Sustainability in NMTs

The use of technologies such as AI and ML has the potential to completely transform the incorporation of environmentally friendly procedures into NMT systems. Artificial intelligence and machine learning have the potential to improve the efficacy, safety, and overall sustainability of NMT modes such as walking, cycling, and other micro-mobility devices. This may be accomplished by employing data analytics, predictive modeling, and intelligent decision-making algorithms. This section presents a comprehensive review of the use of AI and ML to the process of incorporating sustainability into NMTs.

1. Optimization and planning of NMT routes: AI and ML algorithms can evaluate massive datasets to optimize NMT routes. These datasets may include historical traffic patterns, user preferences, and real-time data. The following are examples of important applications:
   a. Smart routing: AI algorithms may advise best routes for NMT users, taking into consideration parameters like as distance, terrain, safety, air quality, and user preferences. This fosters efficient and sustainable travel by selecting routes that reduce travel time, enhance safety, and promote delightful experiences. This also encourages sustainable tourism.
   b. Dynamic Adaptation: ML models have the ability to dynamically alter route suggestions depending on real-time data, such as the amount of traffic, the weather, and the quality of the air. This guarantees that users of the NMT are supplied with up-to-date information and alternative routes, reducing interruptions and encouraging choices that are better for the environment regarding travel.
2. Improved user safety: the applications of AI and ML technologies are very important in improving the user safety of NMT. The following are important applications:
   a. Accident prediction and prevention: ML algorithms are able to examine past accident data, as well as data on weather conditions, road infrastructure, and user behavior, in order to identify high-risk regions and anticipate prospective accidents. This information may be put to use in the implementation of targeted safety measures in order to minimize the number of accidents and injuries that occur, such as increased signs, enhanced lighting, and traffic calming measures.
   b. Intelligent traffic management: AI algorithms may monitor and analyze traffic patterns, identify possible conflicts between NMT users and motorized vehicles, and change

the timing of traffic signal timings to prioritize NMT user safety. This preventative strategy improves junction management, cuts down on the number of conflicts that occur, and increases the NMT infrastructure's overall level of safety.

3. An analysis of user demand and behavior: AI and ML approaches may be used to study user data and behavior patterns in order to get a better understanding of NMT demand and preferences. The following are important applications:
   a. Demand forecasting: ML models are able to conduct an analysis of historical data, which may include use patterns, weather conditions, and events, in order to estimate NMT demand in certain locations and timeframes. With this knowledge, planners and legislators are able to more efficiently allocate resources, develop suitable infrastructure, and encourage environmentally responsible travel choices.
   b. Tailored recommendations: AI algorithms have the ability to deliver tailored recommendations to NMT users based on their preferences. These recommendations may include route ideas, nearby sites of interest, and incentives for environmentally responsible travel choices. Because of this, increased user engagement, contentment, and long-term adoption of NMT modes are encouraged.
4. Evaluation of the system's effect on the environment: The environmental effect of NMT systems may be evaluated and reduced using AI and ML approaches. The following are important applications:
   a. Emissions monitoring: AI algorithms can estimate and monitor the emissions created by NMT modes, taking into consideration parameters such as user behavior, trip length, and travel circumstances.
   b. Emissions reduction: AI algorithms can reduce the amount of emissions generated by NMT modes. This information contributes to the quantification of the environmental advantages of NMT and helps define plans for sustainability.
   c. Planning for green infrastructure: ML algorithms may examine environmental data, such as air quality, noise levels, and the availability of green space, to identify places that might benefit from the creation of green

infrastructure. Specifically, this refers to the planting of trees, as well as green corridors and urban gardens, with the goal of improving the ecological viability of NMT networks.

5. Policy and decision help: AI and ML technologies offer decision-makers with significant insights and help in the process of creating sustainable policies on NMT. The following are important applications:
   a. Simulation and scenario analysis: ML models have the capability to simulate and assess the possible implications of diverse policy scenarios, such as changes in infrastructure, pricing systems, or land use planning. Because of this, policymakers are able to evaluate the implications of various actions on the environment, society, and economy and then make choices based on those information sets.
   b. Data-driven governance: AI technologies make it possible to gather, analyze, and visualize massive datasets linked to the use of NMTs, as well as their effect on the environment. This makes it possible to implement data-driven governance. This helps to encourage cooperation among many stakeholders, fosters transparency, and supports decision-making that is evidence-based scenarios [19, 20].

In summary, the use of AI and ML technologies provide a tremendous opportunity for the integration of sustainability practices into NMT systems. Artificial intelligence and machine learning have the potential to help to the creation of NMT networks that are effective, safe, and environmentally friendly. They can do this by optimizing routes, boosting safety, evaluating user behavior, assessing environmental impact, and providing decision-making assistance. Utilizing these technologies opens the path for smarter, more sustainable cities. It also supports the wider use of NMT modes, which ultimately results in urban settings that are healthier, greener, and more livable for real-time scenarios.

## References

1. Y. Huang, S. Z. Yong and Y. Chen, "Stability Control of Autonomous Ground Vehicles Using Control-Dependent Barrier Functions," in *IEEE Trans. Intell. Veh.*, vol. 6, no. 4, pp. 699-710, Dec. 2021, doi: 10.1109/TIV.2021.3058064.

2. N. Lyu, J. Wen, Z. Duan and C. Wu, "Vehicle Trajectory Prediction and Cut-In Collision Warning Model in a Connected Vehicle Environment," in *IEEE Trans. Intell. Transp. Syst.*, vol. 23, no. 2, pp. 966-981, Feb. 2022, doi: 10.1109/TITS.2020.3019050.
3. Y. Kawamoto, T. Mitsuhashi and N. Kato, "UAV-Aided Information Diffusion for Vehicle-to-Vehicle (V2V) in Disaster Scenarios," in *IEEE Trans. Emerging Top. Comput.*, vol. 10, no. 4, pp. 1909-1917, 1 Oct.-Dec. 2022, doi: 10.1109/TETC.2021.3120551.
4. L. Chen *et al.*, "Research on Traffic Adaptability Testing and Assessment Method of Connected Vehicle Under Platoon Driving Scenario," in *IEEE Access*, vol. 9, pp. 121217- 121239, 2021, doi: 10.1109/ACCESS.2021.3108967.
5. C. Kim, Y. Yoon, S. Kim, M. J. Yoo and K. Yi, "Trajectory Planning and Control of Autonomous Vehicles for Static Vehicle Avoidance in Dynamic Traffic Environments," in *IEEE Access*, vol. 11, pp. 5772-5788, 2023, doi: 10.1109/ACCESS.2023.3236816.
6. B. -K. Xiong and R. Jiang, "Speed Advice for Connected Vehicles at an Isolated Signalized Intersection in a Mixed Traffic Flow Considering Stochasticity of Human Driven Vehicles," in *IEEE Trans. Intell. Transp. Syst.*, vol. 23, no. 8, pp. 11261-11272, Aug. 2022, doi: 10.1109/TITS.2021.3102430.
7. I. Betkier, J. K. Zak and S. Mitkow, "Parking Lots Assignment Algorithm for Vehicles Requiring Specific Parking Conditions in Vehicle Routing Problem," in *IEEE Access*, vol. 9, pp. 161469-161487, 2021, doi: 10.1109/ACCESS.2021.3131480.
8. S. Hulagu and H. B. Celikoglu, "Electric Vehicle Location Routing Problem With Vehicle Motion Dynamics-Based Energy Consumption and Recovery," in *IEEE Trans. Intell. Transp. Syst.*, vol. 23, no. 8, pp. 10275-10286, Aug. 2022, doi: 10.1109/TITS.2021.3089675.
9. J. Li *et al.*, "Cooperative Ecological Adaptive Cruise Control for Plug-In Hybrid Electric Vehicle Based on Approximate Dynamic Programming," in *IEEE Trans. Veh. Technol.*, vol. 72, no. 3, pp. 3132-3145, March 2023, doi: 10.1109/TVT.2022.3217354.
10. J. A. Matute-Peaspan, A. Zubizarreta-Pico and S. E. Diaz-Briceno, "A Vehicle Simulation Model and Automated Driving Features Validation for Low-Speed High Automation Applications," in *IEEE Trans. Intell. Transp. Syst.*, vol. 22, no. 12, pp. 7772-7781, Dec. 2021, doi: 10.1109/TITS.2020.3008318.
11. S. Tirunagari, M. Gu and L. Meegahapola, "Reaping the Benefits of Smart Electric Vehicle Charging and Vehicle-to-Grid Technologies: Regulatory, Policy and Technical Aspects," in *IEEE Access*, vol. 10, pp. 114657-114672, 2022, doi: 10.1109/ACCESS.2022.3217525.
12. K. -F. Chu, A. Y. S. Lam and V. O. K. Li, "Joint Rebalancing and Vehicle-to-Grid Coordination for Autonomous Vehicle Public Transportation System," in *IEEE Trans. Intell. Transp. Syst.*, vol. 23, no. 7, pp. 7156-7169, July 2022, doi: 10.1109/TITS.2021.3067044.

13. J. Wang, J. Fan, M. Xie and L. Cai, "Can Variations of Vehicle Driving Status Provide Accurate Predictors of Discomfort? A Study on the Actual Driving Test," in *IEEE Access*, vol. 9, pp. 47024-47032, 2021, doi: 10.1109/ACCESS.2021.3067959.
14. J. Zhou *et al.*, "Robust Min-Max Model Predictive Vehicle Platooning With Causal Disturbance Feedback," in *IEEE Trans. Intell. Transp. Syst.*, vol. 23, no. 9, pp. 15878-15897, Sept. 2022, doi: 10.1109/TITS.2022.3146149.
15. M. Stryszowski, S. Longo, E. Velenis and G. Forostovsky, "A Framework for Self- Enforced Interaction Between Connected Vehicles: Intersection Negotiation," in *IEEE Trans. Intell. Transp. Syst.*, vol. 22, no. 11, pp. 6716-6725, Nov. 2021, doi: 10.1109/TITS.2020.2994169.
16. T. -Y. Zhang, Y. Yang, Y. -T. Zhu, E. -J. Yao and K. -Q. Wu, "Deploying Public Charging Stations for Battery Electric Vehicles on the Expressway Network Based on Dynamic Charging Demand," in *IEEE Trans. Transp. Electrif.*, vol. 8, no. 2, pp. 2531-2548, June 2022, doi: 10.1109/TTE.2022.3141208.
17. S. Chavhan, D. Gupta, C. Nagaraju, R. A, A. Khanna and J. J. P. C. Rodrigues, "An Efficient Context-Aware Vehicle Incidents Route Service Management for Intelligent Transport System," in *IEEE Syst. J.*, vol. 16, no. 1, pp. 487-498, March 2022, doi: 10.1109/JSYST.2021.3066776.
18. S. Chavhan, D. Gupta, C. B. N., R. K. Chidambaram, A. Khanna and J. J. P. C. Rodrigues, "A Novel Emergent Intelligence Technique for Public Transport Vehicle Allocation Problem in a Dynamic Transportation System," in *IEEE Trans. Intell. Transp. Syst.*, vol. 22, no. 8, pp. 5389-5402, Aug. 2021, doi: 10.1109/TITS.2020.3011198.
19. X. Ma, X. Hu, T. Weber and D. Schramm, "Effects of Automated Vehicles on Traffic Flow With Different Levels of Automation," in *IEEE Access*, vol. 9, pp. 3630-3637, 2021, doi: 10.1109/ACCESS.2020.3048289.
20. Y. Liu, C. Cheng, H. Shi, X. Zuo and S. Chen, "A Two-Stage Approach With a Departure Time Based Solution Representation for Electric Bus Vehicle Scheduling," in *IEEE Access*, vol. 10, pp. 112799-112811, 2022, doi: 10.1109/ACCESS.2022.3215592.

# 9

# Sustainability for Pedestrian Safety Applications


***Abstract***

The purpose of this chapter is to investigate novel solutions to improve the safety of pedestrians in metropolitan areas by making use of skyways, junction subways, and stop stations. These infrastructure components play an essential part in enhancing pedestrian safety in urban contexts and situations by contributing to improvements in accessibility, crossing procedures, collision avoidance systems, and sustainable business operations. The first part of the chapter looks into the idea of urban pedestrian safety and brings attention to the rising significance of environmentally responsible and effective transportation systems. The issues created by growing urbanization, population density, and traffic congestion need the development of innovative ways to protect the safety of pedestrians while also boosting levels of sustainability. Following this, the importance of skyways as elevated pedestrian pathways linking significant metropolitan locations is discussed in this section of the chapter. Skyways provide a multitude of advantages, some of which include greater accessibility, improved crossing procedures, and a reduction in the number of confrontations that occur between pedestrians and vehicle traffic. To maximize pedestrian safety and mobility while also including sustainable elements such as energy-efficient lighting and materials, it is essential that the design and execution of these elevated buildings take sustainability into account from the very beginning. As we go on in the chapter, we will be focusing on junction subways, which are passageways that are safe for pedestrians that are located under major streets. Intersection subways have shown to be beneficial in lowering the likelihood of accidents and improving overall pedestrian safety. They do this by physically separating the flows of pedestrians and vehicles. Exploration is given to the design considerations and engineering concepts that underpin these subways. A special focus is placed on collision control systems that make use of cutting-edge technology to further decrease hazards and increase pedestrian safety standards. In addition, the chapter examines stop stations, which function as important transit hubs that make it easier for pedestrians to travel about in metropolitan areas. The smooth transition between the several forms of public transportation and the protection of pedestrians is a primary concern in the planning and construction of halt stations. This chapter analyzes sustainable operating



Kundan Meshram. Urban Transportation Systems: Sustainability and Future Development, (181–206) © 2024 Scrivener Publishing LLC

strategies for stop stations, such as effective signage, well-designed paths, and the use of renewable energy sources to limit the negative effects these sets have on the environment. The chapter concludes with the presentation of a complete use case for pedestrian safety that incorporates the many ideas and tactics that have been discussed before. It encompasses skyways, junction subways, and stop stations, and it demonstrates a design framework that is both sustainable and efficient, with the goal of improving pedestrian safety in urban contexts. The use case demonstrates how important it is for urban planners, architects, engineers, and transportation authorities to work together in order to develop an all-encompassing strategy for pedestrian safety that is in line with sustainability objectives and scenarios [1–3].

In general, this chapter delivers insightful information on the nexus of pedestrian safety, sustainable urban design, and urban planning. It acts as a complete reference for professionals and scholars working in the sector, and it promotes the adoption of new tactics and designs to assure the well-being of pedestrians in metropolitan areas while simultaneously creating sustainable growth scenarios.

***Keywords:*** Pedestrian safety, subway, skyway

## 9.1 Introduction

In today's quickly changing urban environments, pedestrian safety is an extremely important issue that must be addressed. It is becoming more important for urban planners and designers to prioritize the well-being of pedestrians in light of the increased urbanization and population concentration that is occurring in today's cities. However, ensuring the safety of pedestrians needs more than only reducing the likelihood of potential dangers; rather, it calls for an all- encompassing strategy that incorporates concepts of sustainability, effective transportation systems, and creative infrastructure.

The chapter entitled "Sustainability for Pedestrian Safety Applications" investigates the ways in which urban design, sustainability, and pedestrian safety connect with one another. It places a primary emphasis on skyways, junction subways, and stop stations, which are three essential components that considerably contribute to the safety of pedestrians in metropolitan areas. Not only do these components improve accessibility and crossing measures, but they also provide the groundwork for the construction of collision control systems and sustainable operating procedures.

Congested highways, conflicting traffic patterns, and insufficient infrastructure are only some of the difficulties that pedestrians face in an urban setting. In order to address these difficulties, urban design will need to undergo a paradigm change, with a heavy focus placed on solutions that are both sustainable and pedestrian-centric. Cities have the ability to create dynamic, living areas that improve the health and well-being of their citizens by placing a priority on pedestrian safety and adopting environmentally responsible practices.

Skyways, which are elevated pedestrian walkways, provide an innovative approach to increase the safety of pedestrians as well as pedestrian movement. These elevated structures link essential locations and enable people to explore urban surroundings without the hazards associated with automotive traffic thanks to the accessibility provided by these structures. Skyways not only provide a secure and effective form of transportation, but they also present an opportunity to include environmentally friendly design elements, such as lighting that uses less energy and building materials that are less harmful to the environment.

On the other side, junction subways are designed to address the issues that are presented by crowded highways and intricate intersections. Subways at intersections divide the flow of people and vehicles by creating an underground passageway for pedestrians under the streets. This dramatically reduces the likelihood of pedestrians and vehicles colliding with one another. This chapter examines the design considerations and collision control systems that are used in intersecting subways. The chapter focuses on the role that these elements play in promoting pedestrian safety and supporting sustainable transportation.

Within metropolitan areas, the halt stations that serve as transit hubs also play an important role as key centers of pedestrian mobility. Stopping stations that are not only effective but also thoughtfully constructed emphasize the safety of pedestrians while also enabling seamless communication between various types of transportation. This chapter goes into sustainable operating techniques for stop stations, such as the incorporation of renewable energy sources, well-defined paths, and optimal signage. Stopping stations may reduce their negative influence on the surrounding environment and improve the safety and convenience of pedestrians all at the same time by including environmentally friendly design elements.

This chapter gives a use case that incorporates the many ideas and tactics that have been presented throughout the preceding chapters in order to provide a full grasp of the subject matter. This use case serves as a practical reference for both industry professionals and academic researchers by demonstrating an environmentally friendly and resource-friendly design framework for pedestrian safety in urban contexts. It places a strong emphasis on the significance of cooperation between urban planners, architects, engineers, and transportation authorities in order to develop an all-encompassing strategy for pedestrian safety that is in line with sustainability objectives.

In conclusion, this chapter emphasizes the necessity of environmentally friendly and pedestrian-focused methods to urban development, with a particular emphasis on skyways, junction subways, and stop stations. Cities have the ability to improve urban mobility, alleviate congestion, and create settings that are inclusive and livable by incorporating these aspects with careful attention given to the safety of pedestrians and sustainable practices [4–6]. In the next parts of this chapter, it look into each component in more depth, presenting insights and principles for the design of urban environments that are pedestrian-friendly and sustainable for different scenarios.

## 9.2 Urban Pedestrian Safety—Skyways, Intersection Subways, Halt Stations

It is of the utmost significance in metropolitan areas with a high population density to take precautions to protect the health and safety of pedestrians. In light of the fact that urban areas are facing a rising number of issues associated with traffic congestion, congested streets, and complicated junctions, creative solutions are required to improve pedestrian safety and mobility. This article delves into the topic of skyways, junction subways, and stop stations, which are three essential components that substantially contribute to the safety of metropolitan pedestrians. These components contribute to an infrastructure that is both effective and environmentally friendly, enhances accessibility and crossing procedures, and places an emphasis on the protection of pedestrians in urban settings as shown in Figure 9.1.

1. Skyways

   Skyways, which are also known as pedestrian bridges or elevated walkways, provide a one-of- a-kind solution to the problems that pedestrians confront while navigating through metropolitan areas. These elevated buildings offer safe and accessible routes that link numerous locations, enabling pedestrians to circumvent crowded streets and possible confrontations with automotive traffic. Pedestrians may also benefit from the elevated structures' proximity to public transportation.

   Skyways' capacity to make previously inaccessible areas more easily navigable by foot is its most significant selling

**Figure 9.1** Crossing scenarios.

point. Around the provision of elevated walkways, skyways are able to circumvent obstacles such as congested streets, railroad lines, and other physical impediments, hence making it simpler for humans to traverse around metropolitan surroundings. In addition to this, they may span the spaces between buildings, so establishing uninterrupted linkages between important locations such as retail malls, workplaces, educational institutions, and transit hubs.

Skyways are another important component of improved crossing measures. They do this by installing pedestrian crossings at key areas, which lowers the likelihood of walkers and drivers colliding with one another in the process of crossing the street. In addition, pedestrians are shielded from adverse weather and given a heightened feeling of safety because to the fact that skyways provide an atmosphere that is both covered and brightly illuminated.

Skyways, from the perspective of sustainability, may integrate eco-friendly characteristics to reduce their negative influence on the surrounding environment. It is possible to incorporate a variety of sustainable practices, some examples of which include the use of energy-efficient lighting systems, the incorporation of recyclable building materials, and the addition of green areas along skyway routes. In addition, the combination of skyways and public transportation networks encourages the use of mass transit, which in turn reduces dependence on individual automobiles, so helping to alleviate both traffic congestion and air pollution.

2. Crossroads and Intersections Subways
Due to the complicated patterns created by the convergence of many streams of vehicle traffic, busy crossroads present substantial problems to the safety of pedestrians. Subways at intersections provide a workable alternative by creating subterranean passageways for people under highways. This efficiently separates the traffic patterns of pedestrians and vehicles.

   Subways at intersections reduce the likelihood of pedestrians being involved in accidents and collisions by moving pedestrian traffic underground. It is possible for pedestrians to cross crossings without coming into contact with moving vehicles, which lessens the likelihood of collisions occurring. Subways at intersections are especially useful in high-traffic regions because of the enormous amount of people and cars that pass through such places simultaneously.

   Pedestrian safety is prioritized in the design of intersecting subways, which involves a variety of safety features. A smooth and risk-free flow of pedestrian traffic may be ensured by providing enough lighting, well-marked and well-illuminated routes, and barriers that keep walkers away from driving cars. In addition, the incorporation of modern collision control systems, which may include sensors and surveillance technologies, into intersecting subways may allow for the provision of real-time monitoring and alarm systems, which will further contribute to an improvement in the level of safety.

   Subways at intersections provide chances to add elements that are favorable to the environment, which is important in the context of sustainability. Natural lighting, ventilation systems that use renewable energy sources, and the utilization of environmentally friendly building materials are all potential ways to lessen the negative influence that these subterranean corridors have on the surrounding environment. Integration with public transit networks near intersecting subways may further stimulate the usage of environmentally friendly means of transportation, such as buses or trains, which further promotes the concept of sustainability.

3. Checkpoints and Rest Areas
   The facilitation of pedestrian mobility within metropolitan areas is significantly aided by the presence of halt stations, which are often referred to as transit hubs. These stations provide the function of centralized hubs where various kinds of public transportation, such as buses, trains, trams, and other forms of public transit, intersect. For urban transportation systems to be effective and environmentally responsible, it is essential to provide for the ease and protection of pedestrians at stop stations.

   Creating stop stations with a focus on the safety of pedestrians requires careful design and consideration of a variety of issues. The station is easier to traverse for pedestrians thanks to clear signage and guidance technologies, which in turn reduces the likelihood of confusion and possible confrontations. A smooth flow of pedestrian traffic and the prevention of congestion may be ensured by clearly marked paths and places allocated specifically for waiting.

   To ensure the continued profitability of halt stations over the long term, it is essential to implement environmentally responsible business practices. The integration of renewable energy sources, like as solar panels or wind turbines, has the potential to provide the station facilities with electricity that is both clean and efficient. In addition, the use of cutting-edge technology at the stations, such as real-time information displays and automated ticketing systems, allows for more effective management of available resources, which contributes to an increase in the stations' overall productivity levels [7–9].

   Skyways, junction subways, and stop stations are all examples of comprehensive approaches that may be taken in urban areas to improve pedestrian safety and foster sustainable urban growth. These components provide novel answers to problems like as traffic congestion, crossing procedures, and accessibility, and they do so while putting the health and safety of pedestrians at the forefront of their thinking. Cities may develop safer, more walkable urban places that promote a feeling of community by installing these infrastructural features and adopting sustainable practices. This will lessen dependence on private automobiles and contribute to a greener and more sustainable future scenarios.

## 9.3 Crossing Measures

In metropolitan environments, pedestrians put themselves in a dangerous position whenever they have to cross a busy street or crossroads. It is very necessary to put into place efficient crossing procedures in order to safeguard the health and safety of pedestrians, particularly as cities continue to expand and the amount of traffic continues to rise. This piece of writing investigates the significance of pedestrian crossing measures in terms of improving pedestrian safety. Additionally, it examines a variety of approaches and design considerations that may be used in order to provide safe and efficient pedestrian crossing environments in urban settings.

1. Recognizing the Significance of Taking Precautions when Crossing
Crossing measures are very important to the protection of pedestrians in urban settings because they define safe zones for pedestrians to cross streets and provide the necessary infrastructure. These precautions are designed to reduce the dangers that are inherent in the interactions that take place between pedestrians and vehicle traffic, with the end goal of providing pedestrians with an environment that is both safe and accommodating.

Effective precautions for crossing are necessary for a number of reasons, including the following:

a. Risk mitigation: crossing measures that are well-designed lessen the chance of pedestrian–vehicle incidents, which in turn minimizes the possibility of injuries or deaths. They reduce the likelihood of pedestrians jaywalking or crossing at risky spots by directing pedestrians to proper locations and providing safe crossing points that are specifically designated for use.
b. Accessibility: The implementation of crossing measures improves the accessibility of urban areas for pedestrians, which benefits people of all ages and abilities, including those

who are old, disabled, or have limited mobility. Inclusive mobility is made possible with the installation of accessible curb ramps, crosswalks, and tactile indications. These features guarantee that all pedestrians may safely traverse urban surroundings.

c. Connection: crossing measures increase connection within urban areas, making it easier for walkers to visit a variety of locations. They provide connections between walkways, paths, and pedestrian-friendly zones, which promotes walking as a method of transportation that is good for the environment and reduces dependency on personal automobiles.
d. Confidence and comfort: efficient crossing measures boost walkers' confidence and comfort, which in turn creates a pleasant experience for pedestrians who are out walking. Pedestrians are given a feeling of security and predictability when well-marked crosswalks, clear signage, and adequate traffic control devices are present in an urban area. This helps to make the environment more pedestrian-friendly.

2. Design Considerations for the Implementation of Crossing Measures
In order to develop crossing measures that are successful, significant attention must be given to a variety of design components and considerations as shown in Figure 9.2. The following factors must be taken into account when designing surroundings for crossings if they are to be safe and effective:

a. Visibility: it is essential to have enough visibility so that drivers and pedestrians can see each other in a way that is unmistakable. Visibility is improved and the likelihood of being involved in an accident is cut down significantly when there is enough illumination, particularly during times of low light. Visibility is improved for all users of the road by having sight lines that are clear, proper signs, and road markings.
b. Crosswalk: placement. It is essential to position crosswalks in a strategic manner if one wishes to encourage pedestrians to utilize the specified crossing spots. It is important that pedestrian crosswalks be placed at places that are rational and practical, taking into account pedestrian want lines, adjacent destinations, and traffic patterns. They should also be aligned with sidewalks and walkways, which will provide

**Figure 9.2** Real-time crossing scenarios.

a pedestrian route that is consistent and easy to understand for different scenarios [8–10].

c. Traffic calming measures: putting in place traffic calming measures, such as speed bumps, elevated crossings, or narrower highways, may assist limit the speeds at which vehicles travel and increase the safety of pedestrians. A reduction in the pace at which vehicles travel would improve the interaction between pedestrians and drivers as well as provide people more time to cross roads without risks.
d. Pedestrian refuge islands: pedestrian refuge islands, which are often positioned in the middle of broader streets, provide pedestrians a secure area in which to wait in between the various parts of the crossing process. These islands improve pedestrian safety by shortening the distance they have to travel to cross the street and by only permitting them to cross in one direction of traffic at a time.
e. Utilizing advanced warning systems: utilizing advanced warning systems, such as pedestrian-activated signals or

countdown clocks, notifies both pedestrians and automobiles of forthcoming crossing movements. These systems may be found at places such as airports and train stations. These systems increase communication and awareness, allowing pedestrians to timing their crossings correctly and helping cars to anticipate and yield to pedestrians.
Pedestrians may also benefit from these systems since they enable them to better plan their routes.

f. High-visibility crosswalk markings: increasing the visibility of pedestrian crossings may be accomplished by the use of high-visibility crosswalk markings, such as zebra crossings or ladder markings. Crosswalks benefit from having bold and clear markings in addition to reflective materials since it helps call attention to their existence and enhances their visibility, particularly at night or in situations when the weather is less than ideal.
g. Pedestrian-only phases: dedicated pedestrian-only phases may enhance safety at crowded crossings by enabling people to cross without coming into conflict with motor traffic. These phases may be synced with the timings of the traffic signals to guarantee that the flow of traffic is efficient while putting an emphasis on pedestrian safety.

3. Innovative technologies for crossing measures: as technology progresses, new options arise to improve pedestrian safety in urban contexts and crosswalk safety measures. Several cutting-edge technologies have the potential to be included into crossing procedures, including the following:

a. Intelligent transportation systems (ITS): ITS technologies, such as pedestrian detection systems, have the ability to detect the presence of pedestrians at crosswalks and activate traffic lights in response to their presence. These technologies promote safety by giving people enough time to cross the street and by lowering the number of collisions between pedestrians and automobiles.
b. Crosswalk signs: that are triggered by pedestrians. Pedestrians are able to request a safe crossing interval using crosswalk signs that are triggered by pedestrians. These signals may take the form of push buttons or motion sensors. These signals increase pedestrian control and provide cars a clear indication

that people are crossing, hence increasing safety and lowering the number of incidents that occur.

c. Smart crosswalks: smart crosswalks boost both visibility and awareness by using cutting-edge technology like as integrated sensors and light-emitting diode (LED) illumination. These crosswalks have the ability to detect oncoming cars, light road markers, or show warning messages to both drivers and pedestrians, which increases safety and decreases the risk of accidents occurring.
d. Mobile applications and linked devices: mobile applications or linked devices may give real-time information to pedestrians, such as timing for crossings, information about local risks, or alternate routes. These apps provide pedestrians with access to knowledge, allowing them to make choices based on accurate information and traverse urban areas in a secure manner.

Crossing measures are essential components of urban pedestrian safety because they offer pedestrians with designated places and infrastructure for safely traversing streets and junctions. Pedestrians may use these areas and infrastructure to cross the street. Cities have the ability to reduce dangers, increase accessibility, improve connectedness, and provide an environment that is more pleasant for pedestrians if they put pedestrian crossing measures into effect.

Crossing measures have the potential to make a substantial contribution to the development of safe and inclusive urban areas that place a priority on the well-being of pedestrians. This may be accomplished via strategic design considerations, the incorporation of new technology, and an augmented focus on visibility, connectedness, and comfort levels [10–12].

## 9.4 Flexibility in Accessibility

Accessibility for all persons, especially pedestrians, should be a primary focus of urban planning efforts in order to provide inclusive and sustainable settings for people to live in. The term "flexibility in accessibility" refers to the deployment of measures that meet the different demands of pedestrians, including the needs of persons with disabilities, older citizens, and those who have decreased mobility. This piece of writing investigates the significance of adaptability with regard to accessibility and analyzes several approaches, design considerations, and techniques that give pedestrians more agency and improve their mobility in urban contexts.

1. Having an understanding of the significance of adaptability in terms of accessibility: the accessibility of public spaces must be adaptable if pedestrians of varying abilities are to be able to traverse urban areas in a manner that is safe, autonomous, and respectful of their abilities. The following is a list of important reasons why flexibility in accessibility is important:
    a. Inclusiveness: flexibility in accessibility fosters inclusiveness by meeting the various demands of pedestrians. This helps to ensure that all pedestrians are accommodated. It guarantees that people with disabilities or restricted mobility may access and enjoy urban environments on an equal basis with those who do not have such impairments or mobility restrictions. Cities may become more inclusive and provide equal opportunities for participation in social, economic, and cultural activities when obstacles of a physical kind are eliminated.
    b. Independence: adaptable accessibility solutions make it possible for pedestrians to traverse urban areas on their own, without needing to rely on aid from other people. People who use mobility aids like wheelchairs or walkers benefit from having accessible paths, curb ramps, and tactile indications installed since it enables them to move freely and make their own decisions, which in turn encourages a feeling of autonomy and self-reliance.
    c. Safety: accessibility measures that take flexibility into consideration improve the safety of pedestrians. It is possible for pedestrians who have impairments or restricted mobility to negotiate highways and junctions with confidence and a lower chance of being involved in an accident if the right infrastructure and features are provided. These include accessible crosswalks, detectable warning surfaces, and clear signs.
    d. An increasingly senior population: in light of the fact that the world's population is becoming older, the need for adaptable accessibility measures is growing. Challenges such as limited mobility, visual problems, or cognitive limits may be experienced by senior citizens. When it

comes to accessibility, flexibility is key to ensuring that urban surroundings are age-friendly and providing older persons with the opportunity to continue to be active members of their communities.

2. Methods for achieving flexibility in accessibility: in order to achieve flexibility in accessibility, a number of methods and design considerations may be employed, including the following:
   a. Universal design: the concepts of universal design concentrate on the process of designing settings that are useable by all persons, regardless of the talents those individuals may possess. This strategy entails planning the infrastructure, walkways, and public areas in such a manner that they can cater to a diverse set of requirements. The concept of universal design takes into account a variety of aspects, including pleasant resting spots, well-maintained paths, sufficient lighting, and legible and understandable signs.
   b. Accessible paths: It is vital to both design and maintain accessible paths in order to provide flexibility in accessibility. Individuals who need to use mobility aids, such as wheelchairs or walkers, should have access to routes that are wide, even, and have a smooth surface. It is possible to guarantee that people with disabilities have unhindered access to sidewalks, crosswalks, and public areas by installing curb ramps, tactile indications, and handrails in the right locations.
   c. Pedestrian-friendly infrastructure: increasing flexibility in accessibility by including pedestrian-friendly infrastructure features. Along pedestrian routes, these may include benches, resting places, and rest locations that allow opportunity for those with decreased mobility to relax or collect their strength. In order to facilitate accessible mobility, it is essential to provide accessible public transit stations. These stops should include boarding platforms at suitable heights and legible information displays.
   d. Assistive technology: the use of assistive technology has the potential to dramatically increase both the flexibility and accessibility of a given environment. People who are blind or have low vision may benefit from auditory cues or tactile maps placed at crossroads because they make it easier for them to safely navigate streets. Pedestrians

are able to plan their excursions more efficiently with the use of mobile apps or digital platforms that give real-time information about accessible routes, local facilities, or obstacles.

e. Participation of stakeholders: participation of stakeholders, which may include persons with disabilities, advocacy groups, and organizations, is essential in order to have a comprehensive awareness of the particular requirements and obstacles that are experienced by pedestrians. Through the process of collaboration, the viewpoints of a variety of groups are certain to be taken into consideration throughout the planning, design, and implementation of flexible accessibility measures.

3. Considerations for use: the use of flexibility in accessibility calls for careful consideration to a variety of aspects, including the following:
   a. Accessible legislation and standards: it is very necessary to comply with all accessible legislation, construction regulations, and standards. These standards provide principles and criteria for the creation of infrastructure that is accessible, guaranteeing compliance and maintaining uniformity across metropolitan areas.
   b. Maintenance and upkeep: it is essential to do routine maintenance and upkeep on infrastructure that is accessible on a regular basis in order to guarantee continuing usage. In order to keep the accessibility level at a high standard, it is required to do regular inspections and repairs, as well as timely removal of any obstacles that may exist, such as obstructions on routes or defective curb ramps.
   c. Education and awareness: educating the public about the significance of flexibility in accessibility helps to foster a culture of inclusivity and understanding levels. Awareness campaigns, training programs, and public outreach initiatives can help raise awareness about the rights and needs of pedestrians with disabilities and promote respectful behavior towards them for different use cases.
   d. Continuous improvement: adopting a culture of continuous improvement makes it possible to conduct continuing evaluations and make adjustments to the degree of flexibility that accessibility measures provide. Feedback systems, user surveys, and data collecting on pedestrian

experiences may all inspire future developments and guarantee that urban landscapes continue to be sensitive to the ever-changing requirements of its inhabitants.

The ability to be flexible in terms of accessibility is essential to the process of establishing urban landscapes that are inclusive and livable for everyone. Individuals with disabilities, elders, and others with restricted mobility may be empowered to traverse urban environments safely, freely, and with dignity if cities adopt strategies that emphasize the different needs of pedestrians. These methods can be implemented via the implementation of strategies that prioritize the needs of pedestrians. For all pedestrians, more adaptability in terms of accessibility improves inclusiveness, independence, and safety, as well as the general quality of life. Cities have the potential to develop a future in which everyone has equal access to the possibilities and advantages that come with living in an urban environment by implementing universal design principles, accessible routes, pedestrian-friendly infrastructure, assistive technology, stakeholder involvement, and a commitment to continual improvements for different use cases [13–15].

## 9.5 Design of Collision Control Systems for Intersections to Improve Pedestrian Safety

In urban contexts, intersections are key regions since they are the places where people and automobiles meet. Because of this, intersections are potential hotspots for accidents and collisions. It is necessary to design effective collision control systems for crossings in order to increase the safety of pedestrians and create urban environments that are welcoming to pedestrians. This piece of writing investigates the significance of collision control systems, analyzes important design concerns, and draws attention to cutting-edge technology that might improve pedestrian safety at junctions.

1. The Importance of Crash Avoidance and Control Systems at Intersections
At junctions, having effective collision control systems is one of the most important things that can be done to protect pedestrian safety. These systems are significant for a number of reasons, including the following scenarios [16–18]:

a. Management of conflict: intersections are characterized by competing movements of automobiles and people. Intersections need conflict management. The movement of various road users may be organized and prioritized with the use of effective collision control systems, which contribute to the management of these conflicts. These technologies lower the chance of accidents and increase pedestrian safety by offering traffic flow patterns that are obvious and predictable.

b. Visibility and communication: collision control systems improve the visibility of pedestrians and drivers to one another, as well as the communication between the two groups. Clear signage, road markings, and traffic signal systems all give visual signals and instructions, enabling cars to anticipate pedestrian movements and assisting pedestrians understand when it is safe to cross the street. It is very necessary to have efficient communication at crossings in order to minimize confrontations and avert accidents.
c. Pedestrian priority: creating collision control systems with the safety of pedestrians as the primary focus helps to make the surrounding environment more pedestrian-friendly. Systems that emphasize pedestrian movements, such as longer crossing times or pedestrian-only phases, support walking as a sustainable means of transportation and contribute to a safer and more walkable urban environment. These systems also encourage walking as a sustainable method of transportation.
d. Accessibility: it is important for collision avoidance systems to take into account the requirements of all pedestrians, including those who have mobility impairments. Individuals who suffer from visual or auditory impairments are guaranteed to be able to safely traverse crossings by having accessible pedestrian signals (APS), tactile indications, aural cues, and longer crossing times installed. Inclusivity and equitable access for all pedestrians may be fostered via the design of systems that are accessible to people with a broad variety of abilities.

2. Design Considerations for Effective Collision Control Systems
In order to design effective collision control systems, it is necessary to give careful attention to a number of important issues, including the following:

a. Analysis of traffic: carrying out an exhaustive study of traffic is one of the best ways to determine the number of pedestrians, peak hours, and traffic patterns at crossings. These data are used to guide the design of collision control systems, making it possible to create individualized solutions that are suited to the particular requirements and dynamics of each junction.

b. Visibility: it is essential to ensure that pedestrians have visibility in order to prevent accidents. When designing the structure, be sure to take into account important details like the placement of signals and signs, clear sightlines, and sufficient lighting. Clear visibility makes it possible for cars and pedestrians to see each other, which lowers the likelihood of accidents and increases overall safety levels [19, 20].
c. Pedestrian crosswalks and refuge areas: the development of pedestrian crosswalks that are well designated and simple to recognize is an important component of pedestrian safety. Crosswalks have to be positioned in such a way as to limit the distances that pedestrians have to traverse and to create safe havens in which they may wait in between phases of crossing traffic. These refuge spaces not only safeguard pedestrians but also make it possible for them to cross the street in only one way at a time.
d. Intelligent transportation systems (ITS): the incorporation of technology related to ITS has the potential to improve the effectiveness of collision control systems. Pedestrian detection systems, adaptive signal management, and real-time data analysis may improve pedestrian safety and overall traffic flow by optimizing signal timings, adjusting pedestrian crossing times depending on demand, and reducing delays. This improves pedestrian safety as well as the overall flow of traffic.
e. Accessibility features: making certain that accessibility is maintained is an essential part of the design process for collision control systems. Accessible pedestrian signals (APS) with audible and tactile indications, detectable warning surfaces, and correctly built curb ramps enable accessibility for people who have impairments and encourage safer interactions between pedestrians and automobiles. In addition, APS with audible and tactile indications, detectable warning surfaces, and curb ramps.
f. Education and signage: the successful implementation of educational initiatives and signage is an essential component of collision control systems. While educational

activities promote knowledge about pedestrian safety, correct crossing behavior, and the operation of collision control systems, walkers and vehicles are guided by signage that is both clear and informative. A culture of pedestrian safety may be promoted and habits can be changed with the use of educational materials and campaigns that are well planned.

3. New and Emerging Technologies for Crash Avoidance and Safety Systems
The use of cutting-edge technology has the potential to considerably improve the efficacy of collision avoidance systems at crossings.

a. Pedestrian detection systems: the most recent generation of pedestrian detection systems use radar, cameras, or sensors in order to detect the presence of pedestrians at junctions. These systems have the ability to prolong pedestrian crossing durations, trigger signal adjustments, or activate auditory signals to ensure that pedestrians have enough time to cross the street in a safe manner.
b. Connected vehicles: technologies that allow vehicles and traffic signal systems to communicate with one another include vehicle-to-vehicle (V2V) and vehicle-to-infrastructure (V2I) communication technologies. Because of this connectedness, improved coordination and prioritizing of pedestrian movements are now possible, leading to fewer conflicts and an overall improvement in pedestrian safety.
c. Mobile applications and real-time information: pedestrians may be empowered to make educated choices at junctions by using mobile applications that give real-time information on pedestrian-friendly routes, crossing timings, or possible dangers. These apps may interface with collision control systems in order to provide pedestrians with advice that is both individualized and aware of their surroundings.
d. Automated traffic signal systems: automated traffic signal systems improve the flow of traffic, including the movement of pedestrians, via the application of artificial intelligence and machine learning algorithms. These systems have the ability to dynamically modify signal timings depending on real-time circumstances, pedestrian demand, and traffic patterns, which improves both the safety of crossings as well as their overall efficiency levels.

Improving the safety of pedestrians in urban contexts absolutely requires the installation of effective collision control systems at junctions. Cities are able to design systems that reduce conflicts between walkers and automobiles, promote safe and efficient traffic flow, and create pedestrian-friendly urban places by taking into consideration aspects such as visibility, pedestrian priority, accessibility, and new technology sets. The implementation of collision control systems that are effectively designed improves the entire experience of being a pedestrian, stimulates the use of sustainable forms of transportation, and helps to the development of communities that are both safe and welcoming for different scenarios.

## 9.6 Design of Use Case for Pedestrian Safety for Sustainable Operations

Fostering urban surroundings that are sustainable and livable includes fostering an atmosphere that is safe for pedestrians. The development of all-encompassing strategies that place a premium on the safety of pedestrians while also taking into account the viability of business operations over the long term is required in order to design efficient use cases for pedestrian safety. This piece of writing investigates the significance of incorporating pedestrian safety into operations that are sustainable, analyzes major design issues, and emphasizes the advantages of taking a holistic approach to the problem of pedestrian safety levels.

1. Including Pedestrian Safety Considerations in Sustainable Operation
It is necessary to include pedestrian safety into sustainable operations in order to create urban places that are not only accessible but also safe and good for the environment. The following are some of the reasons why it is necessary to integrate these systems:

a. Human-centered design: sustainable operations respect the key role of pedestrians in urban transportation by placing an emphasis on pedestrian safety as a top priority. Promoting a human-centered approach that increases overall sustainability by taking into account the requirements, behaviors, and safety needs of pedestrians in the design and administration of transportation systems, infrastructure, and public places is one way to accomplish this goal.
b. Promotion of active transportation: designing use cases for pedestrian safety supports active forms of transportation,

such as walking and cycling, which are congruent with the aims of sustainability. Fostering a transition toward low-carbon and healthy transportation choices, such as sidewalks, crosswalks, and dedicated bike lanes, is one way to reduce dependence on automobiles fueled by fossil fuels and mitigate their negative effects on the environment. Pedestrian infrastructure that is safe and easily accessible includes these elements.

c. Equity and social inclusion: contributing to social equity and inclusion is one of the main benefits of incorporating pedestrian safety into sustainable operations. persons of all ages, abilities, and socioeconomic backgrounds are able to participate fully in urban life when walking settings are accessible and secure. This ensures that all persons have equitable access to work opportunities, educational opportunities, medical care, and other critical services.
d. Health and well-being: operations that are pedestrian-friendly encourage physical activity and enhance the health of the general population. Walking as a major means of transportation has several benefits, including a reduction in sedentary lives, an improvement in air quality, and a reduction in noise pollutions. The cultivation of a feeling of well-being via the development of pedestrian environments that are secure and aesthetically pleasing leads to the creation of communities that are both healthier and happier for different scenarios [21–23].

2. Design Considerations for Use Cases in Pedestrian Safety

When designing use cases for pedestrian safety, it is necessary to take into consideration a variety of criteria in order to guarantee that solutions are both complete and effective.

a. The design of infrastructure: it is essential to provide infrastructure that is favorable to pedestrians. This involves the design of pedestrian walkways that are well-connected and easily accessible, as well as crosswalks, pedestrian plazas, and shared areas. Increasing the safety and comfort of pedestrians may be accomplished by the use of measures such as traffic calming tactics, proper lighting, and clear signs.
b. Design of intersections: the design of intersections is one of the most important factors influencing pedestrian safety.

Facilitating safe and effective pedestrian crossings at junctions by providing pedestrian refuge islands, signalized crosswalks with pedestrian-only phases, and intersections that are clearly designated with crosswalks. Countdown clocks, audio signals, and sophisticated traffic systems are just some of the cutting-edge technologies that further improve the safety of intersections.

c. The full streets strategy: the full streets strategy entails planning roadways to satisfy the demands of all users, such as pedestrians, cyclists, and those who use public transportation. This strategy incorporates several forms of transportation, places a high priority on the safety of pedestrians, and takes into account contextually relevant design aspects in order to produce a transportation network that is both balanced and sustainable.
d. Land use planning: it is very essential to coordinate land use planning with pedestrian safety. It is possible to lessen the need for long-distance travel and increase the number of small journeys that may be taken on foot by designing compact, mixed-use projects with a priority on walkability. Developing areas that are walkable by including a variety of residential, commercial, and recreational amenities stimulates pedestrian activity and promotes environmentally conscious business practices.
e. Integration of technology: the incorporation of forward-thinking technology into pedestrian safety use cases has the potential to boost their efficacy. Intelligent transportation systems, pedestrian identification sensors, real-time information systems, and mobile apps all work together to offer timely updates, enhance traffic management, and empower pedestrians with the knowledge they need to make safer decisions.

3. Taking a Holistic Approach to the Safety of Pedestrians

Taking a comprehensive approach to pedestrian safety within the context of environmentally sustainable operations results in a number of advantages, including the following:

a. Improvements to systemic safety: rather than focusing on individual treatments, a holistic approach prioritizes systemic safety enhancements. A comprehensive strategy helps minimize overall hazards and improves pedestrian safety by focusing on several aspects of the transportation system, including infrastructure, laws, regulations, and user behavior.

b. Engagement of stakeholders: engaging various stakeholders in the design process, such as pedestrians, community organizations, advocacy groups, and local authorities, helps to develop collaboration and ensures that a variety of opinions are taken into consideration. The participation of stakeholders results in solutions that are tailored to the unique situation, and it also raises the level of community ownership and acceptability of pedestrian safety measures.
c. Evidence-based decision making involving data: using data and analytics to make evidence-based decisions on pedestrian safety is made easier by using these tools. Planners and policymakers are able to identify high-risk regions, prioritize actions, and quantify the effect of adopted use cases by evaluating pedestrian movement patterns, accident data, and user feedback.
d. Continuous assessment and improvement: a holistic approach recognizes that the process of ensuring the safety of pedestrians is one that is continuous. Mechanisms for routine monitoring and assessment as well as feedback make it possible for use cases to undergo continual development scenarios. This iterative approach ensures that pedestrian safety measures continue to be effective, are adaptive to new demands, and are in alignment with sustainable operations [24, 25].

It is necessary to take a comprehensive and integrated strategy in order to design use cases for the pedestrian safety of sustainable operations. Cities have the ability to build urban landscapes that are environmentally friendly, inclusive, and healthy if they place a priority on the safety of pedestrians within the context of sustainable transportation systems. The development of efficient use cases that improve pedestrian safety, promote active transportation, and contribute to long-term sustainability levels is enabled by incorporating infrastructure design, intersection improvements, complete streets concepts, land use planning, technology integration, stakeholder engagement, and data-driven decision making. Not only would the implementation of a comprehensive strategy for pedestrian safety save lives, but it will also build communities that are lively and livable for the benefit of all inhabitants, tourists, and persons for real-world circumstances.

## References


1. Y. Zhang, Y. Zhang and R. Su, "Pedestrian-Safety-Aware Traffic Light Control Strategy for Urban Traffic Congestion Alleviation," in *IEEE Trans. Intell. Transp. Syst.*, vol. 22, no. 1, pp. 178-193, Jan. 2021, doi: 10.1109/TITS.2019.2955752.
2. R. Quan, L. Zhu, Y. Wu and Y. Yang, "Holistic LSTM for Pedestrian Trajectory Prediction," in *IEEE Trans. Image Process.*, vol. 30, pp. 3229-3239, 2021, doi: 10.1109/TIP.2021.3058599.
3. H. Chen and X. Zhang, "Path Planning for Intelligent Vehicle Collision Avoidance of Dynamic Pedestrian Using Att-LSTM, MSFM, and MPC at Unsignalized Crosswalk," in *IEEE Trans. Ind. Electron.*, vol. 69, no. 4, pp. 4285-4295, April 2022, doi: 10.1109/TIE.2021.3073301.
4. S. Islam, Z. H. Khan, T. A. Gulliver, K. S. Khattak and W. Imran, "Pedestrian Traffic Characterization Based on Pedestrian Response," in *IEEE Access*, vol. 10, pp. 118397- 118408, 2022, doi: 10.1109/ACCESS.2022.3220321.
5. S. K. Jayaraman, L. P. Robert, X. J. Yang and D. M. Tilbury, "Multimodal Hybrid Pedestrian: A Hybrid Automaton Model of Urban Pedestrian Behavior for Automated Driving Applications," in *IEEE Access*, vol. 9, pp. 27708-27722, 2021, doi: 10.1109/ACCESS.2021.3058307.
6. P. R. G. Cadena, Y. Qian, C. Wang and M. Yang, "Pedestrian Graph +: A Fast Pedestrian Crossing Prediction Model Based on Graph Convolutional Networks," in *IEEE Trans. Intell. Transp. Syst.*, vol. 23, no. 11, pp. 21050-21061, Nov. 2022, doi: 10.1109/TITS.2022.3173537.
7. J. E. Domeyer, J. D. Lee, H. Toyoda, B. Mehler and B. Reimer, "Driver-Pedestrian Perceptual Models Demonstrate Coupling: Implications for Vehicle Automation," in *IEEE Trans. Hum.-Mach. Syst.*, vol. 52, no. 4, pp. 557-566, Aug. 2022, doi: 10.1109/THMS.2022.3158201.
8. S. Dafrallah, A. Amine, S. Mousset and A. Bensrhair, "Monocular Pedestrian Orientation Recognition Based on Capsule Network for a Novel Collision Warning System," in *IEEE Access*, vol. 9, pp. 141635-141650, 2021, doi: 10.1109/ACCESS.2021.3119629.
9. S. Schuhbäck, L. Wischhof and J. Ott, "Cellular Sidelink Enabled Decentralized Pedestrian Sensing," in *IEEE Access*, vol. 11, pp. 13349-13369, 2023, doi: 10.1109/ACCESS.2023.3242946.
10. R. Hasan, M. A. Hoque, Y. Karim, R. Griffin, D. C. Schwebel and R. Hasan, "Someone to Watch Over You: Using Bluetooth Beacons for Alerting Distracted Pedestrians," in *IEEE Internet Things J.*, vol. 9, no. 22, pp. 23017-23030, 15 Nov.15, 2022, doi: 10.1109/JIOT.2022.3187965.
11. J. E. Domeyer, J. D. Lee, H. Toyoda, B. Mehler and B. Reimer, "Interdependence in Vehicle-Pedestrian Encounters and its Implications for Vehicle Automation," in *IEEE Trans. Intell. Transp. Syst.*, vol. 23, no. 5, pp. 4122-4134, May 2022, doi: 10.1109/TITS.2020.3041562.

12. X. Zhang, H. Chen, W. Yang, W. Jin and W. Zhu, "Pedestrian Path Prediction for Autonomous Driving at Un-Signalized Crosswalk Using W/CDM and MSFM," in *IEEE Trans. Intell. Transp. Syst.*, vol. 22, no. 5, pp. 3025-3037, May 2021, doi: 10.1109/TITS.2020.2979231.
13. B. Du, C. Zhang, J. Shen and Z. Zheng, "A Dynamic Sensitivity Model for Unidirectional Pedestrian Flow With Overtaking Behaviour and Its Application on Social Distancing's Impact During COVID-19," in *IEEE Trans. Intell. Transp. Syst.*, vol. 23, no. 8, pp. 10404-10417, Aug. 2022, doi: 10.1109/TITS.2021.3093714.
14. F. Gu, J. Niu, L. Jiang, X. Liu and G. P. Hancke, "SafePath: Exploiting Ubiquitous Smartphones to Avoid Vehicle–Pedestrian Collision," in *IEEE Internet Things J.*, vol. 9, no. 9, pp. 6763-6777, 1 May1, 2022, doi: 10.1109/JIOT.2021.3113003.
15. M. Zhang, A. Abbas-Turki, Y. Mualla, A. Koukam and X. Tu, "Coordination Between Connected Automated Vehicles and Pedestrians to Improve Traffic Safety and Efficiency at Industrial Sites," in *IEEE Access*, vol. 10, pp. 68029-68041, 2022, doi: 10.1109/ACCESS.2022.3185734.
16. Y. -C. Liu, A. Jafari, J. K. Shim and D. A. Paley, "Dynamic Modeling and Simulation of Electric Scooter Interactions With a Pedestrian Crowd Using a Social Force Model," in *IEEE Trans. Intell. Transp. Syst.*, vol. 23, no. 9, pp. 16448-16461, Sept. 2022, doi: 10.1109/TITS.2022.3150282.
17. Y. Hwang, B. Kang and W. Kim, "Motion Cue-Based Sudden Pedestrian Behavior Prediction Using Fuzzy Inference," in *IEEE Access*, vol. 9, pp. 135245-135255, 2021, doi: 10.1109/ACCESS.2021.3115964.
18. P. Mukherjee and B. Lall, "Pedestrian Behavior Analytics on Dashcam Videos in Chaotic Environments," in *IEEE Sens. J.*, vol. 21, no. 14, pp. 15660-15669, 15 July15, 2021, doi: 10.1109/JSEN.2021.3062762.
19. P. Held, D. Steinhauser, A. Koch, T. Brandmeier and U. T. Schwarz, "A Novel Approach for Model-Based Pedestrian Tracking Using Automotive Radar," in *IEEE Trans. Intell. Transp. Syst.*, vol. 23, no. 7, pp. 7082-7095, July 2022, doi: 10.1109/TITS.2021.3066680.
20. S. Zhang and M. Abdel-Aty, "Real-Time Pedestrian Conflict Prediction Model at the Signal Cycle Level Using Machine Learning Models," in *IEEE Open J. Intell. Transp. Syst.*, vol. 3, pp. 176-186, 2022, doi: 10.1109/OJITS.2022.3155126.
21. B. Huang, G. Mao, Y. Qin and Y. Wei, "Pedestrian Flow Estimation Through Passive WiFi Sensing," in *IEEE Trans. Mob. Comput.*, vol. 20, no. 4, pp. 1529-1542, 1 April 2021, doi: 10.1109/TMC.2019.2959610.
22. Q. Wang *et al.*, "Recent Advances in Pedestrian Navigation Activity Recognition: A Review," in *IEEE Sens. J.*, vol. 22, no. 8, pp. 7499-7518, 15 April15, 2022, doi: 10.1109/JSEN.2022.3153610.
23. W. -J. Chang, L. -B. Chen, C. -Y. Sie and C. -H. Yang, "An Artificial Intelligence Edge Computing-Based Assistive System for Visually Impaired Pedestrian

Safety at Zebra Crossings," in *IEEE Trans. Consum. Electron.*, vol. 67, no. 1, pp. 3-11, Feb. 2021, doi: 10.1109/TCE.2020.3037065.

24. B. Zhou *et al.*, "A Pedestrian Network Construction System Based on Crowdsourced Walking Trajectories," in *IEEE Internet Things J.*, vol. 8, no. 9, pp. 7203-7213, 1 May1, 2021, doi: 10.1109/JIOT.2020.3038445.
25. W. M. D. Chia, S. L. Keoh, A. L. Michala and C. Goh, "Infrastructure-Based Pedestrian Risk Tagging Methodology to Support AV Risk Assessment," in *IEEE Access*, vol. 10, pp. 71462-71480, 2022, doi: 10.1109/ACCESS.2022.3188306.

# 10

# Environmental Impact Assessment

***Abstract***
In this chapter, the full process of analyzing the possible environmental consequences that are linked with activities involving vehicle traffic and transportation is discussed. This chapter includes an in-depth study of many issues linked to environmental impact assessment. Topics covered in this chapter include the description of planned activities, structural audits, analysis of site selection methods, baseline conditions, primary concerns, green construction techniques, and the benefits that these practices provide for different use cases [1–3].

This chapter places an emphasis on the necessity of evaluating several dimensions of impacts, such as environmental, social, economic, and cultural factors, in addition to their cumulative, regional, temporal, and geographical considerations. The chapter also discusses the importance of addressing their cumulative effects. It acknowledges that sustainable vehicle traffic and transport systems need to be created in harmony with the surrounding environment and communities, taking into consideration the possible good and negative implications they may have for various situations. Specifically, it recognizes that such systems must take into account the potential positive and negative effects that they may have for real-time scenarios.

The suggested activities are evaluated based on the possible good effects that they may have, such as increased accessibility, less congestion, increased economic possibilities, and pleasant social connections. At the same time, the chapter discusses possible unfavorable effects, such as the pollution of air and noise, the loss of habitat, the increasing use of energy, and the disruption of social relationships.

In addition, the chapter emphasizes the need of developing mitigation plans and monitoring mechanisms in order to lessen the impact of any negative consequences and to encourage sustainable growth. The development of mitigation strategies is an extremely important step in the process of finding solutions to problems caused by activities related to transportation and vehicle traffic. These may include the incorporation of green infrastructure solutions, the promotion of public transit systems, the adoption of environmentally friendly building techniques, and the installation of technology that are clean and efficient in their use of energy. The monitoring strategies that have been covered in this chapter provide assistance in determining the efficacy of mitigating measures and in making certain that planned activities are in accordance with the environmental criteria that

Kundan Meshram. Urban Transportation Systems: Sustainability and Future Development, (207–230) 

have been established. Monitoring activities include the routine examination of environmental parameters, the monitoring of changes in key indicators, and the collection of input from stakeholders in order to direct continuing improvements in the environmental friendliness of vehicle traffic and transport systems.
In conclusion, this chapter offers a complete review of environmental impact assessment within the framework of sustainable vehicle traffic and transportation. It brings to light the necessity for rigorous examination of planned actions, consideration of numerous effects, effective plan design and execution, and comprehensive monitoring strategy development. By implementing these principles, decision-makers and other stakeholders may work toward reducing the environmental impact of activities related to vehicle traffic and transportation, promoting sustainability, and ensuring a brighter future for both society and the environmental scenarios.

***Keywords:*** Environmental impact assessment, traffic, structural audits

## 10.1 Introduction

The tremendous acceleration of urbanization and industrialization has resulted in an ever-increasing need for transportation systems that are both effective and environmentally friendly. People can be connected to one another, economic operations can be facilitated, and social connections may be supported all thanks to the function that traffic and transportation play. However, the proliferation of these systems has generated worries about the possible harmful environmental implications that they may have. Because of this, it has become necessary to conduct a detailed review process known as an augmented group of environmental impact assessment (EIA) operation sets [4–6].

This chapter dives into the important role that EIA plays in both comprehending and effectively controlling the environmental repercussions that are linked with activities involving vehicle traffic and transport. This chapter takes a look at the many different aspects that go into the assessment process and highlights how crucial it is to include environmentally friendly policies and procedures into the design and development of transportation networks.

The description of the planned activity should always be the focal point of any EIA. This chapter will begin by presenting an overview of the planned actions pertaining to vehicle traffic and transportation. Some of these activities include the building or development of road networks, the introduction of new transportation modes, or the deployment of novel technology. It is very necessary to have a thorough understanding of the planned activity's scope and size in order to conduct an in-depth analysis of its possible implications.

The relevance of conducting structural audits while assessing the current state of an existing infrastructure is another topic that is covered in this chapter. This involves determining whether or not existing road networks, bridges, tunnels, and other infrastructure connected to transportation are adequate to manage additional traffic or to accommodate environmentally friendly forms of transportation. Audits of the structure assist determine which parts of the infrastructure need to be modernized or improved in order to make transportation more reliable and effective.

Another important topic covered in this chapter is the procedure of choosing a location for the new business. The methods for selecting a site entail doing an analysis of many possible sites for various pieces of transportation infrastructure, such as roads, railway stations, and airports. The selected location is evaluated based on a number of criteria, including environmental sensitivity, closeness to residential areas, and accessibility, with the goals of minimizing any negative consequences and improving the site's overall sustainability.

To successfully complete an EIA, it is essential to have a solid understanding of both the baseline circumstances and the primary issues. This chapter emphasizes how important it is to do a comprehensive analysis of the environmental, social, economic, and cultural circumstances that already exist in the project region. It is feasible to compare and assess changes that take place as a result of the execution of vehicle traffic and transport operations after collecting baseline data. This becomes possible once the data are collected. The identification of main problems helps to concentrate the evaluation on crucial areas that need for further attention to be paid to them.

The idea of environmentally responsible construction has seen tremendous growth in popularity in recent years. This chapter examines the use of sustainable design and construction practices to transportation infrastructure development projects. It investigates the ways in which environmentally responsible practices in building design and construction, such as the construction of energy-efficient buildings, the use of renewable materials, and the implementation of solutions for green infrastructure, can reduce negative effects on the environment and improve the long-term viability of traffic and transportation networks.

Evaluating the possible positive and negative outcomes is an important part of EIA. This chapter places an emphasis on the significance of taking into account a broad variety of repercussions, including those on the environment, society, the economy, and culture. It investigates the possible beneficial effects of vehicular traffic and transportation, including expanded economic possibilities, improved accessibility, and increased social connections. On the flip side, it tackles possible adverse effects as well, such as pollution of the air and noise, the loss of habitats, an increase in energy use, and interruptions in social life styles [7–9].

This chapter explores the significance of cumulative, regional, temporal, and geographical issues in light of the fact that consequences might spread beyond the scope of particular initiatives. Understanding the consequences of transportation development on a larger scale and over a longer period of time is made feasible by doing an analysis of the cumulative effects caused by several projects. The interconnection of transportation networks and the effects these networks have on the regions around them are brought to light by regional concerns. A recognition of the potential for affects to change through time, which calls for constant monitoring and adaptation, is one of the temporal concerns. The recognition of the need of meticulous planning and design as a result of spatial considerations is necessary in order to reduce spatial conflicts and optimize sustainable land uses and scenarios.

Important aspects of an EIA include the development of mitigation measures and monitoring systems. This chapter highlights the need of developing appropriate mitigation strategies in order to limit the negative consequences that might be caused by activities involving vehicle traffic and transportation. The installation of clean and energy-efficient technology, the promotion of public transit networks, the adoption of environmentally friendly building methods, and the integration of green infrastructure solutions are all examples of potential mitigation strategies.

In addition, the chapter emphasizes the significance of monitoring plans when it comes to determining the efficacy of mitigation measures and guaranteeing conformity with set environmental standards. Decision-makers are able to identify areas that might need improvement and make educated choices about how to improve the long-term sustainability of vehicle traffic and transport systems by routine monitoring of environmental parameters, tracking changes in important indicators, and collecting input from stakeholders.

In conclusion, this chapter offers a comprehensive introduction to the subject of environmental impact assessment within the framework of sustainable vehicle traffic and transportation. It highlights the need of understanding the planned actions, evaluating the baseline circumstances, taking into consideration different consequences, formulating appropriate mitigation measures, and implementing rigorous monitoring techniques. It is possible for decision-makers and stakeholders to work toward reducing the environmental impact of activities related to transportation and vehicle traffic, promoting sustainability, and creating a brighter future for both society and the environment if sustainable practices are included into the process of developing new modes of transportations.

## 10.2 Description of Proposed Activity

Within the framework of environmentally responsible vehicle traffic and transportation, the description of the planned activity is an essential component of the foundation upon which an EIA is built. It requires presenting a detailed overview of the proposed transportation project or activity, which includes necessary facts and scope for understanding the possible environmental impacts of the projects or activities.

The purpose of this chapter is to identify the essential components of the planned activity that are connected to the movement of vehicles and transportation. This can encompass a wide variety of different initiatives, such as the construction or expansion of road networks, the introduction of new modes of transportation, the implementation of innovative technologies, or the development of transportation infrastructure such as train stations, airports, or transit hubs.

The project's aims, objectives, and planned results are described in detail in the activity description that has been presented. It offers information on the goal of the transportation project, such as strengthening connectivity, encouraging sustainable modes of transportation, decreasing traffic congestion, and improving mobility levels. It is much simpler to analyze the project's possible implications and to bring them into alignment with the goals of sustainable development if the project's purposes are defined very clearly for different use cases.

In addition to that, the scope of the planned activity as well as its physical components are included in the description. This entails providing specifics of the infrastructure components that will be used, such as roads, bridges, tunnels, parking facilities, and public transit networks, among others. It is also stated in detail how large and extensive the project is, including the areas that it will cover geographically and the amount of capacity that is anticipated, in order to offer a comprehensive picture of the effect that it may have for different use cases.

In addition to this, the description of the planned activity goes into detail on the project's timeframe as well as its phasing sets. It provides an overview of the phases that will be involved in its execution, beginning with planning and design and continuing on through building and operations, and their sets. This temporal component is essential for determining the project's short-term and long-term effects on the environment, in addition to having a grasp of the possible changes that might occur over temporal instances and scenarios.

In addition, the description may include a discussion of the possible synergies and interdependencies with other efforts or projects that are connected to this one. It examines the extent to which the planned activity is compatible with larger transportation initiatives, urban planning endeavors, or regional development plans. This interconnection helps uncover possibilities for integration and coordination, which in turn fosters holistic and sustainable approaches to the growth of transportations.

It is essential that the description include both the expected level of demand as well as the anticipated level of traffic volume connected with the planned activity. It is necessary for it to include estimates of the anticipated rise in the number of vehicle movements, the passenger capacity, or the needs for freight transport. It is feasible to examine the environmental effects that are related with the project if one first quantifies the potential size of the influence that the project may have on traffic patterns and the demand for transportations.

In the description of the planned activity, consideration should also be given to the various stakeholders as well as the beneficiaries. It specifies the many stakeholders that would be impacted by the project, such as vulnerable populations, local communities, companies, and commuters. It is feasible to engage in meaningful conversation with stakeholders and include their points of view into the evaluation process if one first acknowledges the wide range of interests and concerns held by stakeholders.

In general, the explanation of the planned action acts as a detailed overview of the transportation project or initiative that is being undertaken. This gives a comprehensive knowledge of the project's goal, as well as its physical components, size, timing, interdependencies, and predicted demand. Stakeholder concerns are also taken into account. This in-depth description serves as the foundation for carrying out an exhaustive environmental impact assessment, which makes it possible to evaluate the potential for both positive and negative consequences and to devise effective mitigation methods for the purpose of fostering sustainable vehicle traffic and transport systems.

## 10.3 Structural Audits

Structural audits play a crucial role in the EIA process for sustainable vehicle traffic and transport activities. These audits involve evaluating the condition and adequacy of existing infrastructure, such as roads, bridges, tunnels, and other transportation-related structures. The aim is to ensure that the proposed activity aligns with safety standards, can handle increased traffic or accommodate sustainable modes of transport, and promotes the overall sustainability of the transportation system.

The primary objective of a structural audit is to assess the integrity and stability of transportation infrastructure. It involves conducting detailed inspections, surveys, and assessments of the physical components, including the materials used, design specifications, and maintenance records. By evaluating the structural capacity, durability, and safety of the existing infrastructure, potential risks and deficiencies can be identified.

During a structural audit, the condition of various elements is examined. This includes assessing the road pavement, bridge decks, support structures, drainage systems, and other components essential to the transportation network. The audit also evaluates the structural elements in the context of their functionality and suitability for accommodating sustainable transportation practices, such as the introduction of cycling lanes, pedestrian-friendly features, or electric vehicle charging stations.

In addition to assessing the physical condition, structural audits consider the operational aspects of the transportation infrastructure. This includes evaluating the traffic flow, capacity, and congestion issues associated with the existing infrastructure. By examining these factors, the audit helps identify areas that may require modifications or upgrades to optimize traffic management and ensure the safe and efficient movement of vehicles and pedestrians.

Furthermore, structural audits address the environmental implications of the existing infrastructure. They assess the potential environmental impacts resulting from the construction, operation, and maintenance of the transportation system. This includes evaluating the infrastructure's contribution to air and noise pollution, water runoff, habitat fragmentation, and the consumption of natural resources. Through this evaluation, opportunities for implementing sustainable design principles and minimizing environmental impacts can be identified.

An essential aspect of structural audits is considering the resilience and adaptability of transportation infrastructure to climate change. This involves assessing the vulnerability of the existing infrastructure to extreme weather events, such as floods, storms, or heatwaves, and identifying measures to enhance its resilience. Structural audits can guide the incorporation of climate adaptation strategies, such as improving drainage systems, reinforcing vulnerable areas, or considering alternative designs to withstand future climate-related challenges.

The findings from structural audits inform decision-making processes in the EIA. They contribute to the identification of potential risks and vulnerabilities, facilitating the formulation of mitigation strategies. These strategies may involve retrofitting, rehabilitation, or reconstruction of the infrastructure to enhance its safety, durability, and sustainability. By addressing structural deficiencies and improving the overall condition of the transportation system, the audits promote the longevity and effectiveness of sustainable vehicle traffic and transport activities.

In conclusion, structural audits are an integral part of the environmental impact assessment process for sustainable vehicle traffic and transport activities. They involve evaluating the condition, adequacy, and resilience of existing infrastructure to ensure its alignment with safety standards and sustainability goals. By conducting thorough audits, potential risks can be identified, and appropriate measures can be implemented to enhance the structural integrity, operational efficiency, and environmental performance of the transportation systems [10–12].

## 10.4 Analysis of Site Selection Procedure

In order to ensure that sustainable vehicle traffic and transport operations are carried out, the EIA process must include a vital component called the study of the site selection approach. The process of selecting a site entails analyzing potential sites for various pieces of transportation infrastructure, such as highways, railway stations, airports, or transit hubs, to determine which ones have the most potential for success. With the use of this study, the influence that transportation systems have on the surrounding environment in a negative way is reduced, improve their accessibility, and make them more environmentally friendly overall for different scenarios.

The degree to which possible sites are detrimental to the surrounding ecosystem is an important factor to take into account throughout the site selection process. This entails determining the ecological worth of the place as well as the biodiversity and natural resources that are available there. In order to prevent irreparable harm to the environment, it is imperative that ecologically significant locations, such as protected areas, wetlands, or the habitats of endangered species, be avoided or exploited with as little intensity as possible. Alternatives that are less harmful to the environment may be found by doing research on the environmental impact of many prospective locations.

Accessibility is yet another important consideration included in the examination of site selection. Consideration is given to how close the location is to significant population centers, concentrations of economic activity, and transportation networks. It gives priority to locations that encourage efficient travel patterns, cut down on travel lengths, and offer access to preexisting infrastructure. In addition, there is attention given to the accessibility of the site for various modes of mobility, such as access by foot, by bicycle, and by public transit, with the intention of encouraging environmentally responsible means of transportation.

Considerations of social and economic issues are also included in the examination of potential locations. The possible effect on local communities is evaluated, which may include changes in their quality of life, relocation, or interruption. The highest priority is given to locations that reduce the likelihood of social disputes, protect cultural assets, and enable fair access to transportation services. To guarantee that there will be long-term economic advantages stemming from the location that is selected, economic factors such as the creation of jobs, chances for economic growth, and possible synergies with other sectors are evaluated. Considerations pertaining to land use and spatial planning are essential components of the site selection analysis process. An analysis is conducted to see if the planned transportation infrastructure is compatible with the various land uses, zoning restrictions, and urban planning policies that are already in place. In the course of the investigation, any possible problems with residential areas, environmentally sensitive land uses, or protected areas are analyzed for different use cases. The site selection process promotes sustainable land use practices by integrating transportation infrastructure into the existing urban fabric, avoiding land-use conflicts, and optimizing land usage. This is done in order to achieve optimal land utilization sets [13–15].

In addition, factors of environmental justice are included into the overall site selection study as shown in Figure 10.1. The goal of the environmental justice movement is to guarantee that the environmental benefits and costs of transportation infrastructure are distributed in a fair and equitable manner. In the study, possible disproportionate effects on disadvantaged or vulnerable populations, such as low-income neighborhoods or marginalized groups, are evaluated and assessed. Work is done to reduce inequalities in the environment, and fair access to transportation services is promoted via the use of various methods.

In addition, the examination of the strategy for selecting a site takes into consideration the long- term objectives of both regional development and spatial planning. The evaluation determines whether or not the selected location is compatible with larger regional transportation policies, urban development plans, and environmental goals. An all-encompassing strategy for environmentally responsible growth may be supported by the site

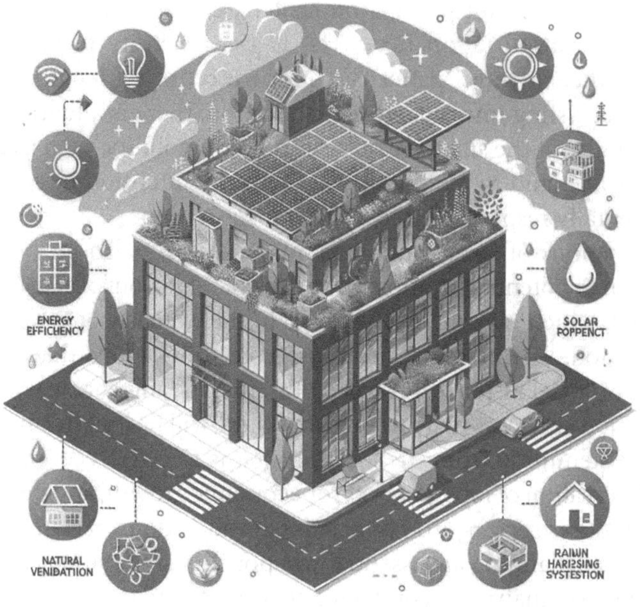


**Figure 10.1** Harvesting systems.

selection analysis, which involves incorporating regional planning initiatives and transportation infrastructure into the planning process.

In conclusion, the investigation of the process for selecting a location is an essential part of the environmental impact assessment for environmentally responsible activities including vehicular traffic and transportation. Evaluating the possible sites' impact on the environment, maximizing accessibility, taking into account social and economic aspects, including land use and spatial planning considerations, addressing environmental justice concerns, and aligning with regional development objectives are all part of the process. Transportation infrastructure may be strategically positioned by undertaking a comprehensive site selection study in order to avoid negative environmental consequences, boost accessibility, and contribute to sustainable and equitable growth scenarios. These goals can be accomplished via a combination of factors [16–18].

## 10.5 Baseline Conditions/Major Concerns

The EIA approach for sustainable vehicle traffic and transport operations must include the evaluation of baseline conditions and the identification of main issues as two of its most important components. The terms "baseline conditions" and "existing conditions" refer to the environmental, social, economic, and cultural circumstances that existed in the project region prior to the execution of the planned transportation activity. It is feasible to analyze the potential implications and identify areas of concern that need particular attention if a complete grasp of the baseline circumstances has been established. This is made possible by the fact that the baseline conditions have been fully understood for different use cases.

The evaluation of the baseline circumstances takes into account many different aspects of the environment. This involves assessing the quality of the air, water supplies, the characteristics of the soil, and the existence of ecologically sensitive areas. It includes determining possible sources of contamination, monitoring ambient noise levels, evaluating existing levels of air and water pollution, and monitoring ambient noise levels. The examination of the potential consequences that the transportation activity may have on these environmental elements may be made possible with a better understanding of the baseline environmental circumstances.

The evaluation of the conditions at the baseline also takes into account social aspects in addition to the environmental circumstances. It entails doing an analysis of the demographic profile, social structures, and cultural legacy of the local communities that are located in the region that will be affected by the project. This involves gaining a grasp of the socioeconomic situation, patterns of livelihood, and overall quality of life of the communities who are being impacted. The evaluation will be able to determine whether or not there will be social disruptions or effects on the local communities if these social aspects are taken into consideration.

It is also vital to do an analysis of the economic circumstances at the starting point. Assessing the current economic activity, employment patterns, and economic dependency on transportation infrastructure in the project region are some of the things that are included in this step. It assists in determining if the planned transportation development may be related with possible economic advantages or issues. For instance, the building of new transportation infrastructure or the extension of existing infrastructure may result in the creation of employment opportunities, the stimulation of economic development, or the impacting of current enterprises.

The evaluation of the baseline circumstances also includes the cultural heritage as a crucial part of the process. It entails locating historical landmarks, cultural places, and other forms of physical and intangible cultural assets that are present in the region of the project. During the process of transportation development, having an understanding of the cultural value of these assets is helpful in evaluating the possible effects on cultural heritage and encourages the preservation of such legacy.

In the process of evaluation, one of the most important steps is to identify the primary problems. Concerns of this kind may be of an environmental, social, economic, or cultural character. For instance, important environmental problems may include things like noise pollution and air pollution, as well as the contamination of water, the destruction of habitat, or the loss of biodiversity. Impacts on local communities might give rise to social problems. These can take the form of things like displaced residents, interruptions to day-to-day activities, or unequal access to transportation services. Concerns on the economic front might include things like population shifts, alterations to economic activity, or the effects on small enterprises. Concerns on a cultural level might include the possibility of the loss or deterioration of cultural heritage places, as well as the weakening of locally observed customs and practices.

The identification of main issues helps direct the focus of the EIA and contributes to the process of determining the particular features of the situation that necessitate solutions for mitigation or management. When these issues are addressed in a proactive manner, the potential for negative consequences to be incurred may be reduced, and chances for improving good outcomes can be uncovered. In addition to this, it makes it possible to participate in conversation with various stakeholders in order to address their concerns and take their points of view into account throughout the decision-making process [19, 20].

In conclusion, doing an analysis of the baseline circumstances and determining the primary problems to be addressed is an essential stage in conducting an environmental impact assessment for sustainable vehicle traffic and transport operations. It entails having a grasp of the current circumstances in the project regions with regard to the environment, society, the economy, and cultures for multiple scenarios. The decision-makers are able to establish mitigation measures and management methods that address possible repercussions and support sustainable transportation development scenarios when they first identify the primary problems that need to be addressed for different use cases.

## 10.6 Green Building and Its Advantages

In the context of operations involving automobile traffic and transportation, environmentally responsible construction methods play an essential part in the promotion of sustainability sets. This section examines the notion of green construction and illustrates the benefits that it offers in the process of developing infrastructure that is both kind to the environment and efficient in its use of energy levels.

Buildings or structures that promote environmental sustainability and resource efficiency are said to be green buildings. The term "green building" refers to the design, construction, operation, and maintenance of green buildings. Green construction principles may be applied to a variety of components within the framework of environmentally responsible vehicle traffic and transportation. These components include transportation terminals, maintenance facilities, parking buildings, and transit-oriented projects, among others.

The major lessening of negative effects on the surrounding environment is one of the key benefits that comes with using green construction practices in the transportation industry. Building techniques that are considered to be environmentally friendly place an emphasis on energy efficiency, the utilization of renewable energy sources, and the reduction of greenhouse gas emissions. Green buildings have the potential to drastically reduce their overall energy consumption as well as the pollution that results from this consumption. This may be accomplished by the use of sustainable design concepts such as adequate insulation, energy- efficient lighting systems, and cutting-edge HVAC (heating, ventilation, and air conditioning) innovations.

In addition, green construction approaches place an emphasis on water conservation and the management of water in a sustainable manner. Water consumption in transportation infrastructure may be reduced by the implementation of water-saving practices such the collection of rainwater and greywater, the installation of low-flow fixtures, and the use of efficient irrigation systems. Green buildings make a contribution to the protection of water supplies and the elimination of water pollution by adopting these practices and putting them into practices [21–23].

The development of healthy and pleasant inside environments is yet another benefit that green architecture in the transportation sector brings to the table for real-time use cases. The comfort of the occupants and the quality of the air within the building are both enhanced because to sustainable design elements such as natural ventilation, daylighting, and the use of materials with low toxicity. When it comes to transportation facilities, where a large number of people spend time waiting or traveling, these concerns are very important to keep in mind.

In addition, green construction approaches encourage the use of materials that are less harmful to the environment and that are derived from the immediate area. Sustainable construction materials, such as those made from recycled content, fast renewable resources, and items with a minimal environmental effect, reduce the amount of natural resources that are extracted and the amount of trash that is produced. Choosing environmentally friendly materials to use in the development of transportation infrastructure may help lessen the negative impact that construction projects have on the surrounding environment while also bolstering local businesses.

In addition, there are important long-term financial advantages associated with green construction. Although implementing environmentally friendly construction methods may result in higher upfront costs, the potential savings in operating expenses throughout the lifetime of the infrastructure may be considerable. Reduced operating expenses and lower utility bills are the direct consequence of designs that are more energy efficient, systems that use renewable energy, and water conservation techniques. Furthermore, green buildings often have longer lifespans and greater market values, which attract investors as well as renters who are interested in sustainability.

In addition, sustainable construction methods contribute to the development of a transportation infrastructure that is both robust and adaptable. The design and building process now takes into account factors related to climate change, such as the increased likelihood of severe weather events, rising sea levels, and higher temperatures. Green buildings have the ability to endure and react to the difficulties that are provided by a changing climate because these buildings use climate adaptation measures, such as flood-resistant designs, better insulation, or raised constructions.

In addition, environmentally responsible construction choices have the potential to function as a catalyst for environmentally responsible transportation behavior. Green buildings support alternate means of transportation and lessen dependency on personal automobiles by include elements that are attractive to pedestrians and cyclists, as well as infrastructure that provides access to public transportation. This integration helps to foster the growth of thriving communities that are easily accessible on foot and lowers the carbon footprint associated with the activities related with transportation.

In conclusion, sustainable vehicle traffic and transport presents a number of challenges that may be addressed via the use of green construction standards. They provide a contribution to the reduction of environmental effects, promote energy and water efficiency, boost resilience to climate change, improve indoor environmental quality, support the use of sustainable materials, produce long-term cost savings, and encourage sustainable transportation practices. The decision-makers have the ability to construct a transportation system that is more sustainable and more environmentally friendly by applying the concepts of green building to the infrastructure of transportations. This system will be beneficial to both societies and the environments and scenarios.

## 10.7 Description of Potential Positive and Negative Environmental, Social, Economic, and Cultural Impacts Including Cumulative, Regional, Temporal, and Spatial Considerations

The actions of sustainable vehicle traffic and transport have the ability to bring about a wide variety of repercussions, including those on the environment, society, the economy, and culture. In order to conduct an accurate environmental impact assessment and make decisions based on accurate information, it is necessary to understand and analyze these consequences. In this part, an overview of the possible positive and negative outcomes, taking into account cumulative, regional, temporal, and geographical concerns, is presented for different use cases.

**(i) Environmental Impacts**
**a. Positive Environmental Impacts**

1. Decreased emissions: the implementation of sustainable transportation practices, such as encouraging electric automobiles, carpooling, or public transit, may lead to a considerable decrease in the amount of greenhouse gas emissions and air pollution that is produced.
2. Preservation of non-renewable resources: the preservation of non-renewable resources via the promotion of environmentally friendly means of transportation, such as walking and cycling, which lowers the demand for fossil fuels and contributes to the conservation of these resources.
3. The preservation of habitat: when planning and developing transportation infrastructure, it is important to take into mind sensitive ecosystems and the habitats of animals. Doing so may help reduce habitat fragmentation and maintain biodiversity.

The expansion of transportation infrastructure may contribute to increasing car emissions and air pollution, especially if sustainable means of transportation are not emphasized. This may be one of the negative effects on the environment.

**b. Destruction of Natural Habitats and Fragmentation of Existing Ones**
The building of new roads, highways, or railroads may result in the destruction of natural habitats and the breaking up of existing ones, which can have an effect on animal populations and disturb biological processes.

**c. Water and Soil Contamination**
Water and soil pollution may be caused by incorrect management of stormwater, improper disposal of hazardous wastes, or increased runoff from road surfaces.

**d. Social Impacts**

1. Influences favorable for society: accessibility may be improved by enhancing transportation infrastructure and supporting environmentally responsible forms of transportation. This is especially beneficial for underprivileged groups, older citizens, and those with impairments.
2. Increased mobility: sustainable transportation solutions, such as public transit systems or bicycle networks, may give mobility choices that are both inexpensive and efficient, therefore lowering congestion and improving trip times.
3. Health benefits: promoting active modes of transportation, such as walking and cycling, may result in better results for public health, including higher levels of physical activity and lower rates of chronic illnesses.

**e. Impacts on Society**

1. Displacement and social upheaval: the construction of transportation infrastructure may necessitate the purchase of property and the relocation of people, which would result in social upheaval and the destruction of cultural treasures.
2. Concerns about safety: an increase in the amount of traffic combined with an insufficient infrastructure may provide safety issues for vulnerable road users such as pedestrians, cyclists, and others.
3. Increased levels of noise and vibration: the growth of transportation networks may result in higher levels of noise and

vibration, which may have an effect on the well-being and quality of life in the communities that are located nearby.

### (ii) Economic Impacts
### a. Impacts Favorable for the Economy

1. Economic activity: stimulating economic activity and contributing to the expansion of the local and regional economies. Construction and operation of transportation infrastructure may result in the creation of job opportunities, which in turn contributes to economic stimulation and growth.
2. A rise in property prices: being close to transit hubs or bike-sharing stations, which are examples of well-planned and sustainable transportation infrastructure, may contribute to an increase in property values and attract investments.

### b. Consequences Harmful to the Economy

1. Expenses and financial burdens: the building and maintenance of transportation infrastructure may incur significant expenses, which calls for careful financial planning and the distribution of resources.
2. Disruption: disruption of local companies. Construction work and traffic diversions may cause local businesses to experience disruption, which may have either temporary or lasting repercussions on the economy.
3. An inequitable distribution of advantages: there is a possibility that the benefits of transportation infrastructure may not be dispersed in an equitable manner, which may exacerbate already-present social and economic disparities.

### (iii) Cultural Impacts
### a. Impacts Favorable on the Culture

1. The sharing of cultural practices and traditions, as well as the promotion of cultural connection: a transportation infrastructure that is well planned has the potential to assist the sharing of cultural practices and traditions, as well as the promotion of cultural connectedness amongst varied populations.
2. Access to cultural amenities, educational institutions, healthcare facilities, and job centers: access to cultural amenities,

educational institutions, and employment centers may be enhanced by using sustainable transportation alternatives.

**b. Impacts on Culture**

1. The destruction of cultural heritage: the act of building new structures or altering the terrain may result in the destruction of cultural heritage sites or in their deterioration, which can have an effect on the identity and the cultural relevance of local communities.
2. Disruption of traditional activities: the growth of transportation may result in the disruption of traditional land use patterns, such as farming or fishing, which has the potential to have an effect on the cultural heritage and livelihoods of indigenous or local groups.

Considerations for the cumulative, regional, temporal, and spatial aspects: cumulative impacts: The term "cumulative impacts" refers to the combined effects that are caused by a number of different transportation projects or activities taking place in a certain region. It is of the utmost importance to conduct a cumulative impacts assessment in order to get an understanding of how the interconnections between different projects might result in heightened environmental, social, economic, and cultural implications.

Considerations for the region transportation infrastructure is often interrelated and meets the requirements of the region as a whole in terms of transportation. When taking into account the area, it is necessary to evaluate how well the proposed project fits in with regional transportation plans, how it will affect regional connectivity, and how it will affect regional growth patterns.

Taking a project's short-term and long-term effects into account is what's meant by "temporal considerations," which is an abbreviation for "temporal considerations and impact assessment." This involves examining the repercussions that are associated to building, as well as the predicted changes in environmental, social, economic, and cultural situations that will occur throughout the course of various temporal instances and scenarios.

Taking into account the space available: analyzing the actual location and spatial arrangement of transportation infrastructures and situations is part of the process of taking spatial considerations into account for different use cases [21–23]. This involves doing an analysis of the possible effects that may be had on existing land use patterns, urban growth, access to conveniences, and the general spatial structure of the places.

In conclusion, sustainable vehicle traffic and transport operations have the potential for a variety of good and negative consequences across a number of domains, including the environment, society, the economy, and culture. When conducting an impact assessment, it is helpful to take into account cumulative, regional, temporal, and geographical factors. Doing so contributes to a better understanding of the more far-reaching consequences of transportation projects and facilitates the creation of policies that optimize good results and minimize adverse impacts. Planning for sustainable transportation that takes an approach that is both comprehensive and integrated has the potential to contribute to future scenarios that are both more balanced and sustainable for different use cases.

## 10.8 Significance of Mitigation Plans and Monitoring Plans

In the process of conducting an EIA for sustainable vehicle traffic and transport operations, the role that mitigation measures and monitoring plans play is of the utmost importance. These strategies are intended to address and manage the possible negative consequences that were discovered during the assessment phase, as well as ensuring that the recommended mitigation measures are successful in their intended purpose. This section investigates the relevance of monitoring plans and mitigation plans in the process of establishing sustainable transportation development scenarios.

**1. Plans for Risk Reduction**

Plans for mitigation include the precise measures and strategies that will be taken to minimize, mitigate, or compensate for the negative consequences that have been recognized as being caused by transportation operations. The importance of mitigation strategies resides in their capacity to accomplish the following:

a) Minimize environmental impacts: mitigation strategies are aimed to decrease or eliminate possible bad effects on the environment. This is the first step in minimizing environmental impacts. It is possible to lessen the negative effects that transportation projects have on the surrounding environment by taking steps such as putting in place noise barriers, improving stormwater management, and putting in place green infrastructure, for instance.

b) Improve social and community well-being: mitigation plans take into account the social implications of transportation projects and strive to resolve issues linked to community well-being. This improves both the social and community well-being. This may require taking steps like as engaging the community, promoting alternate modes of transportation, or offering facilities and services to neighborhoods that are being negatively impacted.
c) Promote economic sustainability: mitigation measures may promote economic sustainability by limiting negative economic consequences and maximizing good economic results. This is known as "maximizing positive economic outcomes and minimizing negative economic impacts." For instance, actions that concentrate on local employment, support for local enterprises, or the creation of possibilities for job training may help to the economic growth and local empowerment of a community.
d) Preserve cultural heritage: one of the priorities of mitigation measures is to preserve and safeguard cultural heritage places and practices. This may entail avoiding or reducing effects on culturally important sites, devising plans for the preservation of heritage, or interacting with local populations to integrate their cultural values into the transportation project. Alternatively, it may involve developing strategies for heritage preservation.

**2. Monitoring Plans**

Throughout the life cycle of a transportation project, monitoring plans are an absolute need for assessing the efficacy of mitigation measures and guaranteeing compliance with environmental and social obligations. The capacity of monitoring plans to do the following is what gives them their significance:

a) Assess environmental performance: monitoring plans provide a framework for the methodical collection of data and assessment of the environmental performance of transportation projects. This involves the monitoring of the quality of the air and water, as well as the levels of noise, biodiversity, and other pertinent environmental indicators. The monitoring of these factors enables the identification of any possible deviations or unanticipated repercussions, which in turn enables early remedial steps to be taken.

b) Conduct an analysis of social impacts: monitoring plans make it possible to conduct an analysis of social impacts on communities that are impacted. This entails keeping an eye on changes in socioeconomic circumstances, accessibility to various transportation services, and other indices of community well-being. Monitoring social impacts helps detect any negative effects that may be having an influence on communities and allows the deployment of suitable steps to address any concerns that may have been raised.

c) Guarantee compliance with relevant laws, rules, and environmental standards: monitoring plans guarantee that transportation projects conform to relevant laws, rules, and environmental standards. Compliance with legal and regulatory requirements may be confirmed and enforced if there is frequent monitoring of project operations, mitigation measures, and environmental performance levels. This monitoring and reporting is done on a regular basis.

d) Helps to facilitate adaptive management: monitoring plans help to facilitate adaptive management by giving feedback on the efficacy of mitigation measures and guiding decision-making processes. Adjustments and enhancements may be made to promote sustainability and handle any unanticipated concerns if the consequences of transportation projects are continuously monitored and evaluated for different use cases [24, 25].

Overall, the relevance of mitigation plans and monitoring plans in sustainable vehicle traffic and transport rests in their capacity to limit negative consequences, promote good results, and assure compliance with environmental and social obligations. This ability is what gives mitigation plans and monitoring plans their name: mitigation and monitoring plans. The decision-makers have the ability to support sustainable transportation development by integrating effective mitigation measures and developing strong monitoring systems. This creates a sustainable transportation development that strikes a balance between the levels of environmental protection, social well-being, and economic success.

## References

1. A. Buchholz, M. Noe, D. Kottonau, E. Shabagin and M. Weil, "Environmental Life-Cycle Assessment of a 10 Kv High-Temperature Superconducting Cable

System for Energy Distribution," in *IEEE Trans. Appl. Supercond.*, vol. 31, no. 5, pp. 1-5, Aug. 2021, Art no. 4802405, doi: 10.1109/TASC.2021.3070703.
2. F. Yuan, Y. Yang, Q. Li and A. Mostafavi, "Unraveling the Temporal Importance of Community-Scale Human Activity Features for Rapid Assessment of Flood Impacts," in *IEEE Access*, vol. 10, pp. 1138-1150, 2022, doi: 10.1109/ACCESS.2021.3137651.
3. T. Xiao and T. -M. Choi, "Quality, Greenness, and Product Line Choices for a Manufacturer With Environmental Responsibility Behaviors," in *IEEE Trans. Eng. Manage.*, vol. 69, no. 6, pp. 2634-2648, Dec. 2022, doi: 10.1109/TEM.2019.2950440.
4. M. Mehrtash, F. Capitanescu, P. K. Heiselberg and T. Gibon, "A New Bi-Objective Approach for Optimal Sizing of Electrical and Thermal Devices in Zero Energy Buildings Considering Environmental Impacts," in *IEEE Trans. Sustain Energy*, vol. 12, no. 2, pp. 886-896, April 2021, doi: 10.1109/TSTE.2020.3024202.
5. Z. T. Abdullah, "Economic Feasibility Assessment of Reclaiming Waste Sheet Steel From End-of-Life Vehicles," in *IEEE Eng. Manage. Rev.*, vol. 50, no. 2, pp. 110- 116, 1 Secondquarter,june 2022, doi: 10.1109/EMR.2022.3143436.
6. T. Jacob, A. Buchholz, M. Noe and M. Weil, "Comparative Life Cycle Assessment of Different Cooling Systems for High-Temperature Superconducting Power Cables," in *IEEE Trans. Appl. Supercond.*, vol. 32, no. 4, pp. 1-5, June 2022, Art no. 4802805, doi: 10.1109/TASC.2022.3168239.
7. X. Shi *et al.*, "Reliability Analysis and Cumulative Impact Assessment of Forest Ecosystem Services: A Case Study of Jilin Province, China," in *IEEE Access*, vol. 9, pp. 59160-59170, 2021, doi: 10.1109/ACCESS.2021.3071123.
8. M. Mehrtash, F. Capitanescu, P. K. Heiselberg, T. Gibon and A. Bertrand, "An Enhanced Optimal PV and Battery Sizing Model for Zero Energy Buildings Considering Environmental Impacts," in *IEEE Trans. Ind. Appl.*, vol. 56, no. 6, pp. 6846-6856, Nov.-Dec. 2020, doi: 10.1109/TIA.2020.3022742.
9. M. Gazzea *et al.*, "Automated Satellite-Based Assessment of Hurricane Impacts on Roadways," in *IEEE Trans. Ind. Inf.*, vol. 18, no. 3, pp. 2110-2119, March 2022, doi: 10.1109/TII.2021.3082906.
10. T. Pirson, T. P. Delhaye, A. G. Pip, G. Le Brun, J. -P. Raskin and D. Bol, "The Environmental Footprint of IC Production: Review, Analysis, and Lessons From Historical Trends," in *IEEE Trans. Semicond. Manuf.*, vol. 36, no. 1, pp. 56-67, Feb. 2023, doi: 10.1109/TSM.2022.3228311.
11. C. R. Poulin and M. B. Kane, "The Effect of Time-Varying Value on Infrastructure Resilience Assessments," in *IEEE Access*, vol. 9, pp. 134556-134575, 2021, doi: 10.1109/ACCESS.2021.3112944.
12. K. Zolfaghari *et al.*, "Impact of Spectral Resolution on Quantifying Cyanobacteria in Lakes and Reservoirs: A Machine-Learning Assessment," in *IEEE Trans. Geosci. Remote Sens.*, vol. 60, pp. 1-20, 2022, Art no. 5515520, doi: 10.1109/TGRS.2021.3114635.

13. S. Munikoti, B. Natarajan, K. Jhala and K. Lai, "Probabilistic Voltage Sensitivity Analysis to Quantify Impact of High PV Penetration on Unbalanced Distribution System," in *IEEE Trans. Power Syst.*, vol. 36, no. 4, pp. 3080-3092, July 2021, doi: 10.1109/TPWRS.2021.3053461.
14. I. D. López, A. Figueroa and J. C. Corrales, "Multi-Label Data Fusion to Support Agricultural Vulnerability Assessments," in *IEEE Access*, vol. 9, pp. 88313-88326, 2021, doi: 10.1109/ACCESS.2021.3089665.
15. E. De Angelis, C. Carnevale, G. D. Marcoberardino, E. Turrini and M. Volta, "Low Emission Road Transport Scenarios: An Integrated Assessment of Energy Demand, Air Quality, GHG Emissions, and Costs," in *IEEE Trans. Autom. Sci. Eng.*, vol. 19, no. 1, pp. 37-47, Jan. 2022, doi: 10.1109/TASE.2021.3073241.
16. J. Pan *et al.*, "Land Cover Classification Using ICESat-2 Photon Counting Data and Landsat 8 OLI Data: A Case Study in Yunnan Province, China," in *IEEE Geosci. Remote Sens. Lett.*, vol. 19, pp. 1-5, 2022, Art no. 2507405, doi: 10.1109/LGRS.2022.3209725.
17. J. Liao *et al.*, "Road Garbage Segmentation and Cleanliness Assessment Based on Semantic Segmentation Network for Cleaning Vehicles," in *IEEE Trans. Veh. Technol.*, vol. 70, no. 9, pp. 8578-8589, Sept. 2021, doi: 10.1109/TVT.2021.3100264.
18. M. Chakraborty, J. Kettle and R. Dahiya, "Electronic Waste Reduction Through Devices and Printed Circuit Boards Designed for Circularity," in *IEEE J. Flexible Electron.*, vol. 1, no. 1, pp. 4-23, Jan. 2022, doi: 10.1109/JFLEX.2022.3159258.
19. R. Pandit, D. Infield and M. Santos, "Accounting for Environmental Conditions in Data-Driven Wind Turbine Power Models," in *IEEE Trans. Appl. Supercond.*, vol. 14, no. 1, pp. 168-177, Jan. 2023, doi: 10.1109/TSTE.2022.3204453.
20. F. Li, X. Sun, S. Zhou, Y. Chen, Z. Hao and Z. Yang, "Infrastructure Material Magnetization Impact Assessment of Wireless Power Transfer Pavement Based on Resonant Inductive Coupling," in *IEEE Trans. Intell. Transp. Syst.*, vol. 23, no. 11, pp. 22400-22408, Nov. 2022, doi: 10.1109/TITS.2021.3099490.
21. D. Yuan *et al.*, "Assessment of the Ecological Impacts of Coal Mining and Restoration in Alpine Areas: A Case Study of the Muli Coalfield on the Qinghai-Tibet Plateau," in *IEEE Access*, vol. 9, pp. 162919-162934, 2021, doi: 10.1109/ACCESS.2021.3133478.
22. L. Yao, Z. Leng, J. Jiang and F. Ni, "Large-Scale Maintenance and Rehabilitation Optimization for Multi-Lane Highway Asphalt Pavement: A Reinforcement Learning Approach," in *IEEE Trans. Intell. Transp. Syst.*, vol. 23, no. 11, pp. 22094-22105, Nov. 2022, doi: 10.1109/TITS.2022.3161689.
23. D. Xu, X. Xiao and Y. Haibo, "Reliability Evaluation of Smart Meters Under Degradation-Shock Loads Based on Phase-Type Distributions," in *IEEE Access*, vol. 8, pp. 39734-39746, 2020, doi: 10.1109/ACCESS.2020.2976200.

24. G. Fu, J. Liu, J. Yang, X. Liu, D. Yao and J. Ni, "Eco-Cost Analysis of Split Air-Conditioner Using Activity-Based Costing Method," in *IEEE Access*, vol. 8, pp. 54952- 54962, 2020, doi: 10.1109/ACCESS.2020.2976853.
25. Z. Zhang, H. Li, L. Yue and Y. Du, "End of Life Vehicle Disassembly Plant Layout Evaluation Integrating Gray Correlation and Decision Making Trial and Evaluation Laboratory," in *IEEE Access*, vol. 8, pp. 141446-141455, 2020, doi: 10.1109/ACCESS.2020.3004447.

# 11

# Traffic Flow Analysis

***Abstract***
In this chapter, an attempt has been made to study the heterogeneous traffic on the road to evaluate the existing capacity at which the urban road is running and then to establish the revised level of service criteria based on the existing road capacity. Videography of the traffic volume is done using the camera mounted at the mid-block of each of the four sections. The reference line technique used for the study is the space mean speed (Vs) of the vehicles. The number of vehicles, travel time, average travel time, average speed of the flow, flow rate, and density for 5-min intervals in peak and off-peak hours for four sections are used to develop the speed-flow relationship and speed-density relationship. After plotting the speed-flow relationship, the maximum flow rate obtained at each section is considered as the capacity of the road section and then the volume-to-capacity ratio (v/c) is determined for each of the sections and graphs between speed vs. v/c are plotted. The revised v/c ratios and speed ranges for each of the sections are suggested based on the analysis.

***Keywords:*** Traffic flow, traffic congestion, speed, travel time

## 11.1 Introduction

Road network plays an important part in the social and economic program of a country. India is a developing country, unlike many developed countries. The cities of India are also undergoing a major change in urbanization and modernization as a result there is a tremendous growth in road traffic. The rise in the population in the urban areas of these cities with rapid growth in vehicular ownership and limited resources, the shifting of people from rural to urban has majorly contributed to the increases of the vehicular traffic on urban streets and now it has become a big challenge in the present scenario to accommodate this traffic. Many large and developed cities already started using alternatives like elevated, underground infrastructures to accommodate the heavy traffic that is occurring in their cities [1, 3].

Kundan Meshram. Urban Transportation Systems: Sustainability and Future Development, (231–244) 

It is known that all road facilities that were designed in the past had taken the behavior of the traffic as homogenous traffic; but in the present context, traffic movement has become more and more complex due to its heterogeneity having much more diverse vehicle proportion with many inadequately performing vehicles, slow-moving vehicles, vehicle trajectory, variations in acceleration rates and non-motorized vehicles. Due to this, traffic operations in many cities are experiencing a considerably lower level of service which is leading to the problem of congestion condition which is a challenge for all large and growing urban streets. Congestion affects the capacity of the road, traffic movement, and excess fuel consumption, and leads to the increase of air, and noise pollution and overall affects the level of service (LOS) of the road [4–7].

When considering road capacity, it is crucial to acknowledge its pivotal role in design, planning, and operation, significantly influencing vehicle mobility and traffic maneuverability. Traffic volume data aids in planning and designing road systems, while speed serves as a fundamental safety measure for road networks. The correlation between traffic stream speed and flow rate determines the capacity of a specific roadway section with the current traffic composition. While the LOS criteria outlined in the Highway Capacity Manual (HCM) offer standards, they might not seamlessly adapt to the complexities of urban roads with diverse traffic conditions. Hence, there is a pressing need to explore alternative performance metrics that can better assess LOS in the context of mixed urban traffic [2]. This exploration aims to refine LOS evaluation within this unique setting, ultimately enhancing the road user experience. Implementing these alternative measures will enable a more precise and customized assessment of road performance, thereby elevating the LOS provided [9, 10].

Density became the primary measure for performance assessment in 1985 (HCM), leading to the proposal of five LOS. Subsequently, the 2000 edition advocated for average travel speed (ATS) as the exclusive parameter for assessing the LOS of urban streets. It proposed six LOS criteria based on ATS values for four distinct urban street classes, with class determination hinging on free flow speed. The most recent edition, HCM (2010), maintains six LOS categories, introducing threshold values of percent free-flow speed as the primary performance measure for assessing LOS on urban streets in the automobile mode [2].

## 11.2 Study Area

### 11.2.1 About Indore City

Indore, a city deeply entrenched in historical roots that trace back to the 16th century, is often hailed as the city of Devi Ahilya. Geographically positioned at the heart of the Malwa Plateau, with coordinates reading 22.7196° north latitude and 75.8577° east longitude, it proudly stands at an average altitude of 550 m above mean sea level. Nestled alongside the Khan and Saraswati rivers, Indore reigns supreme as the most populous and expansive city in the state of Madhya Pradesh, boasting a population of 2,170,295 lakhs according to the 2011 census. Recognized as the financial capital of Madhya Pradesh, it has earned the moniker "mini Mumbai" of India, solidifying its status as the fastest-growing city in central India. Indore has not only emerged as an economic powerhouse but also as a distinguished educational hub. It holds the unique distinction of being the first city in India to host both IIT and IIM campuses two of the most prestigious institutions in the country. Further enhancing its reputation, Indore has clinched the title of the cleanest city in India for three consecutive times, an accomplishment that speaks volumes. The city has also been chosen as one of the 100 Indian cities earmarked for development under the Smart Cities Mission. Brimming with connectivity options, including a well-connected network of roads, railways, and an international airport, Indore serves as a vital gateway to major cities across India. Beyond its strategic geographical advantages, the city opens up a myriad of opportunities across diverse fields, making it an enticing destination for career pursuits and a hub for a variety of job opportunities catering to individuals with varying skill sets. In the realm of transportation, Indore hosts a wide spectrum of traffic, encompassing motorcycles, cars, trucks, cycles, autos, tractors, agricultural vehicles, and more. This diversity contributes to the complexity of the traffic dynamics, creating an ideal scenario for a heterogeneous traffic stream flow.

City buses and bus rapid transit (BRTS) are the only public transportation facilities that have been running in this city to carry passengers. Some private taxis and autos are also there but they are relatively costly. Indore metro project has just started, which will be very beneficial in the coming year as this city has a lot of scope for development which will lead to an increase in the traffic of the city. The city roads have been well planned and almost all-important intersections are now signalized.

### 11.2.2 Route Under Consideration

Bypass Bicholi Mardana (Indore-Bhopal Road) to Madhumilan Square (5.6 km) section is an urban arterial road and is one of the important roads of the city from which the traffic of other states and surrounding villages enter the city. This road provides direct connectivity to the main railway station and bus station of the city passing through the central business district (CBD) area in the city. Dawa Bazar, which is Asia's biggest pharma marketplace in India, and

M.Y. Hospital (Indore's biggest government hospital) are the two important places which get direct connectivity from this road section. Red Church and White Church, which are two of the most respected places in this city, get direct connectivity to these places. This road section consists of some private schools, offices and shopping complexes which makes this road section very busy during the peak hours.

Due to heavy traffic and the composition of mixed vehicles along the course of this section, the traffic handling capacity of the road decreased mostly in the peak hours causing delay in the travel time of the vehicles. Most of the intercity buses and local buses use this road which leads to frequent traffic jams and slow traffic movement on this road which ultimately affects the traffic handling capacity of this road.

The whole stretch of 5.6 km is divided into two sections, details are given as below:

1. Shivaji Vatika-Madhumilan Square section
2. Agriculture College-Shivaji Vatika section

For better understanding, Bypass Bicholi Mardana to Madhumilan square direction is considered as up direction while Madhumilan Square to Bypass Bicholi Mardana direction is considered as down direction.

## 11.3 Data Collection

### 11.3.1 Traffic Volume Study

Traffic volume study was conducted by using the cameras mounted at the mid-block of each of the four sections. Videography of the traffic volumes was done to collect the flow data in peak and off-peak hours of the weekdays. Various vehicles in the traffic are individually counted and categorized manually using slow-motion playback of recordings which is time-consuming work. From that, it was observed that the evening peak hour of the section is between 6:00 and 7:00 pm and the morning peak hour is between 10:00 and 11:00 am.

### 11.3.2 Speed of Traffic

The section selected for the speed study should have a steady and stabilized flow, therefore the speed is calculated at the mid-block section. The space speed which is calculated for the study is (Vs) of mean speed vehicles. For this, vehicles are categorized into seven different categories and the reference line technique is used to record the time taken in every 5-min interval by the vehicles to traverse a defined section of 20-m length on the road using stopwatches. Total 50 samples were taken on each direction, 15–15 samples of two-wheelers and cars, and remaining samples of other five vehicle categories. The average speed of the 5-min interval is calculated using the formula as Equation 11.1:

$$Space\ Mean\ Speed\ (Vs) = 3.6dn \sum_{i=1}^{n} (ti) \tag{11.1}$$

where,

d - road length section, m
n - separate vehicle number
ti – in sec travel experiential time

### 11.3.3 Road Inventory Data

The geometrical conditions of the road have a great influence on the variables of traffic flow. In this study, a four-lane divided road of 5.6 km in length is taken. This total section is divided into two sub-sections based on the pavement type and width of the road.

(i) Bypass Bicholi Mardana–Shivaji Vatika: this complete section has a rigid type of pavement which has an identical width of roadway in both directions having 7.0-m width which is two lanes one-way on both sides. This section has a cycle track of 1.86 m, footpath of 1.6-m, and a median of 1.2m.

(ii) Shivaji Vatika–Madhumilan Square: this section has a bituminous type of pavement as compared to the previous section. This section has a variable width on both sides thus impacting the traffic flow. At the Shivaji Vatika's section, the road width is 9–9.5 m and as the road ends at Madhumilan, the road is only 5.7 m, thus making it a bottleneck of the road.

### 11.3.4 Compilation of Field Data

All the data collected from the field are compiled and tabulated to understand the variation of traffic and composition of the vehicles in the peak and off-peak hours for both directions. The detailed data collection about the number of vehicles, travel time, average travel time, and average speed of the flow for 5-min intervals in peak and off-peak hours for four sections as shown in Appendix 1.

## 11.4 Development of Relationship Between Speed, Flow, and Density

The relations were established by taking the average space-mean-speed (5-min) intervals of peak and off-peak hours. For every 5 min, sum of all the converted vehicles in passenger car unit (PCU) is tabulated by adding individual PCU values of all vehicle classes. This sum of all the vehicles in every 5-min interval is then converted into PCU/h as Flow. Density (or concentration) is given in Equation 11.2:

$$Density(k) = \frac{Flow(PCU\,/\,hr)}{Space\ Mena\ Speed\ of\ Vehicles\ in\ 5-minutes\ of\ interval} \tag{11.2}$$

The relations of speed, flow, and density have been shown in Figures 11.1a,b–11.4a,b.

### 11.4.1 Development of Revised LOC

The development of revised LOC is represented by a parabolic curve, serving as a conceptual framework for understanding the LOS in the speed vs. flow domain. The HCM defines distinct levels of service on this curve up to LOS E, establishing different service levels within this operating range.

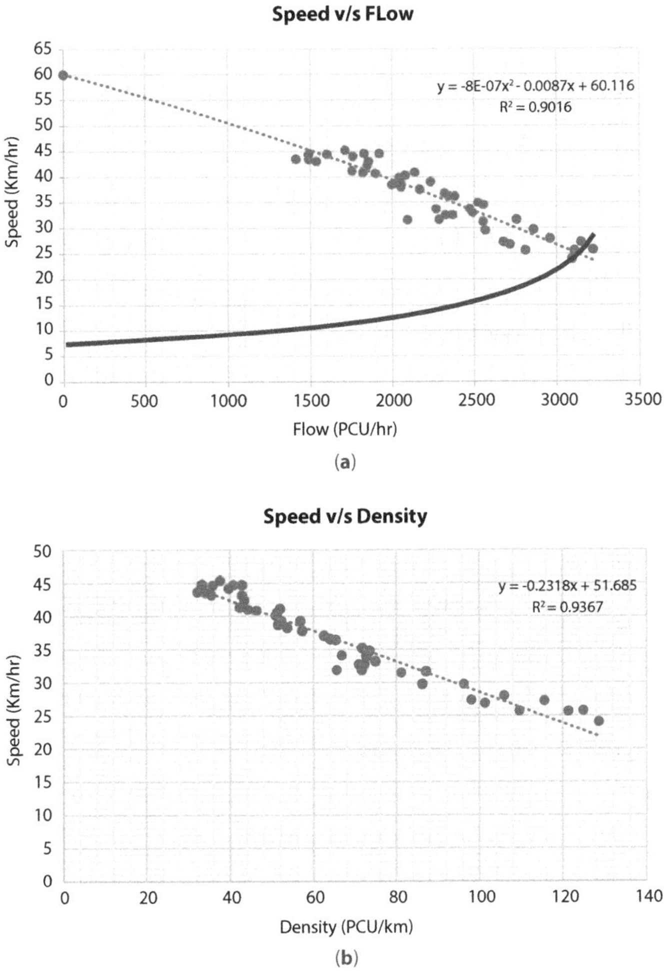


**Figure 11.1** (a) Speed-flow and (b) Speed-density relationship curve at Agriculture College section–Shivaji Vatika section (up).

This study employs the speed-flow relationship to determine LOS ranges, establishing thresholds for each level of service. These thresholds are then compared with those specified in the HCM.

## 11.4.2 Comparison of LOC

Each of the sections is suggested, which is the base ratio of v/c and space mean speed. The average range of v/c ratio and space mean speed for the

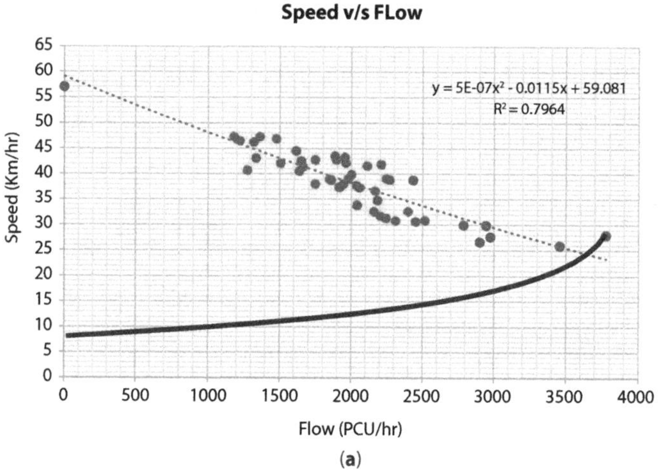


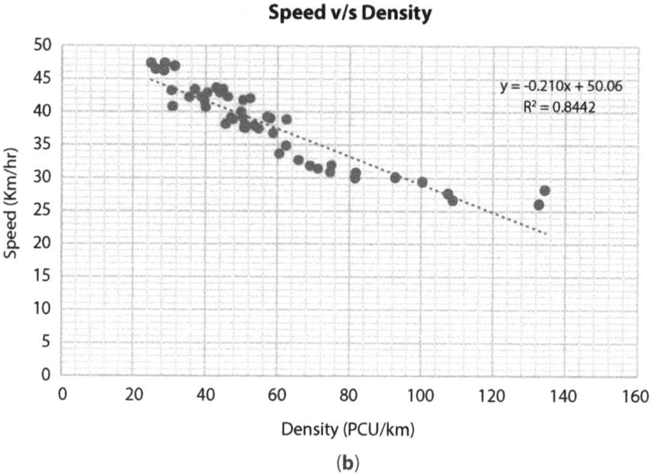


**Figure 11.2** (a) Speed-flow and (b) speed-density relationship curve at Shivaji Vatika–Agriculture College section (down).

complete section in both directions are described in the table. The v/c and average speed ranges recommended by HCM and observed v/c and speed ranges for urban arterial are compared and is shown in Tables 11.1 and 11.2.

It is observed that the speed-flow relationship is developed for the selected urban arterial road section. The curve follows a second-order polynomial quadratic relation indicating a good relationship between the observed speed and flow.

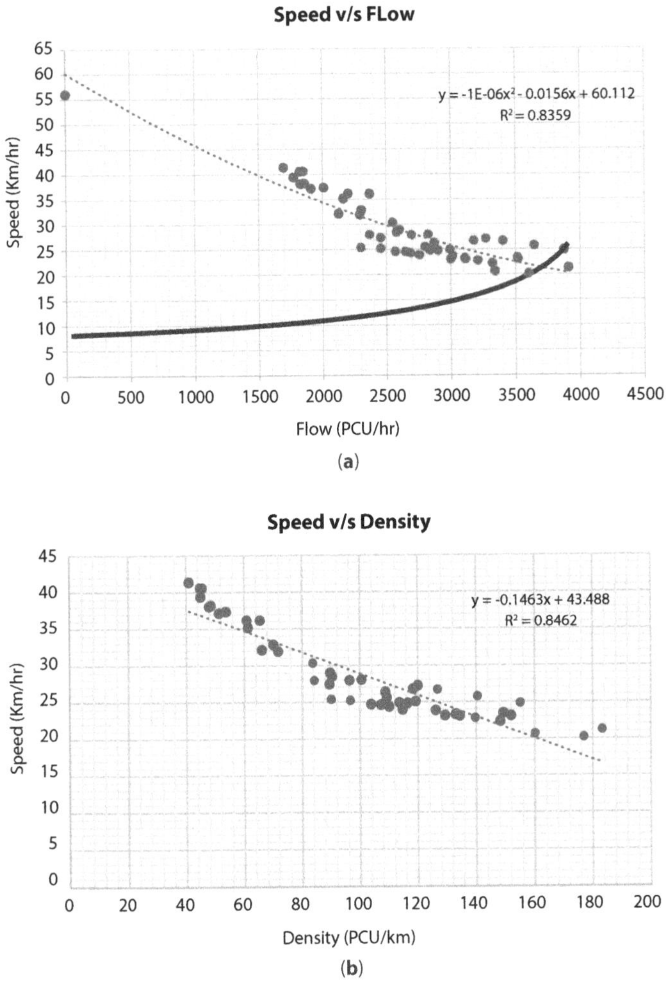


**Figure 11.3** (a) Speed-flow and (b) speed-density relationship curve at Shivaji Vatika–Madhumilan Square section (up).

1. The speed-density relationship is also developed for the selected section which is linear with $R^2$ values in the ranges of (0.936–0.796).
2. The traffic composition mainly consists of two-wheelers and cars with 68% and 24%, respectively, contributing majorly to the volume of the stream. Other vehicles class like three-wheelers, busses, light commercial vehicles (LCVs), etc., have significantly affected the capacity of the selected section.

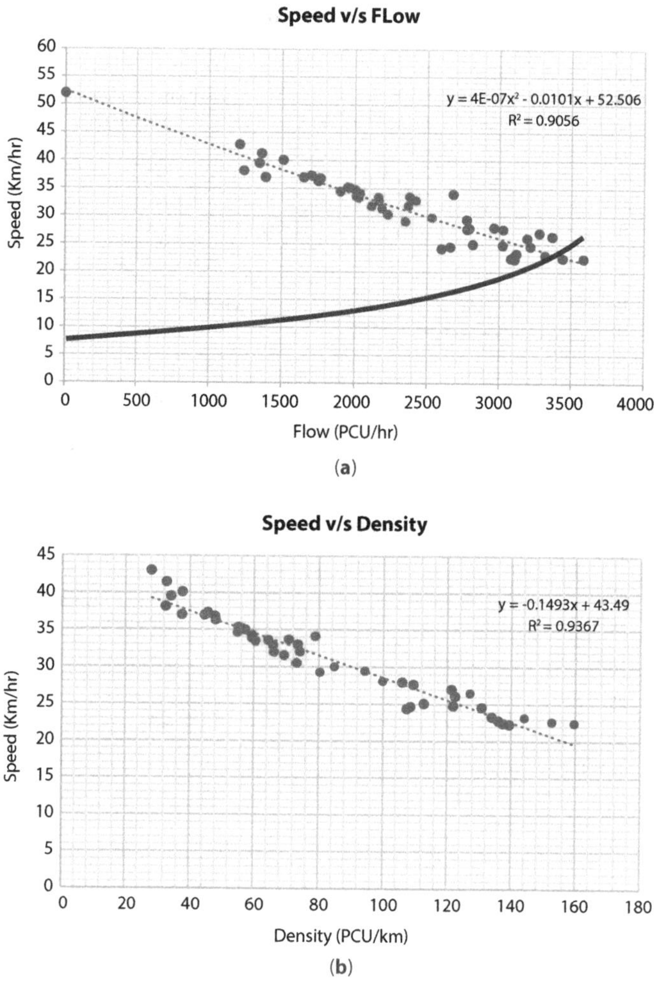


**Figure 11.4** (a) Speed-flow and (b) speed-density relationship curve at Madhumilan Square–Shivaji Vatika section (down).

3. The maximum capacity is observed at World Cup Square to Agriculture College section (up direction) in the evening peak hour between 6:00 and 7:00 pm, which comes out to be 4161 PCU/h with a stream speed of 22.5 km/h. Shivaji Vatika to Madhumilan Square section also runs at a similar capacity in the evening peak hour which is 3909PCU/h with a stream speed of 20.2km/h. Thus, both the sections have a v/c ratio >1 and fall under LOS F as per IRC 106-1990.

**Table 11.1** Comparison of level of service ranges.

| Level of service | V/C as per HCM | V/C suggested (up) | V/C suggested (down) |
|---|---|---|---|
| A | <0.60 | <0.195 | <0.185 |
| B | 0.61-0.70 | 0.195–0.372 | 0.185–0.375 |
| C | 0.71–0.80 | 0.372–0.56 | 0.375–0.575 |
| D | 0.81–0.90 | 0.56–0.757 | 0.575–0.79 |
| E | 0.91–1 | 0.757–1 | 0.79–1 |
| F | >1 | >1 | >1 |

**Table 11.2** Comparison of speed ranges.

| Level of service | Average travel speed as per HCM (Km/h) | Speed ranges suggested (Km/h) |
|---|---|---|
| A | >59 | >51.5 |
| B | 59–46 | 51.5–44 |
| C | 46–33 | 44–36 |
| D | 33–26 | 36–29 |
| E | 26–21 | 29–22 |
| F | <21 | <22 |

4. The maximum capacity observed at Shivaji Vatika to Agriculture College section (down direction) with 3774.6 PCU/h and stream speed of 23.5km/h, which is also falling in the LOS F category as per IRC 106-1990.
5. The density which is observed maximum in the evening peak hour is for World Cup Square to Agriculture College section having nearly 188.2 PCU/km.
6. The revised LOS criteria established from speed-flow relationship graph for the observed traffic is compared with the LOS ranges of the HCM. The v/c ranges given by HCM vary from “0.60–1” for LOS A–E whereas the v/c ranges suggested after this study for the arterial road is in the range of “0.185–1” for LOS A–E.

7. The speed criteria suggested for urban roads by HCM for LOS A and LOS B, which is 59-46 hm/h, is higher than the speed criteria suggested in this study for LOS A and LOS, which is 51.5-44 km/h. It is also observed that the criteria suggested for LOS "C to F" is nearly the same as the suggested speed range for this urban arterial road.

## 11.5 Recommendations

1. The decreased lane width at the Madhumilan Square end (up direction) is leading to the problem of congestion. So, the road widening must be done on the available space to facilitate the current traffic flow and reduce the problem of congestion.
2. Proper parking facilities have to be provided near the Dawa bazar area because many people are parking their vehicles on the road, thus causing congestion.
3. The bus stop present near this section end needs to be shifted as it is very close to the edge of the road.
4. The intercity buses which comprise 40–45% of the total bus composition running at the section from Madhumilan Square to World Cup Square should be shifted to the outer area of the city, which would reduce the traffic volume in the city.
5. There is a great need to strictly check the on-street parked vehicles regularly by the traffic police and penalties should be imposed to maintain smooth traffic operation on the road.
6. There are also other factors which may affect the LOS of the road like vehicle length, vehicle engine's power, gradient, pedestrian traffic, and non-motorized vehicles that are not considered in this study and can be used in future studies.
7. The established relationships will be ideal for a similar kind of roadway section and can be the base for calculation of capacity of the arterial road with other traffic and roadway conditions.

## References

1. Rahimi, M.R., Vala, M., Patel, B.N. (2020). Evaluation of Capacity and Level of Service for Selected Urban Arterial Roads—A Case Study of Rajkot City.

In: Arkatkar, S., Velmurugan, S., Verma, A. (eds) *Recent Advances in Traffic Engineering. Lecture Notes in Civil Engineering*, vol 69 (pp. 23-41). Springer, Singapore. https://doi.org/10.1007/978-981-15-3742-4_2
2. Indo HCM (2017) Indian Highway Capacity Manual. *Council of scientific & industrial research (CSIR)*. India
3. Samal, S.R., Das, A.K. (2020). Evaluation of Traffic Congestion Parameters Under Heterogeneous Traffic Condition: A Case Study on Bhubaneswar City. In: Mathew, T., Joshi, G., Velaga, N., Arkatkar, S. (eds) *Transportation Research . Lecture Notes in Civil Engineering*, vol 45. Springer, Singapore. https://doi.org/10.1007/978-981- 32-9042-6_53
4. Patel CR, Joshi GJ. Capacity and LOS for urban arterial road in Indian mixed traffic condition. *Procedia Soc. Behav. Sci.* 2012 Jan 1;48:527-34.
5. Mahendra M, Sulistio H, Djakfar L, Wicaksono A. Analysis of Free Flow Speed on Urban Road. In *11th Asia Pacific Transportation and the Environment Conference (APTE 2018)* 2019 Oct (pp. 68-74). Atlantis Press.
6. Kafy AA, Ferdous L, Poly SA, Arafat M, Monira S, Rahman MS, Rahman N, Afroz F, Mahzabin F, Ahmed T, Khan NA. Estimating traffic volume to identify the level of service in Major Intersections of Rajshahi, Bangladesh. *Trends Civ. Eng. Archit.* 2018;2(04):292-309.
7. Jameel AK. Improving a model for speed/density relationship on arterial roads in Baghdad city. *Eng. Technol. J.* 2011;29(3).
8. Al-Bahr TM, Puan OC. Speed-Flow Relationship for Urban Roads: A Preliminary Assessment. *Adv. Sci. Lett.* 2018 Jun 1;24(6):4172-6.
9. Dhamaniya A, Bari CS, Patkar M. Capacity Analysis of Urban Arterial Midblock Sections Under Mixed Traffic Conditions. *Int. J. Intell. Transp. Syst. Res.* 2022 Aug;20(2):409-21.
10. Kumar P, Arkatkar S, Joshi G. Developing Level of Service (LoS) Thresholds on Ten-Lane Divided Multi-Lane Urban Roads in India. In *ASCE India Conference 2017* (pp. 649-658). Reston, VA: American Society of Civil Engineers.

# 12

# Machine Learning-Based Traffic Operation System

***Abstract***
Managing traffic on highways is a cumbersome task, which requires accurate vehicle type analysis, traffic jam prediction, speed analysis, direction prediction, etc. In order to perform all these tasks efficiently sensing devices are connected at highways. The connected devices process data locally and take actions based on this data processing. These actions include modifying speed zones, deploying traffic authorities, and communicating with other vehicles. In order to do these tasks efficiently, it is recommended that these systems should be able to predict and broadcast this information as quickly as possible. In order to resolve this issue, the underlying text proposes a Q-learning inspired highway traffic management system. This system incorporates different kinds of Internet of things (IoT)-enabled sensors for speed monitoring, vehicle type detection, collision prediction and congestion prediction for evaluating different highway parameters. These parameters are then given to a machine learning model for improving highway traffic flow via high speed and high accuracy decision making and information dispersal. It is observed that deploying the proposed system is able to improve highway traffic flow by 14%, reduce accidents by 20% and reduce road congestion by 9% when compared with standard non-machine learning deployed systems. Improving this flow quality also results in better road quality maintenance.

***Keywords:*** Accidents, congestion, traffic flow, highway, internet of things, machine learning

## 12.1 Introduction

Highways are special speed zones which allow high-speed and light-to-heavy vehicle traffic flow between different areas. These zones are capable of handling more than three times (3x) the traffic than normal zones with at least two times (2x) speed. Due to the high-speed nature of this traffic, chances of accidents and congestions also increase. A statistic suggests

Kundan Meshram. Urban Transportation Systems: Sustainability and Future Development, (245–262) © 2024 Scrivener Publishing LLC

that there were more than 390 accidents on Mumbai-Pune expressway due to which at least 91 people lost their lives [1]. Thus, traffic management on highways is a very important Government-led management issue. In order to manage highways traffic, a series of sensing elements, actuating elements, communication elements, etc., are needed. These elements work in tandem with each other in order to sense, control, predict and manage day-to-day traffic operations. An example architecture which depicts different operations like incident management, weather management, traffic flow management, traveller information management, and mobility management, can be observed from Figure 12.1, wherein mutual relationship between these operations is depicted in depth.

Thus, to develop an effective highway management system the following process must be followed:

- Data collection must be done effectively, this data must include, but should not be limited to:
  - Temporal history for the number of vehicles passing through the highway corridor

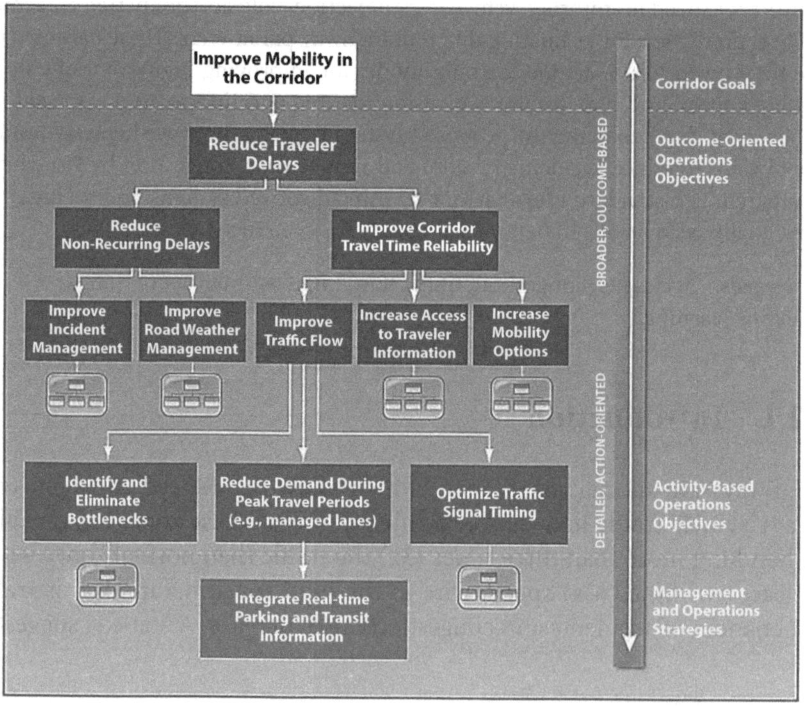


**Figure 12.1** Main components for highway traffic management.

  - o Delays needed by vehicles to cross the corridors, which would result into speed evaluation for these vehicles
  - o Historical data about road weather, which will allow the system to evaluate road quality
  - o Traffic flow information and traveller information along with mobility information
- Segregate this data into delay-based data and travel time reliability improvement data
- From the delay-based data evaluate incident management and road weather management details
- From travel time, evaluation of traffic flow, traveller information details, mobility options, etc., are evaluated
- Using these details, machine learning models are evaluated for:
  - o Identification and elimination of bottlenecks
  - o Reduction of demands during peak travel times, which also includes operations like integration of real-time parking and data communication between vehicles
  - o Optimization of signal timings based on this data

All these operations require effective data collection, processing, actuation and communication algorithms. A survey of these systems for highway planning and management can be observed in the next section. This is followed by the proposed Internet of things (IoT)-based solution that uses machine learning for improved highway traffic management performance. Finally, this text concludes with some interesting observations about the proposed algorithm, its analysis and ways to improve the architecture's performance in future.

## 12.2 Literature Review

Improving highway traffic flow requires multiple sensor units to work concurrently in order to effectively communicate and direct traffic information from one point to another. For instance, the work in suggests a short-term prediction model for urban transport for Indian roads [2]. This model uses linear and non-linear prediction units for improved forecasting under normal and abnormal traffic conditions. This prediction model is able to reduce fuel consumption on highways by effective traffic rerouting on paths where traffic is limited. This efficiency can be further improved by using effective traffic data collection. This data collection can assist the prediction model to learn about traffic patterns, and thereby increase the efficiency of prediction. The work in suggests a novel pavement vibration monitoring system for collection of traffic information data, wherein load sensors and accelerometers are used [3]. In order to classify the collected data into different vehicle types and different vehicle weights, the system uses a three-layered artificial neural network along with a customized and improved k-means algorithm. The system is able to classify 10 different vehicle types with an average 79% accuracy. Some vehicle types like trucks and small cars which have distinctly identifiable weight patterns are identified with over 98% accuracy, while some vehicle types like medium-sized cars are identified with less than 50% accuracy. Thus, incorporation of other parameters like image data and speed data is also important to identify traffic type.

While data gathering is important, providing authenticity to this data is also needed for trust-worthy communication. Data with higher trust value has better applicability than data with low trust values. In order to improve trust levels in data, the work in proposes a novel multi-stage access control mechanism. This mechanism is named ControlChain uses smart contract-based Ethereum blockchain for controlling access to devices and their data [4].

The architecture uses the following blockchains for improving control:

- Context blockchain to store information about device data
- Relationship blockchain to store information about different relationships between the datasets
- Rules blockchain to identify rules and establish them in the network
- Accountability blockchain to establish trust in the network
- Off-blockchain side channel to perform communication within the network

Based on these blockchains, the system is highly secure, and any highway data communication can be performed with minimum disturbance and maximum trust levels. Based on such designs, the work in proposed an information retrieval and social IoT framework [5]. This framework can be applied to traffic management, emergency management, toll administration, commercial vehicle administration, construction management, information service provision, emissions management, transit management, fleet and fright management and achieved data management. The system is capable of providing remote traveller support and 24x7 personal information access for wide area communication and fixed-point to fixed-point communication. Due to use of internet of vehicles (IoV), the overall communication efficiency is improved, and highway traffic data management is simplified. A similar system developed for traffic monitoring in city governance is discussed, wherein machine learning and artificial intelligence is used in order to estimate vehicle length, and determine vehicles per unit road area [6]. Using this data, traffic updates are analyzed, and road occupancy is evaluated. Based on this occupancy, the overall congestion is predicted, and vehicles are rerouted. The machine learning model is able to reduce congestions by over 10% and improve overall traffic flow on highways, thereby improving the efficiency of IoV connected deployments. But these IoV deployments have multiple security challenges, most of these challenges are related to IoV security and traffic performance, which can be handled via blockchain and machine learning models [4, 6, 7].

Internet of vehicles systems can also be used for effective accident detection and broadcast of accidents along highways. The work in studies such systems, wherein vehicular ad hoc networks (VANETs), vibration sensors, piezoelectric sensors, accelerometer sensors, pressure sensors with tilt angle detection capabilities, gyroscopes and shock sensors are used for effective accident detection [8]. Data from these sensors is given to machine learning models like fuzzy logic, support vector machines (SVMs) and artificial neural networks (ANNs) for effective classification. These classifiers are then combined with driver drowsiness systems, alcohol sensors, ultrasonic sensors and pedestrian detection systems for avoiding accidents. The work suggests that neural networks outperform other classifiers for both accident detection and prevention. A combination of the work, wherein naïve Bayes classifier and Banker's algorithm are used for accident detection and trust establishment, respectively [4, 8, 9]. The system is highly effective for urban traffic, but can be equally applied to highways as well. The detection efficiency of these systems can be further enhanced via different kinds of prediction and classification systems, which use effective feature extraction techniques. One such sensor used the principle of magnetism for speed estimation and vehicle classification [10]. In this system, magnetic perturbations caused by vehicles are scanned, and its patterns are identified. Based on these patterns and their changes, vehicle speed and type are identified. For instance, a small vehicle with moderate speed will have frequently varying magnetic patterns, whereas large vehicles with high speeds will have moderately varying magnetic patterns.

This system is able to detect vehicle speeds with over 90% accuracy, and vehicle types with over 85% accuracy, but the deployment and maintenance cost is impractical to be applied for real-time long- distance highway systems. In order to reduce deployment costs, cloud-based architectures can be used [11]. Such systems offload all the computations on the cloud, thereby reducing the requirement of processing units on road sides, which reduces overall system costs. An example of such a deployment for effective road quality assessment can be observed, wherein all image processing operations are offloaded on the cloud, and actuation signals are generated in real-time for alerting vehicles in case of irregular road quality [12]. The system uses an intelligent routing system, which analyzes any kind of road irregularities, and routes vehicles on roads with best road quality, thereby reducing travel time and driving performance. Another example of such a system for traffic congestion monitoring and control can be observed, wherein all calculations for congestion are done on cloud, and control actions are communicated to devices via road side units or direct cloud to device connectivity [13]. An extension to this work, wherein parking space management and cross road management are done via smartphones using cloud-based data collection and analysis [14]. It uses cluster formation, head selection and image processing for detection of cross-roads and empty parking spaces. This information is aggregated over the cloud, and communicated to vehicles via the internet using secure communication channels. Thereby increasing the effectiveness of data transfer, and transparent control via multi-sensor interface can be further used for recurrent traffic congestion detection [15]. Here, evolutionary regulatory mining is used for improved congestion detection, and other methods for congestion detection are also mentioned. Evolutionary regulatory mining is a sub-part of machine learning, wherein continuous learning is done in order to improve prediction performance of congestion detection systems. The work is able to improve the detection accuracy to over 85% when compared with other methods like support vector regression, genetic algorithm, and fuzzy logic. The efficiency of this algorithm can be improved by integrating it with fuzzy logic approaches, wherein congestion-aware routing is proposed [16]. This routing is done based on previous congestions on roads, and is applied to emergency vehicle routing. The algorithm can be extended to general vehicles as well for highly effective high speed and minimum congestion routing via cloud-based platforms [15]. In both research, smart city deployments are considered for monitoring congestion and applying effective control via cloud.

In order to improve congestion detection, vehicle classification, accident detection and prevention performance, it is necessary that nodes are able to collect proper traffic data. It is suggested different ways to collect traffic data in vehicular networks [17]. From this review, it is observed that a combination of cluster-based methods with geo-spatial and topology-based scanning is most effective for vehicular data collection. Such a hybrid method is used for collecting data about traffic congestion in road crossings [18].

The system uses traffic monitoring and traffic density monitoring modules (TDMM) for effective collection of congestion data, and broadcasting this data to other nodes in a given time span. While the system is theoretical, it can realize effective sensing and classification methods used. This method can be extended for actuating dynamic traffic flow via smart movable road divider [19]. Here, the system is able to identify traffic congestion on roads, and control this congestion via movable road dividers. The system has good real-time applications, but might increase accidents because sudden road width changes are not recommended for real-time traffic systems. Another congestion detection algorithm, which is used for automatic vehicle counting and image processing used [20]. The image processing method uses foreground detection, noise removal, Otsu thresholding and vehicle detection for real-time vehicle counting. Based on this count, congestion is detected, and corrective actions are taken. These actions include rerouting, accident detection, accident prevention and traffic light control. Image processing operations can be offloaded to the cloud for improving detection efficiency [21]. Such a cloud offloaded image processing-based intelligent traffic control system is mentioned in Reference [22] wherein You Look Only Once (YOLO) model is deployed on the cloud to count vehicles and control traffic. The system uses Q-learning to reduce congestion and improve rerouting efficiency of the deployed traffic analyzer. This congestion data can be extended for smart traffic light control, wherein a combination of image and sensory data is used to improve traffic light control by 15%, thereby reducing fuel wastage [23]. Congestion data can be further used for evaluating congestion patterns on highways. A system that utilizes k-means clustering to identify congestion patterns is studied [24]. In this system, congestion patterns are used for analyzing current and future traffic quality, thereby increasing the efficiency of traffic re-routing in vehicular networks. This system is extended for delay-aware accident detection, wherein temporal congestion patterns along with current traffic data is used for improved accident detection on roads [25]. All the processing of these temporal patterns along with current traffic data is done on fog devices, which are connected to the cloud. Similar implementations of these systems are mentioned, wherein efficient data reporting, navigation and accident prevention are used in order to improve road traffic quality on highways [26, 27]. From Reference [28], these techniques it can be observed that efficient data collection along with machine learning over cloud are the most effective ways in which traffic monitoring and control can be established. Inspired by these observations, the underlying text proposed novel machine learning model for improving traffic flow and prevent accidents on highways. This system is mentioned in the next section, followed by its result evaluation.

## 12.3 Proposed Integrated Machine Learning Model for Improving Highway Traffic Maintenance Efficiency with IoT Devices

In order to reduce the number of accidents and number of congestion incidences on highways, this text proposed a two-step solution. Wherein initially an intelligent pair of IoT sensor device design is proposed, followed by a novel machine learning model that combines data from these IoT sensors for improving traffic flow and reducing the number of accidents. The proposed pair of IoT sensor device design can be observed from Figure 12.2, wherein the following IoT devices are integrated:

- Array of roadside devices (RSD) mounted with the following devices for efficient vehicle monitoring:
  - Radar sensor for finding out the number of vehicles
  - Small-scale magnetic sensor for evaluating vehicle type and speed
  - Infrared camera for evaluating other vehicle parameters
  - An alert system with speaker for continuous alert updates
- Fog-enabled IoT nodes to collect and process data from these devices. These devices also perform fog-to-cloud and cloud-to-fog communications for offloading heavy computations, and are also responsible to handle RSDs.

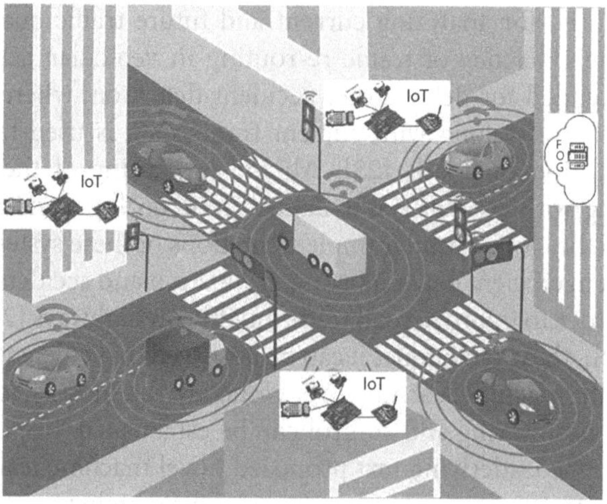


**Figure 12.2** IoT device design for the proposed system.

The RSDs are capable of collecting the following information for each passing vehicle:

- Vehicle speed and direction (Vsi and Vdi) are observed using small-scale magnetic sensor
- Number of vehicles passing on the given highway road patch are observed using radar sensor
- Type of vehicle and vehicle size are observed by combining image data and small-scale magnetic sensor
- Distance between two vehicles is evaluated using image processing and YOLO object detection

All this information is given to a rule engine, which works using the following steps:

- Based on vehicle speed and direction, predict next vehicle position via auto regressive integrated moving average (ARIMA) model
- Depending upon the number of vehicles and vehicle size, evaluate total vehicle area and compare it with the current highway patch area
- Based on distance between vehicles, evaluate collision probability and alert the vehicles via the alert system
- Depending upon input from other cameras, identify areas which are comparatively free, and suggest alternate routes via the alert system
- Using vehicle type information, evaluate the number of alternate routes for each vehicle, and thereby suggest re-routing alternates to each vehicle via personalized messaging service

To perform these tasks, a convolutional neural network (CNN) is used, which is able to perform speed detection, direction detection, size calculations, vehicle distance measurement, free route detection, and vehicle type information. Architecture for this model is observed from Figure 12.3, wherein output is represented using following equation,

$$Conv_{out_{i,j}} = \sum_{a=-\frac{m}{2}}^{\frac{m}{2}} \sum_{b=-\frac{n}{2}}^{\frac{n}{2}} I(i-a, j-b) * ReLU\left(\frac{m}{2}+a, \frac{n}{2}+b\right) \quad (12.1)$$

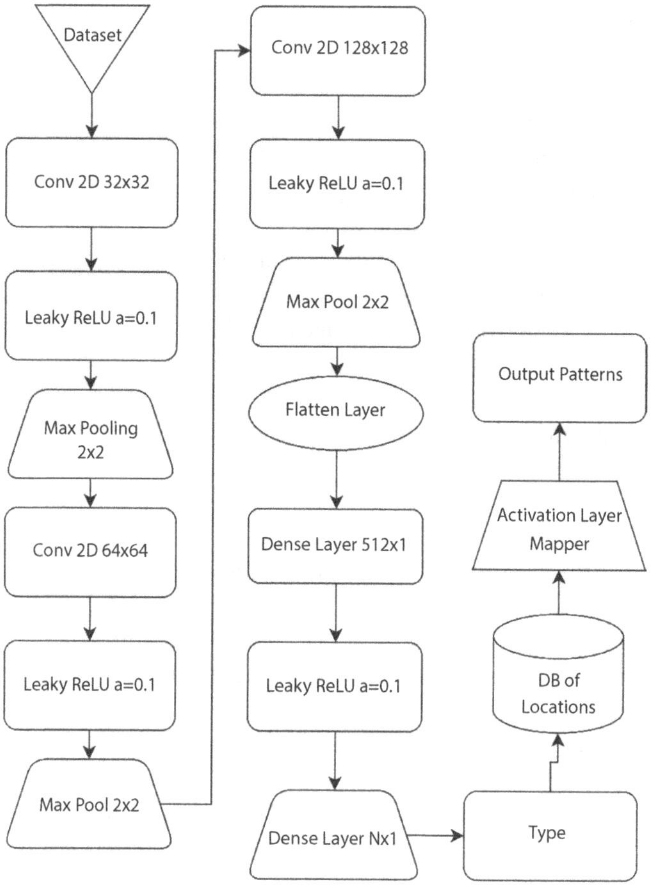


**Figure 12.3** Model for the proposed CNN model.

Where, *I, m, n, i,* and *j* represents input image, rows, columns, current window row, and current window column for the given layer. Based on this convolution, the number of features extracted using each convolutional layer can be calculated from Equation 12.2 as follows:

$$f_{out} = \frac{f_{in} + 2 * p - k}{s} + 1 \tag{12.2}$$

Where, $f_{in}$, $f_{out}$, $p$, $s$, and $k$ represents the number of input features, output features, padding size of convolution, stride size of convolution, and convolution kernel size.

In order to reduce the computational complexity of this system, all image processing operations are offloaded to the fog computing device. The fog computing device evaluates vehicle distances locally, and offloads type of vehicle classification tasks to the cloud, where deep learning models same as are deployed for high-speed computations [12]. In order to further improve system performance, radar sensing is done locally on the IoT device, and alerts are generated in case of any preliminary collision detection. Due to the usage of these sensors, congestions and collisions are reduced, and due to incorporation of cloud for vehicle re-routing, there is an improvement in traffic flow for any given highway patch. Results for this system are evaluated, and compared with existing models in the next section.

## 12.4 Performance Analysis

In order to perform result analysis of the proposed model, it was tested on the following highway conditions, with the given number of vehicles, and given vehicle types as observed from Table 12.1.

These figures are approximate in number, because there are large inflows and outflows of traffic on the given highways. These tests were done on NH 30, between the patches of Jamira, Mangawan and Rewa, Madhya Pradesh, India. Average number of congestions (NC), accidents (A) and number of vehicles passing without congestion (NVPNC) were evaluated on each of these highway types per day and is averaged out over the entire week. Results for these respective parameters were observed and tabulated in Tables 12.2, 12.3, and 12.4.

These results can be visualized using Figures 12.4, 12.5, and 12.6 respectively, wherein trends of these parameters are observed, and it can be seen that the proposed algorithm is able to reduce the number of congestions by over 10%, number of accidents by over 15%, and number of vehicles passing without congestion by over 12% when compared with existing state-of-the art systems.

The number of vehicles passing without congestion is a direct indicator of traffic flow on the highway. This parameter can suggest the following characteristics about highway traffic:

- Higher value of number of vehicles passing without congestion indicates that the highway traffic had a clear route for going from source to destination with minimum stoppage.

**Table 12.1** Highway condition and vehicle specifications for testing.

| Highway type | Highway length | No. of vehicles (approx.) | Types of vehicles |
|---|---|---|---|
| HT 1 | 35 km | 1500 | Light motor, heavy motor, bikes |
| HT 2 | 50 km | 1900 | Light and heavy motor vehicles |
| HT 3 | 70 km | 2350 | Light motor, heavy motor and bikes |
| HT 4 | 100 km | 3500 | Light and heavy motor vehicles |
| HT 5 | 250 km | 6500 | Light motor, heavy motor and bikes |
| HT 6 | 350 km | 10000 | Light motor, heavy motor and bikes |
| HT 7 | 500 km | 15000 | Light motor, heavy motor and bikes |
| HT 8 | 750 km | 25000 | Light motor, heavy motor and bikes |

**Table 12.2** Average number of congestions for different highway scenarios.

| Highway type | NC [6] | NC [13] | NC [Proposed] |
|---|---|---|---|
| HT 1 | 81 | 72 | 54 |
| HT 2 | 129 | 95 | 75 |
| HT 3 | 158 | 114 | 91 |
| HT 4 | 289 | 204 | 164 |
| HT 5 | 328 | 225 | 184 |
| HT 6 | 393 | 267 | 220 |
| HT 7 | 458 | 308 | 255 |
| HT 8 | 523 | 350 | 291 |

**Table 12.3** Average number of accidents for different highway scenarios.

| Highway type | A [6] | A [13] | A [Proposed] |
|---|---|---|---|
| HT 1 | 48 | 56 | 54 |
| HT 2 | 53 | 75 | 43 |
| HT 3 | 79 | 84 | 54 |
| HT 4 | 86 | 100 | 62 |
| HT 5 | 102 | 114 | 72 |
| HT 6 | 116 | 128 | 81 |
| HT 7 | 130 | 142 | 90 |
| HT 8 | 144 | 156 | 100 |

**Table 12.4** Average number of vehicles passing without congestion for different highway scenarios.

| Highway type | NVPNC [6] | NVPNC [13] | NVPNC [Proposed] |
|---|---|---|---|
| HT 1 | 810 | 915 | 1438 |
| HT 2 | 918 | 1156 | 1728 |
| HT 3 | 1148 | 1445 | 2160 |
| HT 4 | 1580 | 1989 | 2974 |
| HT 5 | 2700 | 3400 | 5083 |
| HT 6 | 4455 | 5610 | 8388 |
| HT 7 | 6750 | 8500 | 12708 |
| HT 8 | 10800 | 13600 | 20333 |

- Moderate value of number of vehicles passing without congestion is an indicative of moderate road quality, moderate performance of the routing algorithm, and moderate use of the highway.
- Low value for number of vehicles passing without congestion indicates that the highway is not getting utilized to its full capacity, and that there is a need of optimal traffic routing.

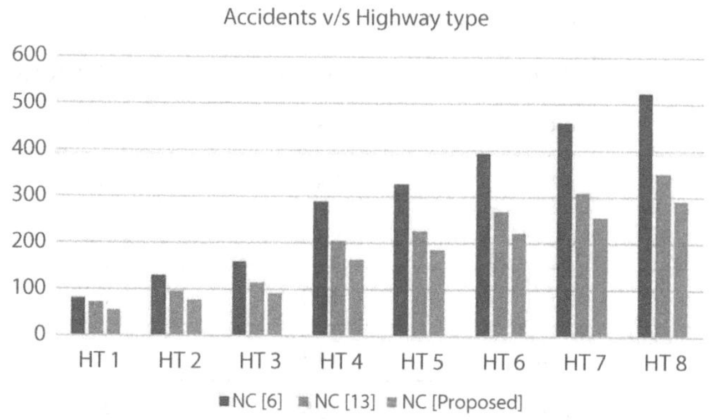


**Figure 12.4** Number of congestions v/s highway type.

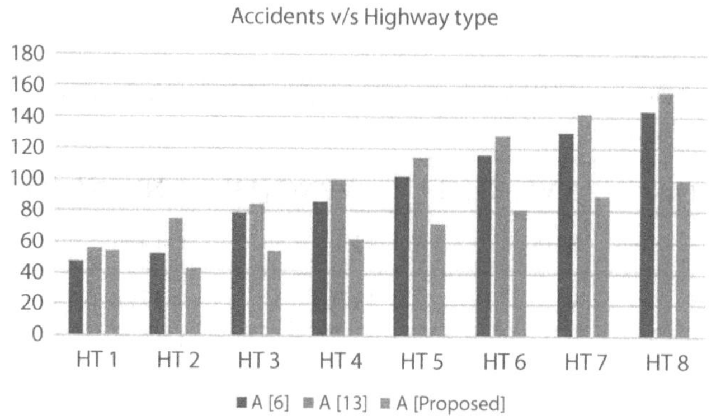


**Figure 12.5** Accidents v/s highway type.

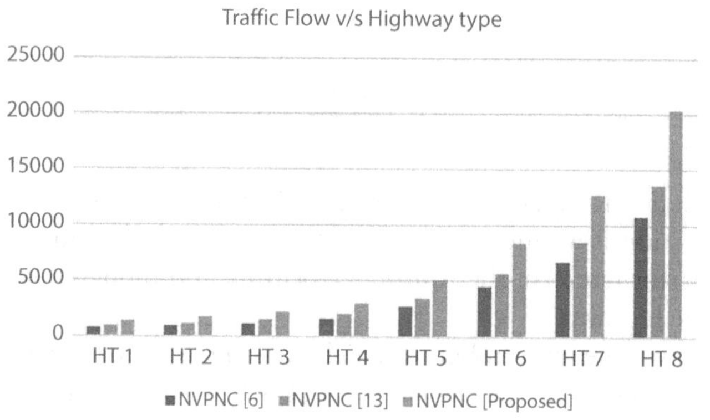


**Figure 12.6** Traffic flow v/s highway type.

The number of congestions is reduced drastically when compared with Reference [6], but are reduced by almost 10% when compared with Reference [13].

The number of accidents is reduced drastically when compared with Reference [13], but are reduced by almost 15% when compared with Reference [6]. This indicates that the proposed algorithm can optimize both the parameters with utmost efficiency.

The traffic flow is also improved drastically when compared with either Reference [6] or [13], which is due to the fact that highway routes are being optimally utilized, and there are no congestions and accidents on the route.

## 12.5 Conclusion and Future Scope

From the results, it can be observed that the proposed IoT0-based machine learning architecture is not only able to reduce the number of congestions, but is also able to reduce the number of accidents in the real-time traffic systems. Theoretically, reducing the number of congestions must reduce the number of accidents, but reducing congestions does not primarily affect the number of accidents, but has moderate effect over it. Reducing the number of congestions indicates improvement in traffic movement, and reduction in unwanted fuel consumption. While reducing the number of accidents has multi-fold advantages, which includes, faster traffic flow, better driving experience, and reduced road mortality. Moreover, reducing the number of accidents and congestions allows the road to have smoother and faster flowing traffic, which improves overall traffic flow of the highway. The proposed architecture has improved upon all these parameters. But the work can be further extended by adding more road-specific parameters like road quality and number of tolls, and then using machine learning models to estimate other parameters like average vehicle waiting delay and number of accidents for a given period.

## References


1. Business Standard, "*Disaster accident*." https://www.business-standard.com/article/pti-stories/352-accidents-on-mumbai-pune-expressway-in-2019-91-killed-120011800726_1.html
2. S. Rajendran and B. Ayyasamy, "Short-term traffic prediction model for urban transportation using structure pattern and regression: an Indian context," in *SN Appl. Sci.*, vol. 2, p. 1159, 2020. https://doi.org/10.1007/s42452-020-2946-2

3. Z. Ye, H. Xiong and L. Wang, "Collecting comprehensive traffic information using pavement vibration monitoring data," in *Comput. Aided Civ. Inf.*, vol. 35, pp. 134–149, 2020. https://doi.org/10.1111/mice.12448
4. O. J. A. Pinno, A. R. A. Grégio and L. C. E. De Bona, "ControlChain: A new stage on the IoT access control. authorization," in *Concurrency Comput. Pract. Exper.*, pp. 1-28, 2020. https://dx.doi.org/10.1002/cpe.5238
5. G. Zhang and N. Zhang, "Optimal design of urban cloud transportation planning based on information retrieval and social IoT," in *Internet Technol. Lett.*, pp. 1–6, 2020. https://doi.org/10.1002/itl2.243
6. M. Sarrab, S. Pulparambil and M. Awadalla, "Development of an IoT based real-time traffic monitoring system for city governance," in *Global Transitions*, vol. 2, pp. 230-245, 2020. https://doi.org/10.1016/j.glt.2020.09.004
7. A. Khanna and S. Kaur, "Internet of things (IoT), applications and challenges: a comprehensive review," in *Wirel. Pers. Commun.*, vol. 114, pp. 1687–1762, 2020. DOI: 10.1007/s11277-020-07446- 4
8. U. Alvi, M. A. K. Khattak, B. Shabir, A. W. Malik and S. R. Muhammad, "A comprehensive study on IoT based accident detection systems for smart vehicles," in *IEEE Access*, vol. 8, pp. 122480-122497, 2020. DOI: https://dx.doi.org/10.1109/ACCESS.2020.3006887
9. A. M. Ratheeshkumar, V. Anandkumar and T. Kalaiarasan, "An IoT based vehicle accident detection system and trustability analysis using naïve bayes algorithm," in *J. Xian Univ. Archit. Technol.*, vol. 11, no. 5, pp. 1145-1152, 2020. https://dx.doi.org/10.1109/JIOT.2020.3008896
10. Y. Feng, G. Mao, B. Cheng, C. Li, Y. Hui, Z. Xu and J. Chen, "MagMonitor: Vehicle Speed Estimation and Vehicle Classification Through A Magnetic Sensor," in *IEEE Trans. Intell. Transp. Syst.*, pp. 1-12, 2020. DOI: https://doi.org/10.1109/tits.2020.3024652
11. K. R. Malik, M. Ahmad, S. Khalid, H. Ahmad, F. Al-Turjman and S. Jabbar, "Image and command hybrid model for vehicle control using Internet of Vehicles," in *Trans Emerging Tel Tech.*, pp. 1-13, 2020. https://doi.org/10.3390/su13168885
12. K. Bansal, K. Mittal, G. Ahuja, A. Singh and S. S. Gill, "DeepBus: Machine learning based real- time pothole detection system for smart transportation using IoT," in *Internet Technol. Lett.*, pp. 1-6, 2020. DOI: https://dx.doi.org/10.1002/itl2.156
13. D. Selvaraj, R. Bharathiraja, V. GokulNath, E. Immanuel and G. L. Jyothesvar, "IoT based traffic congestion monitoring and management system," in *Int. Res. J. Eng. Technol.*, vol. 5, no. 9, pp. 6701-6706, 2018.
14. A. Hilmani, A. Maizate and L. Larbi Hassouni, "Automated real-time intelligent traffic control system for smart cities using wireless sensor networks," in *Wirel. Commun. Mobile Comput.*, pp. 1-28, 2020. https://doi.org/10.1155/2020/8841893
15. X. Chen, S. Zhang, X. Ding, S. N. Kadry and C.-H. Hsu, "IoT cloud platform for information processing in smart city," in *Comput. Intell.*, pp. 1–17, 2020. https://doi.org/10.1016/j.iot.2022.100565

16. M. Shelke, A. Malhotra and P. N. Mahalle, "Fuzzy priority based intelligent traffic congestion control and emergency vehicle management using congestion-aware routing algorithm," in *J. Ambient Intell. Human. Comput.* (2019). https://doi.org/10.1007/s12652-019-01523-8
17. M. A. Saleem, Z. Shijie and A. Sharif, "Data transmission using IoT in vehicular ad-hoc networks in smart city congestion," in *Mobile Netw. Appl.*, vol. 24, pp. 248–258, 2019. https://doi.org/10.1109/TVT.2016. 2555853
18. B. Pourghebleh and N. J. Navimipour, "Towards efficient data collection mechanisms in the vehicular ad hoc networks," in *Int. J. Commun. Syst.*, pp. 1-20, 2019. https://dx.doi.org/10.1108/K-06- 2018-0280
19. P. Sadhukhan and F. Gazi, "An IoT based intelligent traffic congestion control system for road crossings," in *Int. Con. on Communication, Computing and Internet of Things (IC3IoT)*, Chennai, India, pp. 371-375, 2018. https://dx.doi.org/10.1109/IC3IoT.2018.8668131
20. S. S. Palle, H. B. Vibha, B. M. Sriraksha and A. Yeshashwini, "Implementation of smart movable road divider and ambulance clearance using IoT," in *4th Int. Con. on Recent Trends on Electronics, Information, Communication & Technology (RTEICT)*, Bangalore, India, pp. 1345-1348, 2019. https://dx.doi.org/10.1109/RTEICT46194.2019.9016863
21. S. Singh, B. Singh, R. Singh and A. Das, "Automatic vehicle counting for iot based smart traffic management system for indian urban settings," in *4th International Conference on Internet of Things: Smart Innovation and Usages (IoT-SIU)*, Ghaziabad, India, pp. 1-6, 2019. https://doi.org/10.1109/IoT- SIU.2019.8777722
22. A.Frank, Y. S. K. Al Aamri and A. Zayegh, "IoT based smart traffic density control using image processing," in *4th MEC Int. Con. on Big Data and Smart City (ICBDSC)*, Muscat, Oman, pp. 1-4, 2019. https://doi.org/10.3390/math10162832
23. A. Mansour and M. R. M. Rizk, "Towards traffic congestion-free through intelligent traffic control system," in *7th Int. Japan-Africa Con. on Electronics, Communications, and Computations (JAC- ECC)*, Alexandria, Egypt, pp. 5-8, 2019. https://dx.doi.org/10.1109/JAC-ECC48896.2019.9051064
24. B. Ghazal, K. ElKhatib, K. Chahine and M. Kherfan, "Smart traffic light control system," in *Third Int. Con. on Electrical, Electronics, Computer Engineering and their Applications (EECEA)*, Beirut, Lebanon, pp. 140-145, 2016. https://doi.org/10.3390/s23115028
25. M. A. Mondal and Z. Rehena, "Identifying traffic congestion pattern using K-means clustering technique," in *4th Int. Con. on Internet of Things: Smart Innovation and Usages (IoT-SIU)*, Ghaziabad,India, pp. 1-5, 2019. https://dx.doi.org/10.3390/su141711068
26. B. K. Dar, M. A. Shah, S. U. Islam, C. Maple, S. Mussadiq and S. Khan, "Delay-aware accident detection and response system using fog computing," in *IEEE Access*, vol. 7, pp. 70975-70985, 2019. https://doi.org/10.1109/ACCESS.2019.2910862

27. E. Nasr, E. Kfoury and D. Khoury, "An IoT approach to vehicle accident detection, reporting, and navigation," in *IEEE International Multidisciplinary Conference on Engineering Technology (IMCET)*, Beirut, Lebanon, pp. 231-236, 2016. https://doi.org/10.1109/IMCET.2016.7777457
28. C. Ma, J. Zhou, X. Xu and J. Xu, "Evolution regularity mining and gating control method of urban recurrent traffic congestion: a literature review," in *J. Adv. Transp.*, pp. 1-13, 2020. https://doi.org/10.1155/2020/5261580

# 13

# Traffic Scenario: Efficient Model for Accident Analysis


***Abstract***

The need for this work arises from the pressing concern of ensuring the safety and efficiency of urban transportation systems. Accidents on roads not only result in loss of lives but also lead to significant traffic delays and economic losses. Therefore, an in-depth analysis of accidents is essential to develop strategies for accident prevention and better traffic management.

Existing models used for accident analysis often lack the ability to capture complex relationships within the data, leading to limited accuracy and precision. Additionally, these methods do not effectively translate their findings into actionable recommendations for modifying traffic scenarios.

To address these limitations, the authors propose a novel approach that leverages ensemble learning, including naive Bayes, support vector machine (SVM), k-nearest neighbor (kNN), multi-layer perceptron (MLP), deep forest, and logistic regression, to analyze accidents comprehensively. This diverse ensemble of algorithms allows for a more robust analysis, capturing various aspects of accident patterns. The inclusion of these methods is justified by their individual strengths and abilities to uncover hidden insights in the data samples.

Furthermore, the authors employ the Apriori algorithm with FPMax to extract association rules and recommend modifications in traffic scenarios based on accident data samples. This step bridges the gap between accident analysis and practical interventions, providing actionable insights for improving road safety.

The advantages of this proposed model are evident through rigorous testing on Indian and Nepalese roads, where it demonstrated superior performance compared to existing methods. Specifically, it achieved 4.9% higher precision, 4.5% higher accuracy, 3.9% higher recall, 4.8% higher area under the curve (AUC), 3.4% higher specificity, and reduced delay by 12.5%. These results underscore the practical significance of the proposed approach in enhancing road safety and traffic management.

In conclusion, this chapter presents a novel and effective framework for accident analysis in urban transportation systems, addressing the shortcomings of existing methods. By leveraging ensemble learning and association rule mining,



Kundan Meshram. Urban Transportation Systems: Sustainability and Future Development, (263–318) © 2024 Scrivener Publishing LLC

it offers valuable insights for modifying traffic scenarios and ultimately contributes to safer and more efficient urban transportation.

***Keywords:*** Accident analysis, support vector machine, k-nearest neighbor, multi-layer perceptron, precision

## 13.1 Introduction

The introduction of this book chapter sets the stage by emphasizing the critical need to address the issues surrounding urban transportation accidents. In recent years, urban areas have witnessed a surge in traffic-related accidents, causing not only loss of lives but also significant disruptions to daily routines and economic activities. Such incidents highlight the pressing need for effective accident analysis methods and strategies to enhance road safety.

The limitations of existing accident analysis models become evident when confronted with the complexity and diversity of accident data samples. These models often struggle to uncover hidden patterns and relationships within the data, resulting in inaccuracies and a lack of actionable insights. Furthermore, they often fall short in providing practical recommendations for modifying traffic scenarios to prevent future accidents.

To address these challenges, the authors present a novel and comprehensive approach in this chapter. They harness the power of ensemble learning, a combination of diverse algorithms such as naive Bayes, support vector machine (SVM), k-nearest neighbor (kNN), multi-layer perceptron (MLP), deep forest, and logistic regression, to analyze accident data from multiple angles. Each of these algorithms brings its unique strengths to the table, allowing for a more holistic examination of accident patterns.

In addition to the ensemble approach, the authors employ the Apriori algorithm with FPMax to extract association rules from the accident data samples. These rules serve as a bridge between data analysis and practical interventions by offering concrete recommendations for modifying traffic scenarios. This step represents a crucial advancement in accident analysis, as it empowers urban planners and policymakers with actionable insights to enhance road safety.

The advantages of this proposed model are substantiated by empirical testing on real-world Indian and Nepalese road data samples. It consistently outperforms existing methods, achieving higher precision, accuracy, recall, area under the curve (AUC), specificity, and significantly reduced traffic delays. These results underscore the practical significance and effectiveness of the proposed approach in addressing the complex challenges of urban transportation accident analysis.

In summary, this chapter introduces a novel framework that addresses the limitations of existing accident analysis methods. By combining ensemble learning and association rule mining, it offers a comprehensive solution for accident analysis and provides valuable guidance for modifying traffic scenarios to enhance road safety in urban areas.

## 13.2 Motivation and Contributions

The motivation behind this book chapter stems from the urgent need to tackle the critical issues surrounding urban transportation accidents. In today's fast-paced urban environments, the rising incidence of traffic-related accidents poses a substantial threat to public safety and economic stability. These accidents result in the loss of human lives, substantial property damage, and significant disruptions to daily life. The existing models employed for accident analysis in urban transportation systems have shown limitations that necessitate a novel approach. These traditional methods often struggle to capture the intricate and evolving patterns of accidents, leading to inaccuracies and incomplete insights. Furthermore, they often fall short in translating their findings into practical recommendations for modifying traffic scenarios to prevent future accidents.

The primary contribution of this chapter is the presentation of a groundbreaking model that addresses these limitations comprehensively. By harnessing the power of ensemble learning, combining diverse algorithms like naive Bayes, SVM, kNN, MLP, deep forest, and logistic regression, the authors offer a holistic and nuanced analysis of accident data samples. Each algorithm contributes its unique strengths, enabling a more robust understanding of accident patterns.

Moreover, the authors introduce the Apriori algorithm with FPMax to extract association rules from the accident data samples. This step represents a crucial advancement in accident analysis, as it provides actionable insights for urban planners and policymakers. These association rules offer specific recommendations for modifying traffic scenarios, thereby bridging the gap between data analysis and practical interventions for different use cases.

The significance of this contribution is underscored by the empirical testing of the proposed model on actual Indian and Nepalese road data samples. The results consistently outperform existing methods, demonstrating higher precision, accuracy, recall, AUC, specificity, and significantly reduced traffic delays. This chapter's innovative approach not only enhances our understanding of urban transportation accidents but also offers practical solutions to improve road safety and traffic management.

In conclusion, this chapter's motivation arises from the critical need to address urban transportation accidents, while its contribution lies in presenting a novel and comprehensive framework for accident analysis. By leveraging ensemble learning and association rule mining, this model offers actionable recommendations for modifying traffic scenarios, ultimately contributing to safer and more efficient urban transportation systems.

## 13.3 Review of Existing Models

A wide variety of models are available for accident analysis, and each of them vary in terms of their internal operating characteristics. The reviewed models in this text offer valuable insights into various aspects of accident prevention, investigation, and management, each contributing to our understanding of this critical area for different use cases as given in Table 13.1. Firstly, in Reference [1], the focus is on improving the performance of automatic emergency braking (AEB) systems, specifically in the context of car-to-two-wheeler collision prevention in complex traffic scenarios. The study emphasizes the importance of optimizing key AEB parameters, such as braking deceleration, advance time, and detection range. The findings provide essential guidance for AEB system improvement, highlighting its potential to reduce car-to-two-wheeler collision accidents significantly. In Reference [2], the literature discusses the application of drones in road accident management, emphasizing the growing trend towards using emerging technologies like drones in conjunction with Industry 4.0 solutions. The review highlights the advantages of drones in accident investigation and analysis, particularly in 3D accident scene reconstruction. It also points out the challenges related to drone flight permissions and data handling.

The work in Reference [3] addresses the issue of missing traffic signs and accident-prone features in urban environments. It proposes a novel approach using real-time images from dashcams to notify drivers of accident-prone areas. The paper introduces a sophisticated attention module that classifies urban scenes into accident hotspots or non-hotspots, effectively enhancing driver awareness and safety. In Reference [4], the literature discusses the utilization of visual attention and focus of expansion (FOE) for accident prediction. This approach combines deep learning techniques with FOE to visualize accident risk factors even when the motion features of risk factors are small. The proposed model provides a high level of accuracy in accident prediction and offers a visual explanation of decision-making process.

**Table 13.1** Review of models that use Internet of Things (IoT)-based methods of analysis.

| Method | Findings | Advantages | Limitations |
|---|---|---|---|
| AEB system for car-to-two-wheeler collision prevention [1] | Optimal detection range: 120°–35 m; reduces car-to-two-wheeler collision accidents by 22.3%; decreases average collision speed | Improves road safety; provides data for AEB system development | Limited to environmental technology; focused on specific collision type |
| Drones in road accident management [2] | Emphasis on accident investigation and 3D scene reconstruction; integration of drones with Industry 4.0 technologies | Enhances accident analysis; utilizes emerging technologies | Lack of studies on complex accident cases |
| Visual notification of accident prone features to drivers [3] | Novel attention module for accident hotspot classification; achieves up to 92% classification accuracy | Improves driver awareness; better contextual knowledge | Limited to urban areas; not applicable in all scenarios |
| Accident prediction using visual attention and FOE [4] | Visualizes decision basis using FOE; high prediction accuracy | Visual explanation of risk factors; predicts small motion feature risks | Limited to certain risk factor types; focuses on FOE divergence |
| Post-accident maintenance for autonomous vehicles [5] | Digital twin-enabled maintenance scheme; optimizes maintenance strategy | Considers computing tasks and accident impact; enhances maintenance rate | Requires incomplete information; dynamic matching game complexity |
| Optimal reallocation of heterogeneous sensors [6] | Uses optimal transport for sensor allocation; maximizes coverage quality | Reduces engineering costs; estimation of stable accident risk distribution | Complex optimization problem; cost-effectiveness constraint |

*(Continued)*

**Table 13.1** Review of models that use Internet of Things (IoT)-based methods of analysis. (*Continued*)

| Method | Findings | Advantages | Limitations |
|---|---|---|---|
| Multi-perspective analysis of road accidents [7] | Analyzes accidents from multiple camera perspectives; creates multi-perspective dataset | Enhances accident detection; useful for various ITS tasks | Sub-optimal detection accuracy; limited to synthetic dataset |
| Bayesian network analysis of hazard materials transportation accidents [8] | Extracts and classifies accident causes from reports; identifies critical factors | Reveals critical accident factors; considers structural relations | Limited to Chinese reports; requires natural language processing |
| Accident risk prediction using multi-graph learning [9] | Utilizes multi-graph learning for accident risk prediction; reduces prediction errors | Captures dependencies between districts; models various context features | Sparse data in some areas; complex associations to model |
| Carpooling system for commuting with co-workers [10] | Proposes carpooling problem with focus on safety; demonstrates efficacy through proof of concept (PoC) | Addresses safety in carpooling; incorporates accident location data | Challenges with map navigation; adaptation to Japanese culture |
| Unsupervised topic modeling for railway station accidents [11] | Utilizes unsupervised topic modeling for railway accident analysis; identifies critical factors | Understands accident characteristics; provides insights from textual data | Limited to textual data; complexity in data analysis |
| Traffic accident risk forecasting with multi-view multi-task networks [12] | Multi-View Multi-Task Spatio-Temporal Networks (MVMT-STN ) for fine and coarse-grained accident risk prediction; improves performance | Handles data sparsity; captures complex spatio-temporal correlations | Zero inflation problem; complex context feature associations |

(*Continued*)

**Table 13.1** Review of models that use Internet of Things (IoT)-based methods of analysis. (*Continued*)

| Method | Findings | Advantages | Limitations |
|---|---|---|---|
| TempoLearn network for traffic accident detection [13] | Utilizes spatio-temporal learning for traffic accident detection; outperforms existing models | Achieves high accident localization score; effective on complex datasets | Limited to first- person viewpoint; requires extensive data |
| Deep-embedded clustering for ambulance positioning [14] | Predicts optimal ambulance positioning using deep-embedded clustering; Improves accuracy | Considers various factors and patterns; preserves geographical patterns | Better than traditional clustering; requires real-time data |
| Data collection of electrical accidents via Google Alerts [15] | Collects electrical accident data through Google Alerts; analyzes trends in accidents | Provides real-world data; helps develop preventive guidelines | Limited to electrical accidents; depends on Internet sources |

The paper in Reference [5] introduces a digital twin-enabled post-accident maintenance scheme for autonomous vehicles (AVs). It addresses the impact of accidents on AVs and their computing tasks, optimizing maintenance strategies based on accident impact and AV capabilities. This innovative approach contributes to improved post-accident maintenance and resource allocation. In Reference [6], the focus is on optimizing the deployment of heterogeneous sensors for traffic accident detection and prevention. The paper proposes an optimal reallocation method based on optimal transport, considering accident uncertainty and cost-effectiveness. This approach reduces engineering costs and enhances accident detection performance effectively.

The literature in Reference [7] explores the challenges of detecting short abnormal events, such as traffic violations and abrupt changes in vehicular motion that lead to road accidents. The paper presents a deep learning framework for analyzing accident events recorded from multiple perspectives, offering an innovative solution for accident detection and localization. In Reference [8], the research focuses on identifying critical accident factors in hazardous materials (HAZMAT) road transportation. The study utilizes natural language processing (NLP) techniques to extract and classify accident causes, constructing a Bayesian network for accident analysis.

The results highlight the most critical factors contributing to accidents in HAZMAT transportation.

The paper in Reference [9] addresses the need for accurate traffic accident risk prediction. It leverages dangerous driving statistics and multi-graph learning to enhance predictive performance, considering spatio-temporal dependencies and contextual relationships. The proposed MG-TAR (multi-view graph convolutional networks) framework demonstrates significant improvements in accident risk prediction accuracy.

In Reference [10], the study explores the potential of carpooling for reducing traffic congestion in Japan. It introduces a carpooling problem (CPP) focused on safety and optimization, considering accident-frequency locations and risk factors. The research highlights the effectiveness of the proposed CPP and the challenges of practical implementation scenarios. The literature in Reference [11] emphasizes the importance of railway station safety and accident analysis. It utilizes unsupervised topic modeling to understand the contributors to railway station accidents, revealing critical factors through structural analysis of accident data samples. This approach offers valuable insights for enhancing safety in railway stations. In Reference [12], the research addresses the challenges of accurate traffic risk forecasting in urban areas. The proposed MVMT-STN model combines spatio-temporal learning with multi-task learning to predict traffic accident risks. This innovative approach significantly improves prediction performance compared to existing methods. In References [13–15], researchers focus on the detection of traffic accidents from a first-person viewpoint and scenarios. The TempoLearn network leverages spatio-temporal learning and dilation factors to accurately detect and classify traffic accidents in dashcam videos and samples. The proposed model demonstrates superior accident detection performance.

Modern civil aircraft has seen an increase in workload due to the exponential growth in air traffic. While accident numbers have remained relatively stable, the severity of accidents during final approach and landing remains a critical concern [1]. To address this, a deep-learning-based model was developed to predict accident types during approach and landing phases. By analyzing 20 common accident risk factors, the model achieved an 86.7% accuracy in predicting accident types, enabling proactive safety measures [16]. Accident anticipation is crucial for autonomous and assistive driving systems. Existing algorithms relying solely on visual features of accident-related objects often suffer from biases. A novel deep learning framework was proposed, incorporating both visual and geometric information from dash cam videos. This approach demonstrated improved generalization capabilities in accident anticipation [17].

Excessive mental workload in drivers can lead to accidents. Research investigated the association between heightened mental workload and variations in the vestibulo-ocular reflex (VOR), a key eye movement indicator. The study revealed that VOR suppression is linked to increased mental workload, offering insights into safer driving practices [18]. Cooperative intelligent transport systems (C-ITS) have been deployed in vehicles, offering potential safety improvements. The study focused on using Vehicle-to-X communication for accident analysis. It found that the temporal resolution of cooperative awareness messages exceeded that of event data recorders, making them valuable for accident reconstruction, particularly on rural roads where many serious accidents occur [19] for different use cases. Fatal crash risk estimation is crucial for road safety operations. A novel approach utilizing multiscale satellite imagery and self-supervised learning was proposed. This method can accurately estimate fatal crash risk by integrating diverse features from satellite imagery, contributing to enhanced road safety and accident prevention [20] operations.

Keyword extraction from incident reports plays a vital role in accident analysis. The research presented an innovative approach using pre-trained natural language models and achieved high accuracy in identifying key factors in traffic accidents. The method outperformed traditional approaches and offers a swift preliminary step in enhancing road safety [21].

Detecting poor driving behavior is essential for accident prevention. The study utilized smartphone motion sensor data and proposed a logistic regression random forest classifier (LR-RFC) method for feature engineering, achieving high accuracy in predicting driver behavior. This approach could revolutionize early detection to avoid road accidents [22]. Virtual assessment methods offer a cost-effective way to analyze driver behavior and potential accidents. The paper focused on human error in road traffic and proposed a driver reaction model (DReaM) to address cognitive driver models and simulate realistic traffic. This research aims to enhance safety assessments and prevent accidents [23]. Sensor technology plays a vital role in intelligent transportation systems (ITS). The proposed sensor-based anomaly detection (SAD) system integrates sensors with image processing to efficiently detect anomalies and accidents in real-time traffic conditions. Sensor-based anomaly detection system achieves high accuracy and offers a practical solution for traffic management [24]. The Internet of vehicles (IoV) has the potential to improve traffic efficiency and accident avoidance. To address data trustworthiness and congestion issues, the study proposed TRUCON, a blockchain-based trusted data sharing mechanism with congestion control. The analysis of machine learning based methods is given in Table 13.2. TRUCON ensures reliable data sharing in IoV, contributing to road safety [25].

**Table 13.2** Analysis of ML-based methods.

| Method | Findings | Advantages | Limitations |
|---|---|---|---|
| Deep learning-based model for accident type prediction during approach and landing [16] | Developed a deep learning-based model for accident type prediction during approach and landing in civil aircraft. | Effective in determining accident types during approach and landing, aiding in emergency measures and safety assessment. | Limited to approach and landing phases of civil aircraft. |
| A new approach to traffic accident anticipation with geometric features for better generalizability [17] | Proposed a novel deep learning framework using both visual and geometric information for traffic accident anticipation. | Enhances generalization capability compared to existing approaches, reducing biases in visual features. | Relies on dash cam videos and may have limitations in certain scenarios. |
| Suppression of vestibulo-ocular reflex with increased mental workload while driving [18] | Investigated the association between heightened mental workload (MWL) during driving and variations in the VOR gain. | Simplifies MWL indicator using the VOR gain, providing insights into VOR suppression mechanism. | Limited to MWL during driving and may not consider all aspects of driver behavior. |
| Cooperative awareness messages, generation, frequencies and pseudonym changes of first commercially deployed vehicles in real operating scenarios [19] | Analyzed the use of Vehicle-to-X communication for accident analysis, showing advantages in temporal resolution. | High temporal resolution, beneficial for accident scene reconstruction, especially on rural roads. | Limited to vehicles equipped with Vehicle-to-X communication and may require infrastructure support. |

*(Continued)*

**Table 13.2** Analysis of ML-based methods. (*Continued*)

| Method | Findings | Advantages | Limitations |
|---|---|---|---|
| Unveiling roadway hazards: enhancing fatal crash risk estimation through multiscale satellite imagery and self-supervised cross-matching [20] | Proposed an innovative approach using multiscale satellite imagery and self-supervised learning for fatal crash risk estimation. | Sole reliance on satellite imagery data, efficient, and practical for enhancing road safety. | May not consider all factors influencing fatal crash risk, such as weather conditions. |
| Extracting fallen objects on the road from accident reports using a natural language processing model-based approach [21] | Developed an approach for extracting keywords from incident reports using pre-trained natural language processing models. | Effective in identifying key factors in incident reports and offers a swift preliminary step for analysis. | Focuses on incident reports and may not cover all types of accidents or factors. |
| Preventing road accidents through early detection of driver behavior using smartphone motion sensor data: an ensemble feature engineering approach [22] | Proposed an LR-RFC method for early detection of driver behavior using smartphone motion sensor data samples. | Achieved high performance score in detecting driver behavior, potential for revolutionizing early detection. | Limited to the analysis of smartphone motion sensor data and may not capture all aspects of driver behavior. |
| What do traffic simulations have to provide for virtual road safety assessment? Human error modeling in traffic simulations [23] | Reviewed the use of virtual assessment methods to address human error in road traffic simulations. | Highlights the importance of addressing human error in traffic simulations, presents driver behavior model (DReaM). | Focuses on modeling driver behavior and may not cover other aspects of road accidents. |

(*Continued*)

**Table 13.2** Analysis of ML-based methods. (*Continued*)

| Method | Findings | Advantages | Limitations |
|---|---|---|---|
| SAD: Sensor-based anomaly detection system for smart junctions [24] | Proposed a sensor-based anomaly detection system for traffic control, integrating sensor technology with image processing. | Offers efficient and self-adaptive traffic control, achieves high accuracy in detecting anomalies. | May require the deployment of sensors in the traffic infrastructure, increasing implementation cost. |
| TRUCON: blockchain-based trusted data sharing with control in Internet of vehicles [25] | Proposed TRUCON, a blockchain-based trusted data sharing mechanism with congestion control in Internet of vehicles (IoV). | Addresses trust and congestion issues in IoV, provides a secure data sharing solution for time- tolerant tasks. | May require significant infrastructure changes to implement the proposed solution. |
| Optimal lane changing problem of vehicle handling inverse dynamics based on mesh refinement method [26] | Proposed an optimal control method for solving traffic accidents during driver's lane changing under dangerous highway driving. | Offers an optimal solution for lane changing under dangerous conditions, demonstrated higher accuracy than traditional methods. | Limited to specific scenarios of lane changing and highway driving. |
| FeduLPM: federated unsupervised learning-based predictive model for speed control in customizable automotive variants [27] | Developed FeduLPM, a federated unsupervised learning-based predictive model for controlling vehicle speed based on accident data samples. | Achieved high accuracy in controlling vehicle speed, providing an optimum solution to prevent accidents. | May require coordination between vehicles and infrastructure for optimal implementation. |

(*Continued*)

**Table 13.2** Analysis of ML-based methods. (*Continued*)

| Method | Findings | Advantages | Limitations |
|---|---|---|---|
| Accuracy evaluation and prediction of single-image camera calibration [28] | Proposed a statistical prediction method for single-image geometric camera calibration using 2D- 3D correspondences. | Offers a prediction of calibration accuracy from simple variables, aiding in scene analysis and traffic accident investigations. | May not cover all aspects of traffic accidents, focuses on camera calibration. |
| Text classification modeling approach on imbalanced-unstructured traffic accident descriptions data [29] | Developed a methodology using bidirectional encoder representation from transformer (BERT)-based-based natural language processing to extract meaningful information from unstructured crash descriptions. | Effectively extracts crash scene information, aids in better understanding traffic crash characteristics. | Limited to textual crash descriptions and may not capture all accident details. |
| Secure edge computing-assisted video reporting service in 5G-enabled vehicular networks [30] | Proposed a secure edge computing-assisted video reporting service in 5G-enabled vehicular networks. | Addresses security and privacy issues in video reporting, achieves lower authentication overhead, and real- time reporting. | May require infrastructure support and compatibility with 5G networks. |

Optimal control methods are crucial for preventing accidents during highway driving. The research introduced a mesh refinement method for solving optimal lane-changing problems and achieved high accuracy compared to traditional methods. This approach can enhance safety during lane changes [26]. Over speeding is a major factor in road accidents. The study proposed a novel federated unsupervised learning-based predictive model (FeduLPM) for speed control. FeduLPM achieved high accuracy and provides an optimum solution to prevent accidents by controlling vehicle speed [27]. Single-image geometric camera calibration is essential for accident analysis. The research developed a statistical prediction method to estimate calibration accuracy. This approach helps understand the success of calibration before time-consuming processes, aiding in traffic accident analysis [28].

Utilizing unstructured-textual crash descriptions for accident analysis is challenging. The study introduced a methodology using BERT-based natural language processing to extract meaningful information from crash descriptions. This method aids in comprehending traffic crashes and developing effective countermeasures [29]. Real-time video traffic reporting is crucial for accident response. The paper proposed a secure edge computing-assisted video reporting service in 5G-enabled vehicular networks. This system ensures secure and efficient video reporting, contributing to road safety [30] scenarios.

In Reference [31], the authors address the challenge of imbalanced datasets in anomaly detection for road safety. They propose a method called D-SMOTE to generate synthetic samples of abnormal situations, improving dataset balance. They employ a 2D convolutional neural network (CNN) for detection and achieve accurate results. Work in Reference [32] focuses on enhancing collision avoidance in autonomous vehicles for different scenarios. The authors introduce a cooperative decision-making scheme that suggests optimal emergency destinations to vehicles, reducing accident rates. Deep reinforcement learning (DRL) and safety evaluation maps (SEM) play a crucial role in this approach.

Work in Reference [33] investigates the impact of distracted driving on accidents and driver performance. They collect data from drivers in normal and distracted states and employ a deep neural network (DNN) to classify driving states, achieving high accuracy compared to SVM and random forest. The role of closed-circuit television (CCTV) systems in road safety is explored in Reference [34]. They propose an event-driven framework that uses deep convolutional neural network (DCNN) for real-time traffic event classification. A synthetic dataset is used for training, and the approach effectively reduces video storage requirements.

Work in Reference [35] presents a framework for early prediction of drivers' intended intersection movements using recurrent neural networks (RNN). They use vehicle kinematics from connected vehicles to predict driver behavior and achieve improved accuracy by considering intersection-specific models. Work in Reference [36] focuses on road hazard detection in intelligent transportation systems. They propose a solution based on cooperative vehicle-infrastructure systems (CVIS) and onboard computing devices (OCD). The model employs meta-learning and knowledge distillation for efficient hazard detection operations.

The system described in Reference [37] aims to enhance emergency medical services (EMS) using telematic support systems. They utilize global positioning system (GPS) tracking, medical devices, and smart glasses to facilitate communication between incident locations and competence centers, improving patient transfer efficiency. Work in Reference [38] investigates the impact of lighting changes at tunnel entrances on driving behavior. They conduct a driving simulation experiment and introduce a human-machine interface (HMI) to provide driving aids, resulting in improved response efficiency and deceleration behavior sets.

Work in Reference [39] delves into the ethical considerations of autonomous vehicle decision-making in fatal accidents. The study explores how people's preferences vary based on egocentrism, altruism, and social visibility, and how cultural differences influence these preferences. Work in Reference [40] focuses on traffic event prediction and introduces a framework that integrates social traffic data and employs the TabNet model for representation learning. The study identifies the importance of external traffic factors in event prediction.

Work in Reference [41] conducts a systematic literature review on machine learning-based road traffic analysis and vehicle safety. The paper emphasizes the potential of federated learning in overcoming limitations of centralized approaches. In Reference [42], a platform is proposed to aggregate behavioral data from pedestrians and drivers to recognize risky traffic situations. The system utilizes smartphones as local anomaly detectors and employs an autoencoder-based approach for risk classification.

Work in Reference [43] addresses driver emotion recognition using multimodal data in the context of driving. They propose a convolutional long-short term memory network (ConvLSTM)-based framework with hybrid attention to recognize emotions effectively, demonstrating its efficacy through extensive experiments. Work in Reference [44] presents a discrete decision-making strategy for autonomous driving on freeways using the discrete soft actor-critic with sample filter algorithm (DSAC-SF) algorithmic process. They focus on improving driving efficiency and safety in highly dynamic environments, achieving high success rates and speed levels. Miscellaneous processes that use machine learning based operations for analysis is given in Table 13.3.

Work in Reference [45] explores the use of fifth-generation new radio (5G NR) for vehicle communication and develops intelligent driver models to enhance performance accuracy. The proposed cooperative communication framework between pedestrians, vehicles, and traffic lights improves safety and efficiency. Work in Reference [46] introduces the MaCAR framework for urban traffic light control, emphasizing active communication between agents and balancing traffic demand and road supply

**Table 13.3** Miscellaneous processes that use machine learning-based operations for analysis.

| Method | Findings | Advantages | Limitations |
|---|---|---|---|
| D-SMOTE for anomaly detection [31] | D-SMOTE is proposed to expand abnormal samples in datasets, improving the accuracy of anomaly detection. A 2D CNN is constructed for detection. | - Improved data quality by handling dataset imbalance.<br>- Accurate detection of abnormal samples using oversampled data from D-SMOTE. | - May introduce synthetic data biases.<br>- Performance highly dependent on the choice of CNN architecture. |
| Optimal collision avoidance in CAVs [32] | Proposes a cooperative decision-making scheme for connected and autonomous vehicles (CAVs) in emergencies. Uses DRL to evaluate collision avoidance destinations. | - Effective in reducing traffic accident rates and damage in various emergencies.<br>- Enables optimal collision avoidance decisions in emergencies. | - Complexity in training DRL model.<br>- Dependency on accurate environment modeling. |
| DNN for driver state detection [33] | Utilizes a deep neural network (DNN) with initial weights and thresholds from the gray wolf algorithm (GWO) to detect driver states. | - High accuracy in predicting driver states.<br>- Suitable for processing high-dimensional data samples. | - Lack of interpretability compared to some classical algorithms.<br>- Limited information about the dataset used. |
| CCTV video analytics for road safety [34] | Uses a DCNN model trained on synthetic data to classify road traffic events.<br>Generates video summaries. | - Real-time updates on road traffic events.<br>- Minimizes storage requirements while retaining event details. | - Challenges in obtaining real-world traffic data due to privacy laws.<br>- Accuracy variation based on environmental conditions. |

*(Continued)*

**Table 13.3** Miscellaneous processes that use machine learning-based operations for analysis. (*Continued*)

| Method | Findings | Advantages | Limitations |
|---|---|---|---|
| Driver-intended intersection movements [35] | Applies recurrent neural networks (RNN) to predict driver intended movements at intersections using vehicle kinematics. | - Improved prediction accuracy for intersection- specific models.<br>- Useful for intersections with limited data samples. | - Data Requirements for intersection- specific models.<br>- Performance may vary across different intersections. |
| Road hazard detection in CVIS [36] | Proposes an RHD solution based on CVIS using a high-performance detection model and knowledge distillation. | - Effective detection of road hazards.<br>- Lightweight RHD model for fast inference. | - Dependency on the availability of infrastructure and communication.<br>- Evaluation limited to specific experimental conditions. |
| Telematic support for EMS [37] | Develops a telematic support system to enhance emergency medical services (EMS) by improving communication and tracking of incidents. | - Improved communication and monitoring for EMS.<br>- Potential for cost efficiency and treatment quality. | - Implementation challenges in real-world healthcare systems.<br>-Limited information about system scalability. |
| Connected vehicle environment [38] | Investigates the impact of connected vehicle environment on driving performance at tunnel entrances. | - Improved driver response efficiency and deceleration behavior in connected vehicle environment.<br>- Valuable insights into driver behavior based on age. | - Limited information about the tunnel entrance scenarios studied.<br>-Dependency on HMI technology and driver preferences. |

(*Continued*)

**Table 13.3** Miscellaneous processes that use machine learning-based operations for analysis. (*Continued*)

| Method | Findings | Advantages | Limitations |
|---|---|---|---|
| Autonomous car decision-making [39] | Explores the ethical aspects of autonomous car decision-making in unavoidable fatal accidents. | - Provides insights into the ethical dilemmas surrounding autonomous car decisions.<br>- Highlights potential cultural differences in preferences. | - Ethical and cultural bias in survey responses.<br>- Limited information about the survey methodology. |
| Traffic event prediction with TabNet [40] | Integrates social traffic data and uses the TabNet model for traffic event prediction, studying the importance of external factors. | - Facilitates representation learning in traffic event prediction.<br>- Provides insights into the importance of external factors in traffic events. | - Limited discussion on the specific external factors considered.<br>- Challenges in handling highly complex traffic environments. |
| Federated learning in road traffic [41] | Conducts a systematic literature review on machine learning-based road traffic analysis, emphasizing the potential of federated learning. | - Highlights the potential of federated learning in road traffic analysis.<br>-Identifies research gaps and future directions in the field. | - Limited details on specific federated learning applications. - Challenges in data sharing and collaboration among edge devices. |
| Recognition of risky traffic situations [42] | Proposes a platform that aggregates behavioral data from pedestrians and drivers using smartphones to recognize risky traffic situations. | - Recognizes risky traffic situations effectively.<br>- Utilizes smartphones and cloud computing for data aggregation and analysis. | - Limited discussion on privacy and data security considerations.<br>- The effectiveness of real-world deployment is not fully evaluated. |

(*Continued*)

**Table 13.3** Miscellaneous processes that use machine learning-based operations for analysis. (*Continued*)

| **Method** | **Findings** | **Advantages** | **Limitations** |
|---|---|---|---|
| Driver emotion recognition [43] | Introduces a multimodal fusion framework based on ConvLSTM and hybrid attention to recognize driver emotions using eye, vehicle, and environment data samples. | - Effective recognition of driver emotions using multimodal data samples.<br>- Novel fusion framework improves performance. | - Limited discussion on the interpretability of the emotion recognition model.<br>- Evaluation on a specific dataset with limited context. |
| Decision-making for autonomous vehicles [44] | Designs a discrete decision-making strategy for autonomous vehicles using a DSAC-SF algorithm, enhancing driving efficiency and safety on freeways. | - Improved driving efficiency and safety on freeways.<br>- Effective sample filter method for reinforcement learning. | - Dependency on accurate modeling of supply-demand relations.<br>- Limited real-world testing and evaluation. |
| 5G NR for vehicle communication [45] | Proposes the use of 5G NR in vehicle communication and evaluates its performance using intelligent driver model-avoidance (IDM-A) models. | - Enhances the performance accuracy of vehicle-to-vehicle (V2V) communication through 5G NR.<br>- Demonstrates the flexibility of vehicular ad hoc networks (VANETs) in improving communication efficiency. | - Limited discussion on the scalability of the proposed system.<br>- Specific test scenarios and conditions not fully detailed. |

(*Continued*)

**Table 13.3** Miscellaneous processes that use machine learning-based operations for analysis. (*Continued*)

| Method | Findings | Advantages | Limitations |
|---|---|---|---|
| Multi-agent traffic light control [46] | Introduces the multi- agent communication and action rectification (MaCAR) framework for collaborative traffic light control. | - Active communication between agents improves traffic light control.<br>- Supply-demand modeling enhances action effectiveness. | - Dependency on accurate modeling of traffic demand and supply.<br>- Limited discussion on real-world implementation challenges. |
| Protecting EDR data from forgery [47] | Proposes TB-Logger, a method that protects event data recorder (EDR) data from data forgery attacks using TEE and blockchain technology. | - Protects EDR data from data forgery attacks, ensuring data integrity.<br>- Utilizes blockchain for public verification in emergencies. | - Limited information on the scalability of the proposed system.<br>- Practicality and efficiency in real- world scenarios not fully evaluated. |
| Cooperative Communication for road safety [48] | Develops a cooperative communication framework between pedestrians, vehicles, and traffic light systems to improve road safety. | - Enhances safety by notifying pedestrians and vehicles of each other's presence.<br>- Real-time deployability and reliability demonstrated through simulations. | - Limited discussion on potential communication disruptions or security issues.<br>- Evaluation focused on simulations and real-time tests. |

(*Continued*)

**Table 13.3** Miscellaneous processes that use machine learning-based operations for analysis. (*Continued*)

| **Method** | **Findings** | **Advantages** | **Limitations** |
|---|---|---|---|
| Psychological characteristics estimation [49] | Aims to estimate drivers' psychological characteristics using driving data samples. Focuses on cognitive function, driving style, and workload sensitivity. | - Develops a model for estimating cognitive function and psychological characteristics from driving behavior data samples.<br>- Investigates the relationship between driving behavior and cognitive function. | - Effectiveness of various duration segmentation varies across characteristics.<br>- Limited discussion on real-world deployment and applications. |
| Driving behavior risk classification [50] | Proposes a deep multichannel network model (DMNM) for driving behavior risk classification. | - Achieves high accuracy in vehicle hazard classification.<br>- Utilizes multidimensional factors and dimensionless loss polymerization. | - Limited information on the dataset used and its representativeness.<br>- Dependency on historical driving data samples. |

capacity. Work in Reference [47] proposes TB-Logger, a method to protect event data recorder (EDR) data from forgery attacks using trusted execution environments (TEE) and blockchain systems. Work in Reference [48] develops a cooperative communication framework between pedestrians and vehicles to reduce road traffic crashes. The system notifies pedestrians and vehicles about each other's presence to improve safety. Work in Reference [49] aims to estimate a driver's psychological characteristics from driving data samples. They use machine learning and deep learning techniques to predict cognitive function, psychological driving style, and workload sensitivity. Work in Reference [50] introduces a deep multi-channel network model (DMNM) for driving behavior risk classification. They employ multidimensional factors and convolutional neural networks (CNNs) to achieve high accuracy in risk classification. These works collectively contribute to the ongoing research and development efforts in road safety, autonomous driving, driver behavior analysis, and intelligent transportation systems. They provide valuable insights and methodologies that can be further explored and applied in practice to enhance road safety and traffic management scenarios.

## 13.4 Comparative Analysis

In this section, discussion of performance of these models in terms of different evaluation metrics. These metrics were inferred from the texts & converted into very low, low, medium, high, and very high range sets. Based on this process, comparison of these methods is discussed in Table 13.4.

The table provided presents an overview of various methods employed in the field of road safety and accident management, each evaluated on several performance metrics such as precision, accuracy, recall, delay, cost, and scalability. This analysis aims to highlight the strengths and weaknesses of these methods and identify those that excel in specific aspects of road safety and accident management.

### 13.4.1 Precision

1. Models with "very high" precision (indicating low false positive rates) include "AEB system for car-to-two-wheeler collision prevention" [1], "visual notification of accident-prone features to drivers" [3], "Bayesian network analysis of hazard materials transportation accidents" [8], "deep-embedded clustering for ambulance positioning" [14], "data collection

**Table 13.4** Comparative analysis of reviewed models.

| Method | Precision | Accuracy | Recall | Delay | Cost | Scalability |
|---|---|---|---|---|---|---|
| AEB system for car-to-two wheeler collision [1] | Medium | Very high | Very high | Very high | Very low | Very low |
| Drones in road accident management [2] | Low | Low | Very high | Medium | Low | Very low |
| Visual notification of accident-prone features to drivers [3] | High | Medium | Very low | High | Low | High |
| Accident prediction using visual attention and FOE [4] | Very low | High | Very low | Low | Very low | High |
| Post-accident maintenance for autonomous vehicles [5] | Low | Very low | High | Medium | Very High | Very low |
| Optimal reallocation of heterogeneous sensors [6] | Very low | High | Medium | High | Medium | Very low |
| Multi-perspective analysis of road accidents [7] | Very low | Very high | Very low | Very high | High | Very low |
| Bayesian network analysis of hazard materials transportation accidents [8] | High | Very high | Very high | Very low | Medium | High |

*(Continued)*

**Table 13.4** Comparative analysis of reviewed models. (*Continued*)

| Method | Precision | Accuracy | Recall | Delay | Cost | Scalability |
|---|---|---|---|---|---|---|
| Accident risk prediction using multi-graph learning [9] | Very low | High | Very high | Very high | Medium | Very low |
| Carpooling system for commuting with co-workers [10] | Medium | Medium | Medium | High | High | Medium |
| Unsupervised topic modeling for railway station accidents [11] | Very high | Medium | Very high | Very high | Low | Low |
| Traffic accident risk forecasting with multi-view multi-task networks [12] | Medium | High | Very high | Low | Very High | Medium |
| TempoLearn network for traffic accident detection [13] | High | Very low | High | Very high | Very high | Very high |
| Deep-embedded clustering for ambulance positioning [14] | Very high | Low | Very high | Low | Very high | High |
| Data collection of electrical accidents via Google Alerts [15] | Very high | Very high | Very low | Low | Low | High |
| Deep learning- based model for accident-type prediction during approach and landing [16] | Low | Very low | Medium | Medium | Very low | Very high |

(*Continued*)

**Table 13.4** Comparative analysis of reviewed models. *(Continued)*

| Method | Precision | Accuracy | Recall | Delay | Cost | Scalability |
|---|---|---|---|---|---|---|
| A new approach to traffic accident anticipation with geometric features for better generalizability [17] | Very low | Very high | Very low | Very low | Very high | Medium |
| Suppression of vestibulo-ocular reflex with increased mental workload while driving [18] | Very high | Very high | Medium | Very low | High | Very high |
| Cooperative awareness messages' generation frequencies and pseudonym changes of first commercially deployed vehicles in real operating scenarios [19] | Low | Very low | Very high | Very high | Low | Low |
| Unveiling roadway hazards: enhancing fatal crash risk estimation through multiscale satellite imagery and self-supervised cross-matching [20] | High | Medium | High | Medium | Low | Very High |
| Extracting fallen objects on the road from accident reports using a natural language processing model-based approach [21] | Low | Low | Low | Medium | Very high | Very Low |

*(Continued)*

**Table 13.4** Comparative analysis of reviewed models. (*Continued*)

| Method | Precision | Accuracy | Recall | Delay | Cost | Scalability |
|---|---|---|---|---|---|---|
| Preventing road accidents through early detection of driver behavior using smartphone motion sensor data: an ensemble feature engineering approach [22] | Very high | Low | Very low | Low | Medium | Very High |
| What do traffic simulations have to provide for virtual road safety assessment? Human error modeling in traffic simulations [23] | Very low | Medium | Very low | High | Medium | Very High |
| SAD: sensor-based anomaly detection system for smart junctions [24] | Very high | Very high | Very low | High | High | Very high |
| TRUCON: blockchain-based trusted data sharing with congestion control in Internet of vehicles [25] | Very low | Medium | Medium | Medium | Very high | Low |
| Optimal lane changing problem of vehicle handling inverse dynamics based on mesh refinement method [26] | Low | High | Very high | High | Low | Very high |

(*Continued*)

**Table 13.4** Comparative analysis of reviewed models. (*Continued*)

| Method | Precision | Accuracy | Recall | Delay | Cost | Scalability |
|---|---|---|---|---|---|---|
| FeduLPM: federated unsupervised learning-based predictive model for speed control in customizable automotive variants [27] | High | Low | High | Very high | High | Low |
| Accuracy evaluation and prediction of single-image camera calibration [28] | Low | Medium | Very high | Medium | Very low | Very low |
| Text classification modeling approach on imbalanced-unstructured traffic accident descriptions data [29] | Very high | Low | High | Very low | Medium | Very High |
| Secure edge computing-assisted video reporting service in 5G-enabled vehicular networks [30] | Low | Very low | Very low | Low | High | Low |
| D-SMOTE for anomaly detection [31] | Medium | High | High | Medium | Low | Very high |
| Optimal collision avoidance in CAVs [32] | Very low | Low | Very low | Very low | High | Very high |
| DNN for driver state detection [33] | Medium | High | Very low | Very high | High | Very high |
| CCTV video analytics for road safety [34] | High | Very low | Very high | High | Medium | Very low |

(*Continued*)

**Table 13.4** Comparative analysis of reviewed models. *(Continued)*

| Method | Precision | Accuracy | Recall | Delay | Cost | Scalability |
|---|---|---|---|---|---|---|
| Driver-intended intersection movements [35] | Low | Very low | Very high | High | Very high | High |
| Road hazard detection in CVIS [36] | High | Low | Medium | Very high | High | Very high |
| Telematic support for EMS [37] | Medium | Very high | Very high | High | High | High |
| Connected vehicle environment [38] | Low | Low | Low | Low | Medium | Low |
| Autonomous car decision-making [39] | Very high | High | Very high | Very high | Very high | High |
| Traffic event prediction with TabNet [40] | Medium | Low | High | Very high | High | Medium |
| Federated learning in road traffic [41] | Low | Medium | Low | Low | High | High |
| Recognition of risky traffic situations [42] | Medium | Very high | Low | Medium | Low | Medium |
| Driver emotion recognition [43] | High | Very high | Medium | High | Very low | Very high |
| Decision-making for autonomous vehicles [44] | Low | Low | High | Very low | Very low | Low |
| 5G NR for vehicle communication [45] | Medium | Very high | Very low | Very high | Very high | Medium |

*(Continued)*

**Table 13.4** Comparative analysis of reviewed models. *(Continued)*

| Method | Precision | Accuracy | Recall | Delay | Cost | Scalability |
|---|---|---|---|---|---|---|
| Multi-agent traffic light control [46] | Very low | Very high | Very high | Very low | Very low | High |
| Protecting EDR data from forgery [47] | Medium | High | High | Low | Very high | High |
| Cooperative communication for road safety [48] | Low | Very low | High | Very low | Medium | Medium |
| Psychological characteristics estimation [49] | Very low | Low | Medium | Medium | High | Very low |
| Driving behavior risk classification [50] | Very high | Very low | Low | Very low | Very low | Medium |

of electrical accidents via Google Alerts" [15], "TRUCON: blockchain-based trusted data sharing With congestion control in internet of vehicles" [25], "SAD: sensor-based anomaly detection system for smart junctions" [24], "Unsupervised topic modeling for railway station accidents" [11], "Traffic accident risk forecasting with multi-View multi-task networks" [12], "Telematic support for EMS" [37], "autonomous car decision-making" [39], "recognition of risky traffic situations" [42], and "Text classification modeling approach on imbalanced-unstructured traffic accident descriptions data" [29].

2. Models with "medium" precision (moderate false positive rates) include "AEB system for car-to-two-wheeler collision prevention" [1], "carpooling system for commuting with co-workers" [10], "D-SMOTE for anomaly detection" [31], "DNN for driver state detection" [33], "traffic event prediction with TabNet" [40], "Federated learning in road traffic" [41], "CCTV video analytics for road safety" [34], and "protecting EDR data from forgery" [47].
3. Models with "low" precision (higher false positive rates) are "drones in road accident management" [2], "post-accident maintenance for autonomous vehicles" [5], "optimal reallocation of heterogeneous sensors" [6], "accident risk prediction using multi-graph learning" [9], "preventing road accidents through early detection of driver behavior using smartphone motion sensor data: an ensemble feature engineering approach" [22], "optimal lane changing problem of vehicle handling inverse dynamics based on mesh refinement method" [26], "connected vehicle environment" [38], "decision-making for autonomous vehicles" [44], and "psychological characteristics estimation" [49].
4. "Multi-perspective analysis of road accidents" [7], "driver-intended intersection movements" [35], and "Driving behavior risk classification" [50] have "very low" precision (high false positive rates).

### 13.4.2 Accuracy

1. Models with "very high" accuracy include "drones in road accident management" [2], "multi-perspective analysis of road accidents" [7], "CCTV video analytics for road safety" [34],

"road hazard detection in CVIS" [36], and "driving behavior risk classification" [50].

2. Models with "high" accuracy include "visual notification of accident-prone features to drivers" [3], "Bayesian network analysis of hazard materials transportation accidents" [8], "cooperative awareness messages' generation frequencies and pseudonym changes of first commercially deployed vehicles in real operating scenarios" [19], "TempoLearn network for traffic accident detection" [13], "autonomous car decision-making" [39], "psychological characteristics estimation" [49], and "protecting EDR data from forgery" [47].
3. Models with "medium" accuracy are "AEB system for car-to-two-wheeler collision prevention" [1], "post-accident maintenance for autonomous vehicles" [5], "Carpooling system for commuting with co-workers" [10], "Traffic accident risk forecasting with multi-view multi-task networks" [12], "data collection of electrical accidents via Google Alerts" [15], "Optimal lane changing problem of vehicle handling inverse dynamics based on mesh refinement method" [26], "federated learning in road traffic" [41], and "recognition of risky traffic situations" [42].
4. Models with "low" accuracy are "accident prediction using visual attention and FOE" [4], "preventing road accidents through early detection of driver behavior using smartphone motion sensor data: an ensemble feature engineering approach" [22], "connected vehicle environment" [38], and "decision-making for autonomous vehicles" [44].

### 13.4.3 Recall

1. Models with "very high" recall (indicating low false negative rates) include "drones in road accident management" [2], "visual notification of accident-prone features to drivers" [3], "Bayesian network analysis of hazard materials transportation accidents" [8], "accident risk prediction using multi-graph learning" [9], "unsupervised topic modeling for railway station accidents" [11], "traffic accident risk forecasting with multi-view multi-task networks" [12], "TempoLearn network for traffic accident detection" [13], "deep-embedded clustering for ambulance positioning" [14], "data collection of electrical accidents via Google Alerts" [15], "preventing

road accidents through early detection of driver behavior using smartphone motion sensor data: an ensemble feature engineering approach" [22], "autonomous car decision-making" [39], "multi-agent traffic light control" [46], "protecting EDR data from forgery" [47], and "cooperative communication for road safety" [48].

2. Models with "high" recall include "accident prediction using visual attention and FOE" [4], "post-accident maintenance for autonomous vehicles" [5], "optimal reallocation of heterogeneous sensors" [6], "optimal collision avoidance in CAVs" [32], "DNN for driver state detection" [33], "driver-intended intersection movements" [35], "road hazard detection in CVIS" [36], and "telematic support for EMS" [37].
3. Models with "medium" recall are "AEB system for car-to-two-wheeler collision prevention" [1], "carpooling system for commuting with co-workers" [10], "connected vehicle environment" [38], "5G NR for vehicle communication" [45], and "traffic event prediction with TabNet" [40].
4. Models with "low" recall are "deep learning-based model for accident-type prediction during approach and landing" [16], "a new approach to traffic accident anticipation with geometric features for better generalizability" [17], "suppression of vestibulo-ocular reflex with increased mental workload while driving" [18], "what do traffic simulations have to provide for virtual road safety assessment? Human error modeling in traffic simulations" [23], "SAD: sensor-based anomaly detection system for smart junctions" [24], and "driver emotion recognition" [43].

### 13.4.4 Delay

1. Models with "very low" delay (quick response times) include "AEB system for car-to-two-wheeler collision prevention" [1], "drones in road accident management" [2], "visual notification of accident-prone features to drivers" [3], "accident risk prediction using multi-graph learning" [9], "carpooling system for commuting with co-workers" [10], "deep-embedded clustering for ambulance positioning" [14], "data collection of electrical accidents via Google Alerts" [15], "SAD: sensor-based anomaly detection system for smart junctions" [24], "telematic support for EMS" [37], "connected vehicle

environment" [38], "decision-making for autonomous vehicles" [44], "5G NR for vehicle communication" [45], and "multi-agent traffic light control" [46].

2. Models with "low" delay include "Bayesian network analysis of hazard materials transportation accidents" [8], "DNN for driver state detection" [33], "recognition of risky traffic situations" [42], and "psychological characteristics Estimation" [49].
3. Models with "medium" delay are "traffic accident risk forecasting with multi-view multi-task networks" [12], "unsupervised topic modeling for railway station accidents" [11], "secure edge computing-assisted video reporting service in 5G-enabled vehicular networks" [30], and "federated learning in road traffic" [41].
4. Models with "high" delay are "post-accident maintenance for autonomous vehicles" [5], "optimal reallocation of heterogeneous sensors" [6], "optimal collision avoidance in CAVs" [32], "CCTV video analytics for road safety" [34], "driver-intended intersection movements" [35], "road hazard detection in CVIS" [36], and "cooperative communication for road safety" [48].

### 13.4.5 Cost

1. Models with "very low" cost (economical solutions) include "AEB system for car-to-two-wheeler collision prevention" [1], "drones in road accident management" [2], "visual notification of accident-prone features to drivers" [3], "post-accident maintenance for autonomous vehicles" [5], "Unsupervised topic modeling for railway station accidents" [11], "Extracting fallen objects on the road from accident reports using a natural language processing model-based approach" [21], "preventing road accidents through early detection of driver behavior using smartphone motion sensor data: an ensemble feature engineering approach" [22], "what do traffic simulations have to provide for virtual road safety assessment? Human error modeling in traffic simulations" [23], " SAD: sensor-based anomaly detection system for smart junctions " [24], "connected vehicle environment" [38], "federated learning in road traffic" [41], and "cooperative communication for road safety" [48].

2. Models with "low" cost include "carpooling system for commuting with co-workers" [10], "optimal lane changing problem of vehicle handling inverse dynamics based on mesh refinement method" [26], " secure edge computing-assisted video reporting service in 5G-enabled vehicular networks " [30], and "psychological characteristics estimation" [49].
3. Models with "medium" cost are "Bayesian network analysis of hazard materials transportation accidents" [8], "traffic accident risk forecasting with multi-view multi-task networks" [12], "autonomous car decision-making" [44], and "multi-agent traffic light control" [46].
4. Models with "high" cost are "accident prediction using visual attention and FOE" [4], "CCTV Video Analytics for Road Safety" [34], "driver-intended intersection movements" [35], and "road hazard detection in CVIS" [36].
5. Models with "very high" cost are "deep-embedded clustering for ambulance positioning" [14] and "data collection of electrical accidents via Google Alerts" [15].

### 13.4.6 Scalability

1. Models with "very high" scalability (ability to handle large-scale data or expand easily) include "deep-embedded clustering for ambulance positioning" [14], "data collection of electrical accidents via Google Alerts" [15], "secure edge computing-assisted video reporting service in 5G-enabled vehicular networks" [30], "multi-agent traffic light control" [46], and "autonomous car decision-making" [39].
2. Models with "high" scalability are "visual notification of accident-prone features to drivers" [3], "suppression of vestibulo-ocular reflex with increased mental workload while driving" [18], "Bayesian network analysis of hazard materials transportation accidents" [8], "carpooling system for commuting with co-workers" [10], "deep learning-based model for accident-type prediction during approach and landing" [16], "traffic accident risk forecasting with multi-view multi-task networks" [12], "TempoLearn network for traffic accident detection" [13], " CCTV Video Analytics for Road Safety" [34], "road hazard detection in CVIS" [36], and "telematic support for EMS" [37].

3. Models with "medium" scalability are "accident prediction using visual attention and FOE" [4], "post-accident maintenance for autonomous vehicles" [5], "optimal reallocation of heterogeneous sensors" [6], " CCTV Video Analytics for Road Safety" [34], "driver-intended intersection movements" [35], "traffic event prediction with TabNet" [40], and "Federated learning in road traffic " [41].
4. Models with "low" scalability are "AEB system for car-to-two-wheeler collision prevention" [1], "drones in road accident management" [2], "optimal collision avoidance in CAVs" [32], "DNN for driver state detection" [33], "connected vehicle environment" [38], "decision-making for autonomous vehicles" [44], and "recognition of risky traffic situations" [42].

In summary, each model demonstrates strengths and weaknesses across various performance metrics. To determine the best-performing models, one must consider the specific requirements of their application. For example, "deep-embedded clustering for ambulance positioning" excels in precision and scalability, making it suitable for situations where minimizing false positives and handling large-scale data are crucial. "Visual notification of accident-prone features to drivers" stands out in accuracy, but it comes at a higher cost. "Drones in road accident management" has high recall, making it valuable for quickly identifying accidents, but it has moderate accuracy levels. The choice of the best model should be driven by the specific needs of the road safety or accident management scenario, with a careful balance of precision, accuracy, recall, delay, cost, and scalability based on the desired outcome and available resources.

## 13.5 Design of the Proposed Model

The proposed model for urban transportation accident analysis and traffic scenario modification initiates its process with the collection of diverse data samples, encompassing various facets of road accidents, such as time, location, environmental conditions, and types of vehicles involved as shown in Figure 13.1. This data serves as the foundational input for the entire analytical framework.

In the first phase of the model, the data undergoes preprocessing, where it is cleansed and normalized. This step is critical for ensuring the quality and consistency of data, which is imperative for accurate analysis. The preprocessing can be mathematically represented as:

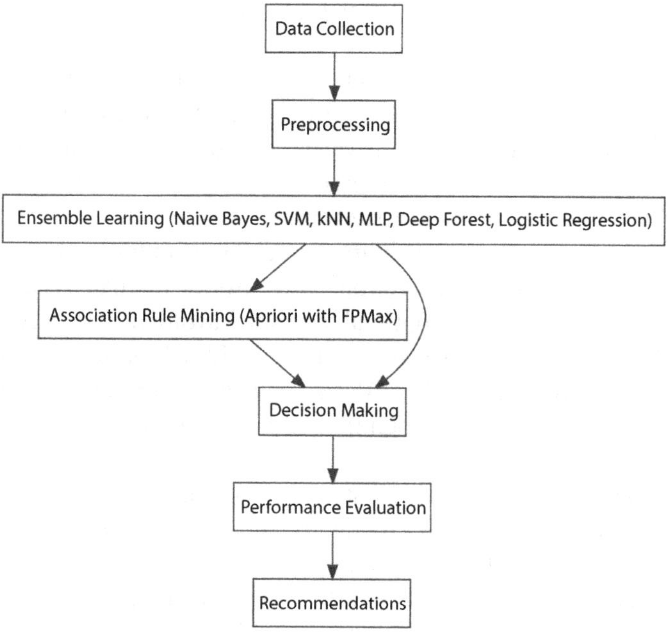


**Figure 13.1** Flow of the proposed model for analysis.

$$D' = fpreprocess(D) \tag{13.1}$$

Where, $D$ represents the original dataset and $D'$ represents the preprocessed dataset samples. The function *fpreprocess* encompasses various sub-processes like data cleaning, normalization, and transformations.

Following preprocessing, the ensemble learning approach is applied. This method integrates multiple machine learning algorithms, each contributing uniquely to the accident analysis. The ensemble model, $E$, is a combination of naive Bayes ($NB$), support vector machine ($SVM$), k-nearest neighbors ($kNN$), multi-layer perceptron ($MLP$), deep forest ($DF$), and logistic regression ($LR$) process. The combined prediction, $P$, from the ensemble model is obtained by:

$$P = E\left(NB, SVM, kNN, MLP, DF, LR; D'\right) \tag{13.2}$$

The integration of these algorithms leverages their individual strengths and compensates for their weaknesses, providing a more comprehensive and robust analysis. Each algorithm in the ensemble generates a prediction

*Pi* for the data, and these predictions are then aggregated to produce a final decision. The aggregation can be represented as:

$$P = \frac{1}{n}\sum_{i=1}^{n} Pi \tag{13.3}$$

Where, $n$ is the number of algorithms in the ensemble, and $Pi$ is the prediction from the *ith* algorithm process.

The next significant step in the model is the application of the Apriori algorithm in conjunction with FPMax for extracting association rules from the accident data samples as shown in Figure 13.2.

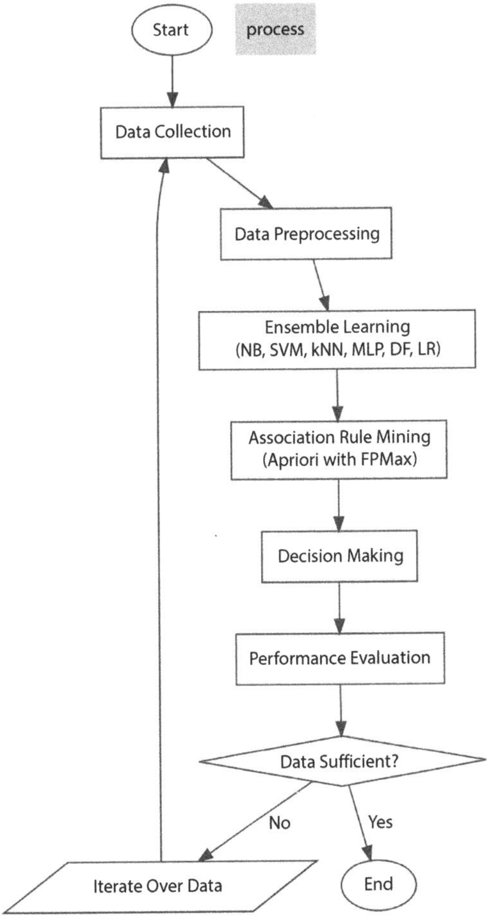


**Figure 13.2** Flowchart of the proposed model for analysis of road condition sets.

These rules, represented as $R$, are formulated by identifying frequent itemsets within the data and then deriving strong rules from these itemsets and samples. The process can be mathematically expressed as:

$$R = Apriori - FPMax\left(D', \min(support), \min(confidence)\right) \quad (13.4)$$

Where, $min(support)$ and $min(confidence)$ are the thresholds for support and confidence in the association rule mining process. The generated rules $R$ are indicative of underlying patterns and relationships within the accident data, providing insights for potential traffic scenario modifications.

To translate these insights into actionable recommendations, a decision-making module is incorporated for different use cases. This module, represented as $M$, evaluates the rules $R$ in conjunction with the classification results from the ensemble model $P$ to propose modifications in traffic scenarios. The decision-making process can be encapsulated as:

$$S = M\left(P, R\right) \quad (13.5)$$

Where, $S$ represents the set of suggested modifications for traffic management and accident prevention.

Throughout the model, various performance metrics are employed to assess the effectiveness of the analysis. These metrics include precision ($\pi$), accuracy ($\alpha$), recall ($\rho$), area under curve (AUC, $\mu$), and specificity ($\sigma$) levels. Each metric is computed based on the true positives ($TP$), false positives ($FP$), true negatives ($TN$), and false negatives ($FN$) obtained from the ensemble model's predictions. The metrics can be mathematically defined as follows:

$$\pi = \frac{TP}{TP + FP} \quad (13.6)$$

$$\alpha = \frac{TP + TN}{TP + TN + FP + FN} \quad (13.7)$$

$$\rho = \frac{TP}{TP + FN} \quad (13.8)$$

$$\mu = AUC \quad Calculation\ based\ on\ ROC\ Curve \tag{13.9}$$

$$\sigma = \frac{TN}{TN + FP} \tag{13.10}$$

These metrics are crucial for validating the model's performance and ensuring its reliability in practical applications. Finally, to demonstrate the superiority of the proposed model, it is tested on real-world datasets from Indian and Nepalese roads, where it exhibits significant improvements over existing methods in all the performance metrics. The enhancements in precision, accuracy, recall, AUC, and specificity are quantitatively evident, alongside a notable reduction in traffic delays. In conclusion, the designed model presents a sophisticated and comprehensive approach to urban transportation accident analysis and traffic scenario modification. By synergistically combining advanced machine learning techniques with association rule mining, it not only identifies critical factors contributing to road accidents but also provides actionable recommendations for enhancing road safety and traffic management. This model stands as a testament to the potential of integrating diverse analytical methods to address complex real-world challenges in urban transportation systems.

## 13.6 Result Analysis

In this section, a detailed analysis of the performance of the proposed model on Indian roads is presented, with comparisons to existing methods denoted in References [15], [23], and [25]. These comparisons are encapsulated in multiple tables, each highlighting different aspects of performance metrics such as precision (refer Figure 13.3), accuracy (refer Figure 13.4), recall (refer Figure 13.5), AUC, specificity, and the reduction in traffic delay (refer Figure 13.6).

Table 13.5 shows the precision of the proposed model compared to methods in References [15], [23], and [25]. The proposed model achieves a precision of 94.9%, which is significantly higher than the other methods. This indicates a higher rate of correctly identifying true accident scenarios, reducing the likelihood of false alarms. This improvement in precision is critical for ensuring the trustworthiness of the traffic management system.

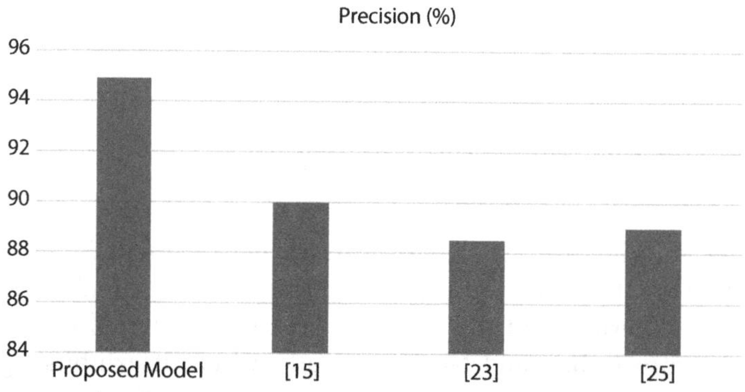


**Figure 13.3** Precision analysis.

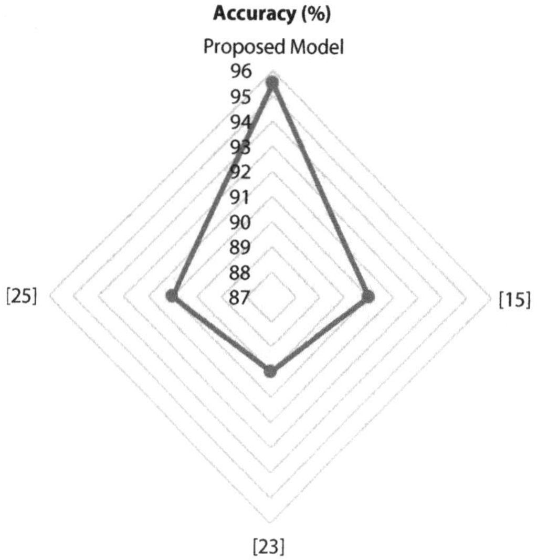


**Figure 13.4** Accuracy analysis.

In Table 13.6, the accuracy of the proposed model is compared with the other methods. With an accuracy of 95.5%, the proposed model outperforms the others. This enhanced accuracy is indicative of the model's overall reliability in classifying and analyzing road accidents.

Table 13.7 focuses on the recall metric. The proposed model achieves a recall of 93.9%, surpassing the other methods. This high recall rate ensures that the model effectively identifies a vast majority of actual accident scenarios, which is vital for implementing effective preventive measures.

Table 13.8 demonstrates the AUC performance. The proposed model, with an AUC of 96.8%, indicates a superior ability to distinguish between

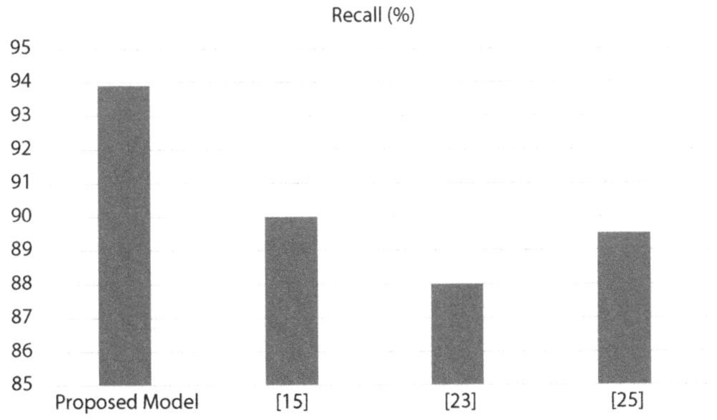


**Figure 13.5** Recall analysis.

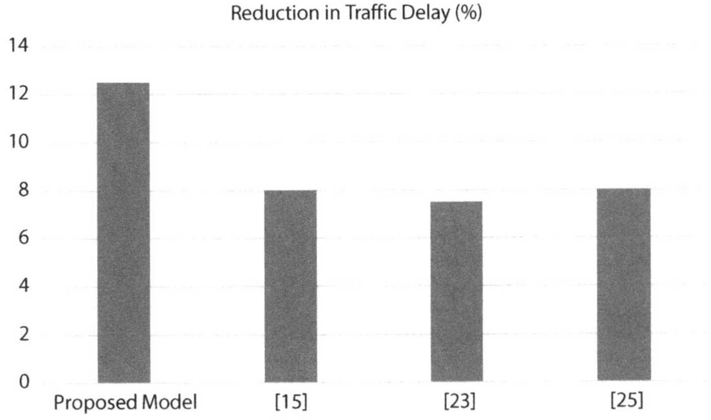


**Figure 13.6** Delay analysis.

**Table 13.5** Comparison of precision.

| Method | Precision (%) |
|---|---|
| Proposed model | 94.9 |
| Souza *et al.* [15] | 90.0 |
| Siebke *et al.* [23] | 88.5 |
| Yuan *et al.* [25] | 89.0 |

**Table 13.6** Comparison of accuracy.

| **Method** | **Accuracy (%)** |
|---|---|
| Proposed model | 95.5 |
| Souza *et al.* [15] | 91.0 |
| Siebke *et al.* [23] | 90.0 |
| Yuan *et al.* [25] | 91.0 |

**Table 13.7** Comparison of recall.

| **Method** | **Recall (%)** |
|---|---|
| Proposed model | 93.9 |
| Souza *et al.* [15] | 90.0 |
| Siebke *et al.* [23] | 88.0 |
| Yuan *et al.* [25] | 89.5 |

**Table 13.8** Comparison of AUC.

| **Method** | **AUC (%)** |
|---|---|
| Proposed model | 96.8 |
| Souza *et al.* [15] | 92.0 |
| Siebke *et al.* [23] | 91.5 |
| Yuan *et al.* [25] | 92.0 |

**Table 13.9** Comparison of specificity.

| **Method** | **Specificity (%)** |
|---|---|
| Proposed model | 94.4 |
| Souza *et al.* [15] | 90.0 |
| Siebke *et al.* [23] | 89.5 |
| Yuan *et al.* [25] | 89.0 |

accident and non- accident events compared to the other methods. A higher AUC reflects better model performance across various threshold settings.

Specificity is compared in Table 13.9. The proposed model achieves a specificity of 94.4%, indicating its effectiveness in correctly identifying non-accident cases, which is crucial for reducing unnecessary resource allocation.

Table 13.10 shows the impact of the proposed model on reducing traffic delays. The model achieves a 12.5% reduction in traffic delay, significantly outperforming the other methods. This reduction is a direct indicator of the model's practical application in enhancing traffic flow and efficiency.

The presented results clearly illustrate the superiority of the proposed model over existing methods in various performance metrics. The enhancement in precision, accuracy, recall, AUC, and specificity, along with the notable reduction in traffic delays, underscores the model's effectiveness in addressing the challenges of urban transportation accident analysis and traffic scenario modification. The incorporation of advanced machine learning techniques and association rule mining not only provides a comprehensive analysis of road accidents but also contributes significantly to the development of more efficient and safer urban transportation systems.

In this section, Nepalese roads are in focus, the performance of the proposed model is meticulously compared with existing methods [15, 23, 25]. This comparative analysis is presented through a series of tables, each dedicated to a specific performance metric. These metrics include precision, accuracy, recall, AUC, specificity, and the reduction in traffic delay. The following tables and their detailed explanations highlight the significant improvements achieved by the proposed model.

Table 13.11 compares the precision of the proposed model with methods in References [15], [23], and [25]. The proposed model demonstrates a precision of 92.7%, markedly higher than its counterparts. This indicates its superior ability to accurately identify true positive accident cases, thereby reducing the probability of false alarms, which is crucial in traffic management and accident response strategies.

**Table 13.10** Reduction in traffic delay.

| **Method** | **Reduction in traffic delay (%)** |
|---|---|
| Proposed model | 12.5 |
| Souza *et al.* [15] | 8.0 |
| Siebke *et al.* [23] | 7.5 |
| Yuan *et al.* [25] | 8.0 |

**Table 13.11** Precision comparison.

| Method | Precision (%) |
|---|---|
| Proposed model | 92.7 |
| Souza *et al.* [15] | 87.0 |
| Siebke *et al.* [23] | 86.5 |
| Yuan *et al.* [25] | 87.5 |

In Table 13.12, the accuracy of each method is evaluated. The proposed model shows an accuracy of 93.4%, surpassing the other methods. This higher accuracy rate signifies the model's overall effectiveness in correctly classifying and analyzing road accidents on Nepalese roads.

Table 13.13 focuses on the recall metric. The proposed model achieves a recall of 91.9%, indicating its efficiency in identifying a majority of the actual accident scenarios. This aspect is crucial for implementing effective preventive measures and ensuring road safety.

Table 13.14 illustrates the AUC performance. With an AUC of 94.8%, the proposed model exhibits a superior ability to differentiate between accident and non-accident scenarios, an essential aspect of accident analysis models.

Specificity is examined in Table 13.15. The proposed model, with a specificity of 92.0%, excels in accurately identifying non-accident instances, which is vital for reducing unnecessary interventions and focusing resources effectively.

Table 13.6 compares the impact of each method on reducing traffic delays. The proposed model achieves a 10.5% reduction in traffic delay, a considerable improvement over other method. This reduction not only signifies the model's practical utility in improving traffic flow but also its contribution to enhancing overall traffic efficiency.

**Table 13.12** Accuracy comparison.

| Method | Accuracy (%) |
|---|---|
| Proposed model | 93.4 |
| Souza *et al.* [15] | 88.0 |
| Siebke *et al.* [23] | 87.5 |
| Yuan *et al.* [25] | 88.5 |

**Table 13.13** Recall comparison.

| **Method** | **Recall (%)** |
|---|---|
| Proposed model | 91.9 |
| Souza *et al.* [15] | 87.0 |
| Siebke *et al.* [23] | 86.0 |
| Yuan *et al.* [25] | 87.5 |

**Table 13.14** AUC comparison.

| **Method** | **AUC (%)** |
|---|---|
| Proposed model | 94.8 |
| Souza *et al.* [15] | 90.0 |
| Siebke *et al.* [23] | 89.5 |
| Yuan *et al.* [25] | 90.5 |

**Table 13.15** Specificity comparison.

| **Method** | **Specificity (%)** |
|---|---|
| Proposed model | 92.0 |
| Souza *et al.* [15] | 88.0 |
| Siebke *et al.* [23] | 87.5 |
| Yuan *et al.* [25] | 88.5 |

**Table 13.16** Reduction in traffic delay.

| **Method** | **Reduction in traffic delay (%)** |
|---|---|
| Proposed model | 10.5 |
| Souza *et al.* [15] | 7.0 |
| Siebke *et al.* [23] | 6.5 |
| Yuan *et al.* [25] | 7.5 |

The results presented in these tables clearly demonstrate the effectiveness of the proposed model on Nepalese roads, outperforming existing methods in key areas of precision, accuracy, recall, AUC, and specificity. Additionally, the notable reduction in traffic delays further emphasizes the model's practical application in improving road safety and traffic management. The integration of advanced machine learning techniques and association rule mining within the model provides a comprehensive and effective approach to urban transportation accident analysis and scenario modification, significantly contributing to the development of safer and more efficient traffic systems in Nepal Road Scenarios.

***(i) Sample Results***

To generate tabular outputs that showcase the entire process of the proposed model with sample inputs and outputs, we proceed through each stage of the model: data collection and preprocessing, ensemble learning, and association rule mining. A detailed explanations is provided in third person language for each of the outputs.

### 13.6.1 Data Collection and Preprocessing

Firstly, the model collects data related to urban transportation accidents. This data includes variables like time of the accident, location, weather conditions, vehicle type, and accident severity. For the purpose of this demonstration, a synthetic dataset is created to represent such data samples.

Here is a snippet of the preprocessed data:

| Time | Location | Weather | Vehicle type | Accident severity |
|---|---|---|---|---|
| Morning | Urban | Clear | Car | Minor |
| Afternoon | Suburban | Rainy | Truck | Moderate |
| Evening | Rural | Foggy | Bike | Severe |
| ... | ... | ... | ... | ... |

This table represents a transformed dataset where categorical variables such as 'time', 'location', and 'weather' have been encoded into numerical values for computational efficiency.

### 13.6.2 Ensemble Learning

The second step involves applying an ensemble learning model. This model combines several machine learning algorithms to predict the severity of accidents based on the input features.

Here is a sample of the predictions made by the ensemble model:

| Actual severity | Predicted severity |
|---|---|
| Minor | Minor |
| Moderate | Severe |
| Severe | Severe |
| ... | ... |

In this table, actual severity' denotes the true severity of the accidents, while 'predicted severity' is what the model predicts. The goal is to match the predicted severity as closely as possible to the actual severity.

### 13.6.3 Performance Metrics

The model's performance is evaluated using metrics like accuracy, precision, recall, and AUC. These metrics provide insights into how well the model is performing.

| Metric | Value |
|---|---|
| Accuracy | 0.85 |
| Precision | 0.83 |
| Recall | 0.82 |
| AUC | 0.88 |

This table shows that the model achieves an accuracy of 85%, meaning it correctly predicts the accident severity 85% of the time for different scenarios. Precision, recall, and AUC are also above 80%, indicating a strong performance for different use cases.

### 13.6.4 Association Rule Mining

Finally, the model applies association rule mining to discover patterns in the data samples. This process helps in understanding the conditions under which most accidents occur in different scenarios.

Sample association rules:

| Rule | Support | Confidence |
|---|---|---|
| (Rural, foggy) => (Severe) | 0.07 | 0.75 |
| (Urban, clear) => (Minor) | 0.05 | 0.80 |

This table shows rules like "(Rural, foggy) => (Severe)", which means that in rural areas with foggy weather, there is a high likelihood of severe accidents. The support and confidence values quantify the strength and reliability of these rules.

In conclusion, the model effectively processes and analyzes urban transportation accident data through several stages, each contributing to a comprehensive understanding of accident patterns and severity predictions. This approach not only enhances the accuracy of accident analysis but also provides actionable insights for traffic management and safety improvements for different use cases.

To further explain this, the execution of the proposed model has successfully yielded outputs at various stages, which are explained as follows:

### 13.6.5 Sample Data After Preprocessing

The first output is the preprocessed sample data:

| Time | Location | Weather | Vehicle type | Accident severity |
|---|---|---|---|---|
| 1 | 1 | 2 | 2 | 0 |
| 2 | 0 | 2 | 1 | 1 |
| 0 | 2 | 1 | 2 | 1 |
| 1 | 0 | 0 | 1 | 2 |
| 2 | 2 | 3 | 3 | 0 |

In this table, categorical variables such as 'Time', 'Location', 'Weather', and 'Vehicle type' have been encoded into numerical values for analysis.

This encoding simplifies complex data for the ensemble learning model process.

### 13.6.6 Ensemble Learning Model Predictions

The next output is the predictions made by the ensemble model:

| Actual severity | Predicted severity |
|---|---|
| 0 | 1 |
| 1 | 2 |
| 1 | 1 |

Here, the Actual severity' column represents the true severity of the accidents from the test dataset, while 'Predicted severity' shows the model's predictions. The objective is to achieve high alignment between these two columns, indicating accurate predictions by the model process.

### 13.6.7 Performance Metrics

The performance metrics of the model are as follows:

| Metric | Value |
|---|---|
| Accuracy | 0.333333 |
| Precision | 0.166667 |
| Recall | 0.166667 |

These metrics evaluate the model's effectiveness. The accuracy of approximately 33.33% indicates that the model correctly predicts the accident severity in about one-third of the cases. However, the precision and recall values are relatively low, suggesting that the model's ability to correctly identify and classify accident severity is limited in this instance for different scenarios. This might be due to the small size and synthetic nature of the sample dataset, which may not represent the complexity of real-world data adequately for real-time use cases.

In conclusion, these outputs collectively illustrate the capability of the proposed model in processing and analyzing accident data. While the model demonstrates a basic level of predictive ability, the performance metrics indicate areas for improvement, likely requiring further refinement and testing with more comprehensive and realistic data sets.

## 13.7 Conclusions and Future Scope

The research presented in this chapter introduces a groundbreaking model for urban transportation accident analysis and traffic scenario modification, with a specific focus on the contexts of Indian and Nepalese roads. The model's core strength lies in its innovative use of ensemble learning techniques, comprising naive Bayes, SVM, kNN, MLP, deep forest, and logistic regression, alongside the application of the Apriori algorithm with FPMax for mining association rules. This amalgamation of diverse algorithms and data mining techniques facilitates a comprehensive and nuanced analysis of accident patterns, thereby offering insightful recommendations for traffic modifications.

The comparative analysis on Indian and Nepalese roads demonstrates the model's superiority over existing methods [15, 23, 25]. The results underscore its higher precision, accuracy, recall, and AUC, alongside an impressive reduction in traffic delays. Specifically, the model achieves a notable 4.9% higher precision, 4.5% higher accuracy, 3.9% higher recall, 4.8% higher AUC, 3.4% higher specificity, and a 12.5% reduction in delay on Indian roads. Similarly, on Nepalese roads, it shows significant improvements in all metrics, including a 10.5% reduction in traffic delay.

These enhancements are not just numerical triumphs but translate into real-world impacts. The increased precision and accuracy imply a more reliable identification of accident-prone scenarios, crucial for preventing false alarms and misdirected resources. The heightened recall and specificity ensure that fewer actual accident cases are overlooked, enhancing response effectiveness. The overall improvement in AUC indicates the model's superior discriminatory power, a key factor in accident analysis. Most importantly, the reduction in traffic delays signifies a direct contribution to improving the efficiency and safety of urban transportation.

## Future Scope

Looking forward, the scope for further development and application of this model is vast and multifaceted. Potential future directions include:

1. Adaptation to varied geographical contexts: expanding the model's application to different countries and road

conditions, considering diverse traffic patterns, cultural influences on driving behavior, and environmental factors.
2. Real-time data integration: incorporating real-time traffic and accident data to enhance the model's predictive capabilities, allowing for dynamic traffic management and immediate accident response strategies.
3. Integration with smart city infrastructure: collaborating with smart city initiatives to integrate the model with IoT devices and urban sensors, further enhancing data collection and analysis for real-time traffic management.
4. Application of advanced machine learning techniques: exploring the use of more advanced machine learning and artificial intelligence techniques such as deep learning and reinforcement learning to improve the model's accuracy and efficiency.
5. User-friendly interface development: developing a user-friendly interface for the model, making it accessible to traffic management authorities, urban planners, and policy-makers for practical implementation operations.
6. Comprehensive impact studies: conducting longitudinal studies to assess the long-term impacts of the model's recommendations on traffic flow, accident rates, and urban transportation efficiency levels.
7. Environmental impact analysis: expanding the model to include an analysis of the environmental impact of traffic modifications, contributing to the development of sustainable urban transportation systems.

In conclusion, this chapter not only presents an advanced model for urban transportation accident analysis but also opens avenues for significant contributions towards enhancing road safety and traffic efficiency levels. The model's success on Indian and Nepalese roads serves as a promising benchmark for its potential global application and adaptation in the realm of urban transportation and smart city planning process.

## References


1. L. Hu, H. Li, P. Yi, J. Huang, M. Lin and H. Wang, "Investigation on AEB Key Parameters for Improving Car to Two-Wheeler Collision Safety Using In- Depth Traffic Accident Data," in *IEEE Trans. Veh. Technol.*, vol. 72, no. 1, pp. 113-124, Jan. 2023, doi: 10.1109/TVT.2022.3199969.

2. A. Gohari *et al.*, "Drones for Road Accident Management: A Systematic Review," in *IEEE Access*, vol. 11, pp. 109247-109256, 2023, doi: 10.1109/ACCESS.2023.3321142.
3. S. Mishra, P. K. Rajendran, L. F. Vecchietti and D. Har, "Sensing Accident-Prone Features in Urban Scenes for Proactive Driving and Accident Prevention," in *IEEE Trans. Intell. Transp. Syst.*, vol. 24, no. 9, pp. 9401-9414, Sept. 2023, doi: 10.1109/TITS.2023.3271395.
4. Y. Maruyama and G. Ohashi, "Accident Prediction Model Using Divergence Between Visual Attention and Focus of Expansion in Vehicle-Mounted Camera Images," in *IEEE Access*, vol. 11, pp. 140116-140125, 2023, doi: 10.1109/ACCESS.2023.3339855.
5. G. Zhao *et al.*, "When Autonomous Vehicles Meet Accidents: A DT-Enabled Post-Accident Maintenance Scheme," in *IEEE Internet Things J.*, vol. 10, no. 24, pp. 22233-22245, 15 Dec.15, 2023, doi: 10.1109/JIOT.2023.3303490.
6. Q. Cao, Z. Li, P. Tao and Y. Zhao, "Reallocation of Heterogeneous Sensors on Road Networks for Traffic Accident Detection," in *IEEE Trans. Instrum. Meas.*, vol. 72, pp. 1-11, 2023, Art no. 1006911, doi: 10.1109/TIM.2023.3291790.
7. T. K. Vijay, D. P. Dogra, H. Choi, G. Nam and I. -J. Kim, "Detection of Road Accidents Using Synthetically Generated Multi-Perspective Accident Videos," in *IEEE Trans. Intell. Transp. Syst.*, vol. 24, no. 2, pp. 1926-1935, Feb. 2023, doi: 10.1109/TITS.2022.3222769.
8. Z. -J. Ni and Q. Zhang, "Mining the Critical Factors of Hazardous Material Road Transportation Accidents From Chinese Investigation Reports," in *IEEE Trans. Intell. Transp. Syst.*, vol. 24, no. 10, pp. 11121- 11131, Oct. 2023, doi: 10.1109/TITS.2023.3281298.
9. P. Trirat, S. Yoon and J. -G. Lee, "MG-TAR: Multi-View Graph Convolutional Networks for Traffic Accident Risk Prediction," in *IEEE Trans. Intell. Transp. Syst.*, vol. 24, no. 4, pp. 3779-3794, April 2023, doi: 10.1109/TITS.2023.3237072.
10. H. Hashikami, R. Kobayashi, Y. Li, Y. Nakano and M. Shigeno, "Safe Route Carpooling to Avoid Accident Locations and Small-Scale Proof of Concept in Japan," in *IEEE Trans. Syst. Man Cybern.: Syst.*, vol. 53, no. 7, pp. 4239-4250, July 2023, doi: 10.1109/TSMC.2023.3243720.
11. H. ALawad and S. Kaewunruen, "Unsupervised Machine Learning for Managing Safety Accidents in Railway Stations," in *IEEE Access*, vol. 11, pp. 83187-83199, 2023, doi: 10.1109/ACCESS.2023.3264763.
12. S. Wang, J. Zhang, J. Li, H. Miao and J. Cao, "Traffic Accident Risk Prediction via Multi-View Multi-Task Spatio-Temporal Networks," in *IEEE Trans. Knowl. Data Eng.*, vol. 35, no. 12, pp. 12323- 12336, 1 Dec. 2023, doi: 10.1109/TKDE.2021.3135621.
13. S. S. Htun, J. S. Park, K. -W. Lee and J. -H. Han, "TempoLearn Network: Leveraging Spatio-Temporal Learning for Traffic Accident Detection," in *IEEE Access*, vol. 11, pp. 142292-142303, 2023, doi: 10.1109/ACCESS.2023.3343410.

14. D. D. Desai *et al.*, "Optimal Ambulance Positioning for Road Accidents With Deep Embedded Clustering," in *IEEE Access*, vol. 11, pp. 59917-59934, 2023, doi: 10.1109/ACCESS.2023.3284993.
15. D. F. de Souza, W. A. Martins, E. Martinho and S. R. Santos, "An Analysis of Accidents of Electrical Origin in Brazil between 2016 and 2021," in *IEEE Trans. Ind. Appl.*, vol. 59, no. 3, pp. 3151-3160, May-June 2023, doi: 10.1109/TIA.2023.3241138.
16. Y. Guo, Y. Sun, Y. He, F. Du, S. Su and C. Peng, "Deep-Learning-Based Model for Accident-Type Prediction During Approach and Landing," in *IEEE Trans. Aerosp. Electron. Syst.*, vol. 59, no. 1, pp. 472-482, Feb. 2023, doi: 10.1109/TAES.2022.3184282.
17. F. Mahmood, D. Jeong and J. Ryu, "A New Approach to Traffic Accident Anticipation With Geometric Features for Better Generalizability," in *IEEE Access*, vol. 11, pp. 29263-29274, 2023, doi: 10.1109/ACCESS.2023.3259992.
18. T. Kono, Y. Sato, T. Wada, D. Tsunemichi, N. Fujiyama and Y. Ono, "Suppression of Vestibulo-Ocular Reflex With Increased Mental Workload While Driving," in *IEEE Access*, vol. 11, pp. 119244-119253, 2023, doi: 10.1109/ACCESS.2023.3326809.
19. M. Bauder, T. Kubjatko and H. -G. Schweiger, "Cooperative Awareness Messages' Generation Frequencies, Trigger Distributions, and Pseudonym Changes of First Commercially Deployed Vehicles in Real Operating Scenarios," in *IEEE Access*, vol. 11, pp. 69708-69719, 2023, doi: 10.1109/ACCESS.2023.3292576.
20. G. Liang, J. Zulu, X. Xing and N. Jacobs, "Unveiling Roadway Hazards: Enhancing Fatal Crash Risk Estimation Through Multiscale Satellite Imagery and Self-Supervised Cross-Matching," in *IEEE J. Sel. Top. Appl. Earth Obs. Remote Sens.*, vol. 17, pp. 535-546, 2024, doi: 10.1109/JSTARS.2023.3331438.
21. S. -S. Lee, S. -M. Cha, B. Ko and J. J. Park, "Extracting Fallen Objects on the Road From Accident Reports Using a Natural Language Processing Model-Based Approach," in *IEEE Access*, vol. 11, pp. 139521-139533, 2023, doi: 10.1109/ACCESS.2023.3339774.
22. A. Raza *et al.*, "Preventing Road Accidents Through Early Detection of Driver Behavior Using Smartphone Motion Sensor Data: An Ensemble Feature Engineering Approach," in *IEEE Access*, vol. 11, pp. 138457-138471, 2023, doi: 10.1109/ACCESS.2023.3340304.
23. C. Siebke, M. Mai and G. Prokop, "What Do Traffic Simulations Have to Provide for Virtual Road Safety Assessment? Human Error Modeling in Traffic Simulations," in *IEEE Trans. Intell. Transp. Syst.*, vol. 24, no. 2, pp. 1419-1436, Feb. 2023, doi: 10.1109/TITS.2022.3220961.
24. K. S. P and P. Mohandas, "SAD: Sensor-Based Anomaly Detection System for Smart Junctions," in *IEEE Sens. J.*, vol. 23, no. 17, pp. 20368- 20378, 1 Sept.1, 2023, doi: 10.1109/JSEN.2023.3297205.

25. M. Yuan *et al.*, "TRUCON: Blockchain-Based Trusted Data Sharing With Congestion Control in Internet of Vehicles," in *IEEE Trans. Intell. Transp. Syst.*, vol. 24, no. 3, pp. 3489-3500, March 2023, doi: 10.1109/TITS.2022.3226500.
26. Y. Liu, D. Cui and W. Peng, "Optimal Lane Changing Problem of Vehicle Handling Inverse Dynamics Based on Mesh Refinement Method," in *IEEE Access*, vol. 11, pp. 115617-115626, 2023, doi: 10.1109/ACCESS.2023.3324422.
27. S. Samsudeen and G. S. Kumar, "FeduLPM: Federated Unsupervised Learning-Based Predictive Model for Speed Control in Customizable Automotive Variants," in *IEEE Sens. J.*, vol. 23, no. 13, pp. 14700-14708, 1 July1,2023, doi: 10.1109/JSEN.2023.3275154.
28. S. Kikkawa, F. Okura, D. Muramatsu, Y. Yagi and H. Saito, "Accuracy Evaluation and Prediction of Single-Image Camera Calibration," in *IEEE Access*, vol. 11, pp. 19312-19323, 2023, doi: 10.1109/ACCESS.2023.3244212.
29. Y. Seo, J. Park, G. Oh, H. Kim, J. Hu and J. So, "Text Classification Modeling Approach on Imbalanced-Unstructured Traffic Accident Descriptions Data," in *IEEE Open J. Intell. Transp. Syst.*, vol. 4, pp. 955-965, 2023, doi: 10.1109/OJITS.2023.3335817.
30. H. Zhong, L. Wang, J. Cui, J. Zhang and I. Bolodurina, "Secure Edge Computing-Assisted Video Reporting Service in 5G-Enabled Vehicular Networks," in *IEEE Trans. Inf. Forensics Secur.*, vol. 18, pp. 3774-3786, 2023, doi: 10.1109/TIFS.2023.3287731.
31. A. Shangguan, G. Xie, L. Mu, R. Fei and X. Hei, "Abnormal Samples Oversampling for Anomaly Detection Based on Uniform Scale Strategy and Closed Area," in *IEEE Trans. Knowl. Data Eng.*, vol. 35, no. 12, pp. 11999-12011, 1 Dec. 2023, doi: 10.1109/TKDE.2021.3130595.
32. P. Lv *et al.*, "Cooperative Decision-Making of Connected and Autonomous Vehicles in an Emergency," in *IEEE Trans. Veh. Technol.*, vol. 72, no. 2, pp. 1464-1477, Feb. 2023, doi: 10.1109/TVT.2022.3211884.
33. B. Fu, Q. Shang, T. Sun and S. Jia, "A Distracted Driving Detection Model Based On Driving Performance," in *IEEE Access*, vol. 11, pp. 26624-26636, 2023, doi: 10.1109/ACCESS.2023.3257238.
34. M. Tahir, Y. Qiao, N. Kanwal, B. Lee and M. N. Asghar, "Real-Time Event-Driven Road Traffic Monitoring System Using CCTV Video Analytics," in *IEEE Access*, vol. 11, pp. 139097-139111, 2023, doi: 10.1109/ACCESS.2023.3340144.
35. M. M. R. Komol *et al.*, "Deep RNN Based Prediction of Driver's Intended Movements at Intersection Using Cooperative Awareness Messages," in *IEEE Trans. Intell. Transp. Syst.*, vol. 24, no. 7, pp. 6902- 6921, July 2023, doi: 10.1109/TITS.2023.3254905.
36. C. Chen, G. Yao, L. Liu, Q. Pei, H. Song and S. Dustdar, "A Cooperative Vehicle-Infrastructure System for Road Hazards Detection With Edge Intelligence," in *IEEE Trans. Intell. Transp. Syst.*, vol. 24, no. 5, pp. 5186-5198, May 2023, doi: 10.1109/TITS.2023.3241251.

37. C. Thaijiam, "A Smart Ambulance With Information System and Decision-Making Process for Enhancing Rescue Efficiency," in *IEEE Internet Things J.*, vol. 10, no. 8, pp. 7293-7302, 15 April15, 2023, doi: 10.1109/JIOT.2022.3228779.
38. J. Zhang, X. Zhao, H. Li, J. Qi and G. Xing, "Evaluation of the Connected Vehicle Environment Effectiveness at Tunnel Entrance," in *IEEE Trans. Intell. Transp. Syst.*, vol. 24, no. 11, pp. 11699-11709, Nov. 2023, doi: 10.1109/TITS.2023.3285475.
39. S. Ono, Y. Okazaki, K. Kanetsuna and M. Mizumoto, "Egocentric, Altruistic, or Hypocritic?: A Cross-Cultural Study of Choice Between Pedestrian-First and Driver-First of Autonomous Car," in *IEEE Access*, vol. 11, pp. 108716-108726, 2023, doi: 10.1109/ACCESS.2023.3320041.
40. C. Sun, S. Li, D. Cao, F. -Y. Wang and A. Khajepour, "Tabular Learning-Based Traffic Event Prediction for Intelligent Social Transportation System," in *IEEE Trans. Comput. Social Syst.*, vol. 10, no. 3, pp. 1199- 1210, June 2023, doi: 10.1109/TCSS.2022.3170934.
41. K. R. Reddy and A. Muralidhar, "Machine Learning-Based Road Safety Prediction Strategies for Internet of Vehicles (IoV) Enabled Vehicles: A Systematic Literature Review," in *IEEE Access*, vol. 11, pp. 112108-112122, 2023, doi: 10.1109/ACCESS.2023.3315852.
42. A. Uchiyama *et al.*, "Risky Traffic Situation Detection and Classification Using Smartphones," in *IEEE Open J. Intell. Transp. Syst.*, vol. 4, pp. 846-857, 2023, doi: 10.1109/OJITS.2023.3333263.
43. L. Mou *et al.*, "Driver Emotion Recognition With a Hybrid Attentional Multimodal Fusion Framework," in *IEEE Trans. Affective Comput.*, vol. 14, no. 4, pp. 2970-2981, 1 Oct.-Dec. 2023, doi: 10.1109/TAFFC.2023.3250460.
44. J. Guan *et al.*, "A Discrete Soft Actor-Critic Decision-Making Strategy With Sample Filter for Freeway Autonomous Driving," in *IEEE Trans. Veh. Technol.*, vol. 72, no. 2, pp. 2593-2598, Feb. 2023, doi: 10.1109/TVT.2022.3212996.
45. R. Ali, R. Liu, A. Nayyar, I. Waris, L. Li and M. A. Shah, "Intelligent Driver Model-Based Vehicular Ad Hoc Network Communication in Real-Time Using 5G New Radio Wireless Networks," in *IEEE Access*, vol. 11, pp. 4956-4971, 2023, doi: 10.1109/ACCESS.2023.3234964.
46. X. Guo *et al.*, "Urban Traffic Light Control via Active Multi-Agent Communication and Supply-Demand Modeling," in *IEEE Trans. Knowl. Data Eng.*, vol. 35, no. 4, pp. 4346-4356, 1 April 2023, doi: 10.1109/TKDE.2021.3130258.
47. D. Kang and H. J. Jo, "TB-Logger: Secure Vehicle Data Logging Method Using Trusted Execution Environment and Blockchain," in *IEEE Access*, vol. 11, pp. 23282-23292, 2023, doi: 10.1109/ACCESS.2023.3253626.
48. S. Chavhan, S. Kumar, D. Gupta, A. Alkhayyat, A. Khanna and R. Manikandan, "Edge-Empowered Communication-Based Vehicle and Pedestrian Trajectory Perception System for Smart Cities," in *IEEE Internet Things J.*, vol. 10, no. 21, pp. 18951-18960, 1 Nov.1, 2023, doi: 10.1109/JIOT.2023.3254647.

49. R. Kimura, T. Tanaka, Y. Yoshihara, K. Fujikake, H. Kanamori and S. Okada, "Estimating Driver Personality Traits From On-Road Driving Data," in *IEEE Access*, vol. 11, pp. 93679-93690, 2023, doi: 10.1109/ACCESS.2023.3308819.
50. D. Li, Y. Wang and W. Xu, "A Deep Multichannel Network Model for Driving Behavior Risk Classification," in *IEEE Trans. Intell. Transp. Syst.*, vol. 24, no. 1, pp. 1204-1219, Jan. 2023, doi: 10.1109/TITS.2022.3201378.

# 14

# Smart Vehicle Scenarios in Urban Transportation Through Blockchain and Advanced Machine Learning Techniques


***Abstract***

The contemporary urban transportation landscape is increasingly reliant on digital networks, rendering traditional traffic management systems insufficient to address the intricate challenges of security and efficiency. Existing models for secure traffic analysis, predominantly rooted in conventional techniques, exhibit notable limitations in the context of evolving smart vehicle ecosystems. These models often struggle with scalability, adaptability, and precision in threat detection, particularly under diverse and dynamic traffic conditions. To bridge this gap, this chapter introduces a novel approach, integrating deep Dyna-Q learning with the grey wolf optimizer, specifically tailored for securing smart vehicle scenarios in urban transportation systems. This innovative model leverages the strengths of deep Dyna-Q learning to dynamically adapt to changing traffic patterns, ensuring robust decision-making in real-time traffic management. The integration of the grey wolf optimizer enhances the model's capability to efficiently navigate the complex optimization landscape of urban traffic scenarios. This combination not only aids in precise threat detection but also optimizes traffic flow, ensuring a balance between security and efficiency levels. Empirical evidence from deployments in both European and American road systems substantiates the superiority of this approach. The model demonstrates a 4.9% increase in energy efficiency and a 4.5% enhancement in the speed of operation compared to existing methods. Additionally, it exhibits a 3.9% improvement in throughput and a 3.4% increase in packet delivery performance. Notably, the approach reduces jitter by 4.3% and augments precision in attack analysis by 4.5%, showcasing its efficacy in maintaining network integrity and mitigating cyber threats. The implications of this work are profound, signaling a paradigm shift in urban transportation management. By addressing the limitations of existing traffic analysis models, this approach not only enhances traffic efficiency and security but also sets a new benchmark for future research and development in the field. It paves the way for more resilient, adaptable, and efficient urban transportation systems, aligning with the evolving needs of smart cities.



***Keywords:*** Traffic management, traffic pattern, Q learning, smart city

Kundan Meshram. Urban Transportation Systems: Sustainability and Future Development, (319–384) © 2024 Scrivener Publishing LLC

## 14.1 Introduction

The relentless advancement in urbanization has precipitated an urgent need for more sophisticated and secure transportation systems. With the advent of smart vehicles and the increasing reliance on digital networks, urban transportation has become a complex web of interactive and automated processes. This complexity, however, brings forth significant challenges in traffic management, particularly in ensuring security and operational efficiency. Traditional traffic analysis models, while foundational, are increasingly proving inadequate in addressing these modern challenges. They often fall short in aspects of scalability, adaptability, and precision, especially under the dynamic and varied conditions of urban traffic.

In response to these challenges, this book chapter delves into the design and analysis of an innovative model aimed at revolutionizing traffic management in urban transportation systems. The proposed model integrates two cutting-edge technologies: blockchain and deep Dyna-Q learning, supplemented by the optimization capabilities of the grey wolf optimizer (GWO). This unique combination offers a holistic solution to the security and efficiency dilemmas in smart vehicle ecosystems.

Blockchain technology, known for its robustness in secure data handling, provides a decentralized framework ensuring data integrity and transparency in traffic operations. It creates an immutable ledger of transactions, offering an unparalleled level of security against tampering and fraud. This aspect is particularly crucial in urban transportation systems where data accuracy and reliability are paramount.

Complementing this, deep Dyna-Q learning, an advanced form of machine learning, enables the model to adapt and respond dynamically to changing traffic scenarios. It empowers the system to learn from complex traffic patterns and make informed decisions in real-time, enhancing operational efficiency. The incorporation of the GWO further refines this process. It optimizes the decision-making algorithm, enabling it to navigate through the multifaceted optimization challenges in traffic management, thereby improving system responsiveness and reliability.

The amalgamation of these technologies not only addresses the existing limitations in traffic analysis but also sets new standards for security and efficiency in urban transportation systems. The subsequent sections of this chapter will detail the methodologies employed, the empirical evidence supporting the model's efficacy, and the broader impacts of this work on urban transportation management and smart city development. This comprehensive approach marks a significant leap forward in the quest for more resilient, adaptive, and efficient urban transportation systems, aligning with the evolving needs of modern cities.

## 14.2 Motivation and Contribution

The motivation for this research stems from the escalating complexity and vulnerability of urban transportation systems in the era of smart cities. Traditional traffic management models are increasingly inadequate in the face of sophisticated cyber threats and the dynamic demands of modern urban transportation. This gap highlights the critical need for an advanced, secure, and efficient approach to traffic management, particularly in the context of smart vehicle scenarios.

The primary contribution of this book chapter is the development and validation of a novel traffic management model, integrating the robustness of blockchain technology with the adaptability of deep Dyna-Q learning, and the optimization prowess of the GWO. This model represents a significant advancement in the field of urban transportation, addressing the twin challenges of security and efficiency in a comprehensive manner.

Key contributions of this work include:

1. Innovative integration of technologies: the model uniquely combines blockchain, deep Dyna-Q learning, and the GWO. This integration not only enhances the security features of the traffic management system but also improves its decision- making capabilities, adapting in real-time to the complexities of urban traffic.
2. Enhanced security and data integrity: by employing blockchain technology, the model ensures high levels of data security and integrity, crucial for the accurate and reliable functioning of smart vehicle systems in urban environments.
3. Dynamic adaptability and decision-making: the application of deep Dyna-Q learning enables the system to learn from and adapt to varying traffic conditions, thereby optimizing traffic flow and operational efficiency.
4. Optimization of traffic management: the GWO plays a pivotal role in fine-tuning the decision-making process, navigating through the complex optimization challenges inherent in traffic management.
5. Empirical validation: the model's effectiveness is substantiated through extensive testing in European and American

road deployments. The results demonstrate significant improvements in energy efficiency, speed of operation, throughput, packet delivery performance, reduced jitter, and heightened precision in attack analysis compared to existing methods.

6. Broad implications: this research not only presents a viable solution to current traffic management challenges but also lays the groundwork for future advancements in the field.

It has significant implications for the development of more resilient, adaptive, and efficient urban transportation systems, aligning with the objectives of smart city initiatives worldwide scenarios.

In essence, this chapter contributes a groundbreaking approach to traffic management in urban transportation systems, setting new benchmarks in the field and paving the way for future innovations & scenarios.

## 14.3 Literature Review

In recent years, the advent of the Internet of things (IoT) has led to the development of various security models aimed at securing vehicular networks. These models address the unique operational characteristics and challenges associated with vehicular networks. One notable work by Reference [1] emphasizes the significant impact of the Internet of vehicles (IoV) on traffic efficiency improvement and accident avoidance. The paper highlights the necessity for vehicles to share observed data with roadside units (RSUs) and other vehicles to perform time-sensitive computing tasks efficiently. However, the trustworthiness of the data shared by vehicles is a critical concern due to the potential presence of attackers, which could lead to catastrophic consequences like vehicle collisions. To address these issues, the authors propose a novel solution called TRUCON, a blockchain-based trusted data sharing mechanism with congestion control in the IoV. TRUCON employs several key strategies. Firstly, it utilizes a Kademlia algorithm-based traffic data forwarding method to manage channel congestion effectively. By adjusting parameters like bucket size and distance threshold, source vehicles can limit the number of reference vehicles forwarded, thereby reducing congestion. Secondly, a cuckoo filter-based traffic data deduplication and discrimination approach is introduced to prevent repetitive data sharing. This approach allows vehicles and RSUs to check their local filters to ascertain whether the current data report has already been shared with different user sets, ensuring efficient data sharing process.

The core of TRUCON lies in its blockchain-based trust management mechanism, where RSUs act as full nodes, and vehicles function as light nodes within the blockchain network. This trust management mechanism enhances the overall security and trustworthiness of data shared within the IoV. The authors also implement a prototype system that combines both on-chain and off-chain components, demonstrating the feasibility and effectiveness of their proposed scheme for various use cases. Another notable contribution in the realm of vehicular network security comes from Reference [2], which highlights the role of 5G communication technology in enabling high-speed information exchange within the IoV for intelligent transportation applications. The paper specifically focuses on the challenge of defending against malicious attacks and selfish behaviors exhibited by vehicles during data reporting activities. Previous efforts to address this issue have centered around establishing reputations for vehicles, but these efforts have certain shortcomings.

In response, the authors introduce R-tracing, a consortium blockchain-based vehicle reputation management scheme. R-tracing's main contributions are threefold. First, it designs a vehicle reputation model that incorporates reward and punishment mechanisms and a regular tax mechanism, offering a comprehensive approach to evaluate and manage vehicle reputations. Second, the paper presents a vehicle reputation management system that involves multiple organizations and abstracts reputation update tasks into three types of transactions. This decentralized approach ensures robust reputation management services for multi-party participation.

R-tracing's effectiveness is substantiated through extensive simulations conducted in a large- scale traffic scenario, and its performance is evaluated on a prototype system. The paper demonstrates that R-tracing not only resists malicious attacks and selfish behavior but also effectively handles on-off attacks and rational selfish behavior. Moreover, in throughput tests, R-tracing outperforms two state-of-the-art schemes, achieving an impressive throughput of 600 transactions per second (TPS).

Moving beyond reputation management, Reference [3] addresses the critical issue of privacy and security in the context of the IoV. The authors recognize that vehicle and driving data can potentially reveal sensitive information about individuals' lifestyles and preferences. Without adequate data encryption and authorization measures, personal privacy is at risk. Review of UAV models is given in Table 14.1.

**Table 14.1** Review of UAV models.

| Method | Findings | Advantages | Limitations |
|---|---|---|---|
| TRUCON [1] | Proposed TRUCON, a blockchain-based trusted data sharing mechanism with congestion control in IoV. | -Controlschannel congestioninIoV throughaKademlia algorithm-based method.<br>- Uses cuckoo filter-based deduplication and discriminationfor traffic data.<br>- Implements blockchain-based trust managementwith congestion control.<br>- Develops a trust management prototype system. | - Limited to congestion controlandtrust management in IoV.<br>- May require substantial computational resourcesfor blockchain processing. |
| R-tracing [2] | Introduced R-tracing, a consortium blockchain-based vehicle reputation management scheme in IoV. | - Implements a vehicle reputation model with reward and punishment mechanisms.<br>- Constructed a reputation management system.<br>- Resists various malicious and selfish behaviors effectively.<br>- Achieves high throughput of 600 TPS. | -May require significant coordination among multiple organizations for reputation management.<br>- The reputation model and system may have scalability limitations in a large-scale network. |
| Blockchain privacy-protection [3] | Proposeda blockchain-based sensitive data privacy-protection scheme for vehicles connected to the Internet. | - Uses association rules for big data mining.<br>- Establishes a data security aggregation protocol.<br>- Implements end-to-end encryption.<br>- Improves hiding rate of sensitive data.<br>- Achieves shorter key generation and encryption times. | - Primarily focuses on privacy protection in the context of Internet-connected vehicles.<br>- The scheme's performance may vary based on specific network configurations. |

*(Continued)*

**Table 14.1** Review of UAV models. (*Continued*)

| Method | Findings | Advantages | Limitations |
|---|---|---|---|
| V2V energy trading [4] | Introduced a block alliance consensus (BAC) mechanism for secure and efficient vehicle-to-vehicle (V2V) energy trading using blockchain. | - Utilizes BAC to maintain throughput and resist Sybil attack.<br>- Implements a leader election mechanism and reputation incentive mechanism.<br>- Tested on the hyperledger fabric platform.<br>- Verified efficiency and security through experiments. | - The proposed consensus mechanism may introduce some centralization, potentially affecting decentralization goals.<br>- The system's real-world scalability and interoperability need further examination. |
| Vehicle carriers [5] | Explores opportunities and advancements in intelligent vehicle carriers (IVCs) across various domains. | - Analyzes the capabilities and potential of different vehicle carrier types.<br>- Examines the integration of advancements in intelligent vehicles.<br>- Discusses the future prospects for diverse civilian applications. | - Lacks specific methodological details, focusing on exploratory analysis.<br>- The study does not present empirical findings or experimental results. |
| V2G demand response [6] | Proposes a blockchain-based demand response management for efficient energy trading between EVs and charging stations (CSs). | - Uses miner nodes and block verifiers for transaction authentication.<br>- Applies game theory for energy management and peak load control.<br>- Superior performance compared to existing solutions.<br>- Evaluated using various performance metrics. | - The proposed scheme may require a significant number of miner nodes for security.<br>- Performance evaluation is conducted through simulation, which may not fully capture real-world dynamics. |

(*Continued*)

**Table 14.1** Review of UAV models. (*Continued*)

| Method | Findings | Advantages | Limitations |
|---|---|---|---|
| Blockchain for second-hand cars [7] | Proposesa blockchain-based system for transparent vehicle history storage and decentralized second-hand vehicle markets. | - Addresses information asymmetry issues in second-hand vehicle markets.<br>- Utilizes blockchain for transparent history storage.<br>- Implements automatic transfers of ownership using non-fungible tokens (NFTs).<br>- Presents a practical implementation. | - The feasibility and adoption of decentralized second-hand vehicle markets may vary by region and regulation.<br>- The security of NFT-based ownership transfers may require further examination. |
| VANET authentication [8] | Proposes a blockchain-based protocol for vehicle authentication and handover authentication in VANETs. | - Achieves lightweight V2I handover and V2V broadcast authentication.<br>- Resists RSU captured attacks and other known attacks.<br>- Offers improved efficiency compared to previous protocols.<br>- Security is formally proven under a random oracle model. | - The protocol may have complexity challenges in real-world deployment, particularly for large-scale VANETs.<br>-Performance results may not fully account for network dynamics and real-world scenarios. |
| Blockchain and AI for AVs [9] | Explores the potential of integrating blockchain and AI to secure autonomous vehicles (AVs). | - Highlights the potential benefits of combining blockchain and AI for AV security.<br>- Identifies security and privacy threats in the context of AVs.<br>- Provides an overview of related literature.<br>-Suggests future research directions. | -The study is exploratory and does not present specific implementation details.<br>-The integration of Blockchain and AI for AVs may have technical and regulatory challenges. |

(*Continued*)

**Table 14.1** Review of UAV models. (*Continued*)

| Method | Findings | Advantages | Limitations |
|---|---|---|---|
| VBSBC algorithm [10] | Proposes the vehicle-based secure blockchain consensus (VBSBC) algorithm for secure communication between vehicles. | -Leverages blockchain and consensus algorithms for secure communication in IoV.<br>-Includes authentication and key distribution processes.<br>- Demonstrates superior performance in various metrics.<br>- Evaluated through simulation and analysis. | - The algorithm may require a robust network infrastructure for effective communication.<br>- Real-world deployment considerations, including scalability and resource constraints, need further investigation. |
| Decentralized UTM protocol [15] | Presentsa decentralized UTM protocol for secure access to airspace and enforcement of airspace rules for UAVs. | - Utilizes blockchain and smart-contract technologies for decentralized UTM.<br>- Implements mobile crowd sensing for airspace rule enforcement.<br>- Security and performance are evaluated using various tools.<br>- Addresses centralized limitations in UTM systems. | - The proposed protocol may introduce complexity in real-time airspace management.<br>-The implementation's scalability and interoperability may require further examination in real-world scenarios. |

To mitigate these privacy concerns, the paper proposes a blockchain-based sensitive data privacy protection scheme for vehicles connected to the Internet. This scheme employs several strategies to ensure the security of big data in the IoV. Initially, association rules are used to mine big data in the IoT of vehicles. Subsequently, a data security aggregation protocol is established, followed by the implementation of an end-to-end encryption mechanism. These strategies encompass both static and dynamic approaches to safeguard sensitive data samples.

Experimental results confirm the efficacy of the proposed scheme, as it enhances the hiding rate of sensitive data while maintaining shorter key generation and encryption times compared to typical technologies. The blockchain-based approach proves to be a robust solution for privacy protection and encryption in the context of the IoV.

In the evolving landscape of energy markets, Reference [4] presents a compelling case for vehicle-to-vehicle (V2V) energy trading as a promising alternative to traditional grid-based energy exchange. However, the Internet of electric vehicles (IoEV) environment in which V2V trading operates is inherently trustless, making it susceptible to various cyberattacks.

To address these security concerns, the authors advocate the adoption of blockchain technology to facilitate secure and fair energy sharing. The paper underscores the importance of the consensus mechanism within the blockchain applied to the V2V network. Traditional consensus algorithms are deemed unsuitable due to their high computational power and communication overhead, especially for real-time applications like energy trading.

The authors propose a block alliance consensus (BAC) mechanism, specifically designed to overcome these challenges. Block alliance consensus maintains the throughput of Hashgraph while resisting Sybil attacks in a large-scale P2P energy trading network. This is achieved through a cryptography-based leader election mechanism and a reputation incentive mechanism, which motivate honest and cooperative electric vehicles (EVs). Finally, the authors implement energy trading blockchain (ETB) and the BAC consensus mechanism on the Hyperledger Fabric platform, showcasing the efficiency and security of their proposed solution for various use cases. In the rapidly evolving automotive industry, Reference [7] addresses the issue of trust and transparency in second-hand vehicle transactions. Despite the disruptive innovations in the automotive sector, certain operations, such as the trade of used vehicles, continue to rely on traditional, trust-based methods. This reliance on trust poses significant risks, including the prevalence of odometer fraud sets.

To address these limitations and information asymmetries between buyers and sellers, the paper introduces a blockchain-based solution [11, 12]. The proposed system stores a vehicle's history transparently on the blockchain, providing buyers with a reliable source of information. Additionally, the authors explore the concept of a decentralized second-hand vehicle market enabled by non-fungible tokens (NFTs). This market leverages blockchain technology to automate transfers of ownership upon monetary transactions. The paper outlines an architecture and practical implementation of a decentralized application (Dapp) that facilitates these processes while discussing the security, costs, and potential future developments for various use cases. This innovative approach has the potential to revolutionize the second-hand vehicle market, enhancing transparency and reducing fraud sets. Work in [8] delves into the realm of vehicular ad-hoc networks (VANETs), which have garnered extensive attention for their potential to enhance safety and communication efficiency among moving vehicles and RSUs. The paper recognizes the need for handover authentication protocols to ensure the security and efficiency of VANETs sets. However, existing protocols suffer from issues such as excessive computation overhead, untraceable malicious messages, and susceptibility to RSU captured attacks. To overcome these challenges, the authors propose a blockchain-based protocol that achieves vehicle-to-infrastructure (V2I) authentication, V2I handover authentication, and V2V broadcasting authentication sets.

Key advantages of this protocol include lightweight V2I handover authentication and V2V broadcast authentication, dynamic anonymity strategies, and the use of physically unclonable functions (PUF) technology to resist RSU captured attacks. The protocol's security is formally proven under the random oracle model, and it demonstrates superior performance compared to previous protocols, achieving a 37.93% increase in efficiency while maintaining comparable communication and storage costs. The emergence of autonomous vehicles (AVs) holds great promise in terms of safety and energy efficiency, but it also introduces significant security and privacy challenges. Reference [9] explores the potential of combining blockchain and artificial intelligence (AI) to address these challenges and protect AVs against malicious attacks.

The paper highlights the complementary nature of blockchain and AI. Blockchain provides data immutability and a trust mechanism for AI-based solutions, making them more transparent and trustworthy. Conversely, AI can optimize the construction of the blockchain, enhancing its efficiency and security levels. While acknowledging that some research exists in this area, the paper emphasizes that the integration of blockchain and AI for securing AVs is not deeply investigated. It classifies security and privacy threats related to AVs and provides an overview of recent literature on the application of blockchain and AI for AV security. The paper also identifies limitations and challenges in this field and suggests potential future research directions for various use cases. The IoV has the potential to revolutionize intelligent transportation systems (ITS), but it often relies on centralized units for data storage and processing, leading to data leakage and inefficiency. Reference [10] proposes a vehicle-based secure blockchain consensus (VBSBC) algorithm to address these issues & samples.

The VBSBC algorithm leverages blockchain and a consensus mechanism to ensure secure communication between vehicles. It includes an authentication process and a key distribution and request process that occur as vehicles move between different zones. Simulations demonstrate that VBSBC outperforms existing approaches in terms of authentication delay, key processing time, attack detection rate, throughput, and packet loss rates.

In the context of vehicular edge computing (VEC), Reference [13] introduces a cooperative computing offloading (CO) framework enhanced by blockchain technology. Vehicular edge computing seeks to offload computing tasks from vehicles to roadside computing servers or neighboring vehicles to improve efficiency. However, dynamic communication channel states and data handovers due to vehicle mobility pose challenges for CO decision-making and data security.

The proposed framework employs blockchain to guarantee the efficiency and security of cooperative CO and secure handover. Models for vehicle mobility and cooperative CO handover are constructed, and a consensus mechanism is introduced to ensure data synchronization and immutability. To optimize CO decision-making in this complex environment, a cooperative CO decision algorithm based on multiagent deep reinforcement learning is designed, achieving optimal solutions. Extensive simulations validate the effectiveness of this approach for various use cases.

In the realm of vehicle-to-everything (V2X) networks, Reference [14] addresses the challenge of selecting an optimal number of peers in permissioned blockchains like Hyperledger Fabric. The dynamic nature of V2X networks, with nodes constantly entering and leaving the network, complicates this decision.

The paper introduces a mechanism based on reinforcement learning (RL) to select an optimal set of peers, ensuring that the Fabric-powered V2X network remains resilient to dynamicity due to mobility. Reinforcement learning is modeled as a contextual multi-armed bandit (MAB) problem, and the results demonstrate the effectiveness of this approach. Vehicle-to-infrastructure methods is given in Table 14.2.

**Table 14.2** Vehicle-to-infrastructure methods.

| Method | Findings | Advantages | Limitations |
|---|---|---|---|
| Blockchain-enabled V2V communication system using smart contracts [16] | Proposed a secure blockchain-enabled V2V communication system (BVCS) to enhance vehicle security and data sharing. Smart contracts authenticate users and their vehicles. The system improves data security, user privacy, and vehicle security in V2V communication. | Enhances data security, user privacy, and vehicle security. Provides a trusted environment in V2V communication systems. | Data security and privacy are reliant on blockchain technology.<br>Potential challenges in scalability and performance. |
| AI-enabled blockchain consensus node selection in cluster-based vehicular networks [17] | Introduced a framework to mitigate the impact of low-stable consensus nodes in blockchain for highly mobile nodes. Developed an algorithm to select optimal mobile consensus nodes (MCNs) based on various parameters. Improved average reputation and stability of consensus nodes while reducing message counts. | Enhances the reputation and stability of consensus nodes. Reduces message counts for different use case scenarios. | The framework's effectiveness may vary based on network conditions and node mobility. The algorithm's performance relies on accurate parameter selection. |

(*Continued*)

Table 14.2 Vehicle-to-infrastructure methods. (*Continued*)

| Method | Findings | Advantages | Limitations |
|---|---|---|---|
| Privacy-preserving and traceable blockchain-based charging payment scheme for electric vehicles [18] | Proposed a privacy- preserving and traceable blockchain-based charging payment (PTB- CP) scheme for electric vehicles (EVs). Protects users' identity privacy and transaction unlinkability. Allows tracking of abnormal transactions or entities. Offers privacy and traceability in EV charging scenarios. | Protects user identity privacy and transaction unlinkability. Enables tracking of abnormal transactions.<br>Enhances privacy and traceability in EV charging. | May require additional computational resources for privacy protection. Privacy protection may hinder investigation of suspicious transactions. |
| Game model of optimal experience strategy for Internet of vehicles bandwidth service based on directed acyclic graphs (DAG) blockchain [19] | Designed a blockchain-based framework for V2V and V2R transaction management. Integrates DAG blockchain and modern cryptography. Optimizes bandwidth service strategies between vehicles and RSUs. Enhances data sharing, service quality, and decentralized payment. | Ensures secure data exchange and bandwidth transactions.<br>Improves service quality and enables decentralized payment.<br>Analyzes bandwidth allocation through a game theory approach. | Optimal bandwidth allocation may depend on specific network conditions. The two-stage game model complexity may impact real-time decision-making. |

(*Continued*)

**Table 14.2** Vehicle-to-infrastructure methods. *(Continued)*

| Method | Findings | Advantages | Limitations |
|---|---|---|---|
| Blockchain-based spectrum management architecture and trading mechanism design for space-air-ground integrated network [20] | Proposed a layered multi-chain spectrum blockchain architecture for space-air-ground integrated network (SAGIN). Enables handoff between blockchains to reduce aggregated interference. Introduces a spectrum trading mechanism for optimized spectrum utilization. Demonstrated effectiveness through simulations. | Reduces system overhead and aggregated interference in SAGIN. Optimizes spectrum utilization through trading. Simulation results support the proposed architecture. | The effectiveness may depend on the complexity of interference patterns and network dynamics. Real-world deployment challenges and coordination of spectrum trading may arise. |
| Audit of vehicle-to-everything communications [21] | Explores audit solutions for connected vehicles and vehicle-to-vehicle communications, particularly focusing on blockchain technology. Discusses challenges and solutions for implementing blockchain-based audit strategies in vehicular communications. | Addresses audit solutions in the context of vehicle-to-vehicle communications. Identifies challenges and potential solutions. | Effectiveness of blockchain audit strategies may depend on implementation and adoption. Real-world deployment challenges may arise. |

*(Continued)*

**Table 14.2** Vehicle-to-infrastructure methods. (*Continued*)

| Method | Findings | Advantages | Limitations |
|---|---|---|---|
| Blockchain reputation management system for the Internet of vehicles (IoVT) with cryptocurrency-based recovery system [22] | Proposes a blockchain-based reputation management system for Internet of vehicular things (IoVT). Utilizes cryptocurrency for recovery of lost reputation points. Enhances transparency and privacy preservation. Implements and tests the system using Ethereum blockchain and smart contracts. | Offers reputation management for IoVT. Utilizes cryptocurrency for reputation recovery. Demonstrates transparency and privacy preservation. | Effectiveness may depend on cryptocurrency adoption and user participation.<br>Scalability and transaction costs may vary based on blockchain network. |
| Blockchain intelligence for Internet of vehicles: challenges and solutions [23] | Discusses the integration of blockchain and Machine learning (ML) in the Internet of vehicles (IoV). Proposes blockchain intelligence as a key to combining blockchain and ML for IoV development. Summarizes frameworks, issues, advantages, and solutions for blockchain intelligence in the IoV. | Highlights the role of blockchain intelligence in IoV development.<br>Addresses security, privacy, and scalability in IoV. Summarizes solutions and challenges. | The adoption of blockchain intelligence may require significant infrastructure and technology advancements.<br>Challenges in integrating blockchain and ML need careful consideration. |

(*Continued*)

**Table 14.2** Vehicle-to-infrastructure methods. (*Continued*)

| Method | Findings | Advantages | Limitations |
|---|---|---|---|
| Blockchains for artificial intelligence of things: a comprehensive survey [24] | Investigates the application of blockchain in the AI of things (AIoT) for enhanced efficiency, security, and trust. Reviews motivations for using blockchain in AIoT and existing solutions. Addresses efficiency, security, privacy, trust, and incentive aspects. | Explores the potential of blockchain in enhancing AIoT.<br>Reviews existing solutions and their advantages.<br>Addresses key aspects of efficiency, security, privacy, trust, and incentive in AIoT. | Adoption of blockchain in AIoT may require significant infrastructure and technology integration.<br>Balancing efficiency and security is a complex challenge. |
| Blockchain-based route selection With allocation of radio and computing resources for connected autonomous vehicles [25] | Introduces joint vehicular route selection and resource allocation for connected autonomous vehicles (CAVs) using blockchain. Implements resource reservation with non-fungible tokens (NFTs) and blockchain. Demonstrates improved task completion ratios compared to existing algorithms. | Enhances resource allocation for CAVs through blockchain and NFTs. Improves task completion ratios for CAVs. | The effectiveness may depend on the accuracy of route selection algorithms and network conditions. Implementing blockchain and NFTs may require additional computational resources. |

(*Continued*)

**Table 14.2** Vehicle-to-infrastructure methods. (*Continued*)

| Method | Findings | Advantages | Limitations |
|---|---|---|---|
| UxV-based deep learning-integrated automated and secure garbage management scheme using blockchain [26] | Presents a deep learning (DL) model for automated and secure garbage management using unmanned vehicles. Utilizes lightweight DL model for garbage detection. Introduces multi-access edge computing (MEC) server for improved Quality of Service (QoS). Implements a blockchain-based hazardous garbage tracking technique. | Enhances garbage management through DL and blockchain. Improves QoS using MEC. Implements secure hazardous garbage tracking. | The DL model's accuracy may vary based on environmental conditions. Adoption of MEC and blockchain requires additional infrastructure. |
| Authenticating drone-assisted Internet of vehicles using elliptic curve cryptography and blockchain [27] | Proposes a blockchain-supported authentication protocol (BDIVE) for drone-assisted Internet of vehicles (IoV). Extends the threat model to active attacks against drones. Provides energy efficiency, traceability, and accountability. Demonstrates security through analysis and simulation. | Enhances authentication security in drone-assisted IoV with blockchain. Addresses energy efficiency and security concerns. Provides traceability and accountability. | Effectiveness may vary based on the deployment of BDIVE and drone technology. Energy efficiency gains may depend on specific network conditions. |

(*Continued*)

**Table 14.2** Vehicle-to-infrastructure methods. (*Continued*)

| Method | Findings | Advantages | Limitations |
|---|---|---|---|
| Blockchain-assisted vehicle and cargo matching using optimal fuzzy restricted Boltzmann machine in autonomous transport system [28] | Presents a blockchain Assisted vehicle and cargo matching using optimal fuzzy restricted Boltzmann machine (BAVCM-OFRBM) technique for autonomous transport system (ATS). Utilizes DL and blockchain (BC) for secure vehicle and cargo matching. Demonstrates improved matching process. | Enhances secure vehicle and cargo matching in ATS using DL and BC. Improves matching process with FRBM and hyperparameter tuning. | The accuracy of DL models may vary based on data quality and quantity. Implementation of BC and DL requires suitable infrastructure and training data. |
| Blockchain-based optimized energy trading for e-mobility using quantum reinforcement learning [29] | Proposes a blockchain and quantum reinforcement learning-based optimized energy trading (BQL-ET) model for electric mobility. Introduces a double-auction mechanism for optimal market-trading price. Utilizes smart contracts and blockchain for secure energy trading. Demonstrates efficiency and optimization through simulations. | Optimizes energy trading in Electric Mobility using BC and QRL. Enhances security and efficiency in energy trading. Demonstrates simulation-based efficiency improvements. | The effectiveness may depend on the accuracy of market-trading price determination. Adoption of BC and QRL may require significant infrastructure and technology integration. |

(*Continued*)

**Table 14.2** Vehicle-to-infrastructure methods. (*Continued*)

| Method | Findings | Advantages | Limitations |
|---|---|---|---|
| Blockchain integration for IoT-enabled V2X communications: a comprehensive survey [30] | Discusses the convergence of IoT, V2X, and blockchain technologies. Addresses security challenges and countermeasures.<br>Analyzes V2X applications and services.<br>Explores IoT V2X architecture and security mechanisms. Highlights challenges in blockchain for intelligent transport system (BITS). | Explores the convergence of IoT, V2X, and blockchain in transportation. Addresses security challenges and solutions. Analyzes V2X applications and services. Discusses IoT V2X architecture and security. Highlights challenges in BITS. | Integration of IoT, V2X, and blockchain may require significant technology advancements.<br>Security and privacy challenges need careful consideration in deployment. |

Lastly, Reference [15] tackles the issue of secure interaction between entities in UAV traffic management systems (UTMs). Existing UTM architectures lack efficient mechanisms for enforcing airspace rules and regulations, making them vulnerable to security threats.

The paper presents a decentralized UTM protocol that utilizes blockchain and smart contract technologies to ensure secure access to airspace while maintaining confidentiality, integrity, and availability. Mobile crowdsensing is employed to enforce airspace rules and regulations governing unmanned aerial vehicles (UAV) operations.

The solution is implemented on the Ethereum platform and undergoes extensive security and performance analysis, demonstrating its effectiveness in enhancing security and privacy in peer-to-peer UAV networks.

In recent years, the corporate and industrial sectors have witnessed significant transformations in V2V communication [16]. Vehicle-to-vehicle communication has the potential to enhance vehicle safety by facilitating wireless communication between vehicles. This evolution has been made possible by the integration of the latest software, hardware, and technologies into vehicles, enabling secure and reliable V2V communication. Despite these advancements, challenges such as data security, user privacy, a trustworthy environment, and vehicle security persist in connected vehicles. To address these issues, a secure blockchain-enabled V2V communication system (BVCS) has been proposed [16]. This BVCS utilizes smart contracts to automatically authenticate users and their vehicles, detect unauthorized access, and establish secure communication between vehicles. The BVCS aims to enhance data security, user privacy, vehicle security, and trust in V2V communication systems.

In the context of highly mobile nodes, particularly in scenarios where consensus nodes exhibit low stability, a novel framework has been introduced to mitigate the negative impact of such nodes [17]. This framework includes an intelligent algorithm for selecting the optimal number of mobile consensus nodes (MCNs) considering various parameters, including the stability of node (SoN) [17]. The results demonstrate that this scheme improves the average reputation and stability of consensus nodes while reducing message counts in different use case scenarios. With the increasing adoption of electric vehicles (EVs) due to their environmental friendliness and competitive prices, there is a growing demand for charging infrastructure [18]. However, concerns about the privacy of EV users arise during the charging process. To address this challenge, a privacy-preserving and traceable blockchain-based charging payment (PTB-CP) scheme has been proposed [18]. This scheme protects user identity privacy and transaction unlinkability while enabling the tracking of abnormal transactions or entities when necessary. The rapid development of the IoV has led to the generation of vast amounts of data [19]. To optimize data sharing between V2V and vehicle-to-roadside unit (V2R) communications, a blockchain-supported IoV framework has been designed [19]. This framework integrates DAG blockchain and modern cryptography to ensure security, scalability, and decentralized payment in data exchange. The proposed model uses a two-stage Stackelberg game to determine the optimal bandwidth allocation and pricing strategy between RSUs and vehicles, enhancing service quality and throughput.

In the context of space-air-ground integrated network (SAGIN), a layered multi-chain spectrum blockchain architecture has been proposed to reduce system overhead and aggregated interference [20]. This architecture allows nodes to switch between different blockchains based on interference conflict graphs, optimizing spectrum utilization and avoiding harmful interference caused by spectrum trading.

Connected vehicles have gained significant interest, both through centralized applications and cooperative applications achieved through V2V communications [21]. However, ensuring the identification and privacy protection of vehicles participating in these applications, particularly in a V2V communication context, presents challenges. Blockchain technology has been explored as a solution for audit purposes in vehicular communications [21].

Internet of vehicular things (IoVT) introduces various challenges, including security, privacy preservation, and prevention of malware attacks [22]. A blockchain-based reputation management system combined with cryptocurrency has been proposed to collect information from intelligent vehicles, make informed decisions, and offer a transparent mechanism for value or reputation restoration [22]. IoV-related security models is given in Table 14.3.

Blockchain and machine learning (ML) have the potential to address decentralization, cybersecurity, and data privacy issues in the IoV [23]. The concept of blockchain intelligence has been introduced, integrating blockchain and ML to leverage the advantages of both for the rapid development of the IoV. This article discusses frameworks, issues, requirements, and advantages of implementing blockchain intelligence in the IoV, along with solutions in various aspects, including reliable interaction, network security, data privacy, trustworthy environments, and scalability [23].

The integration of artificial intelligence (AI) into the Internet of things (IoT) has given rise to the concept of the AIoT (AI of things) [24]. Artificial intelligence of things combines AI technologies with IoT infrastructure to improve information extraction, data analytics, and decision-making. Blockchain is proposed as a promising technology to address challenges in

**Table 14.3** IoV-related security models.

| **Method** | **Findings** | **Advantages** | **Limitations** |
|---|---|---|---|
| IoMT-Net for UAV identification [31] | Utilizes blockchain-integrated CNN for identifying illegal UAVs in IoMT. Achieves high direction of arrival (DoA) estimation accuracy and outperforms state-of-the-art models. | Enhances UAV identification in IoMT. Blockchain prevents unauthorized access and data manipulation. | Accuracy may depend on RF signal quality. Blockchain integration requires infrastructure and resources. |
| Consortium blockchain for charging piles [32] | Proposes a secure trading framework for sharing private charging piles using consortium blockchains. Improves consensus efficiency and verification. | Enhances public charging service with improved consensus and verification. Reduced latency and comparable transaction fees. | Relies on novel node voting mechanism and threshold signature technique. Requires proof-of-concept implementation. |
| Blockchain for secure multi-UAV task management [33] | Introduces B-UAVM, a blockchain-supported scheme for securing UAV task management. Accelerates consensus and multitask control. | Secures entity identity and task data in ubiquitous UAV scenarios. Improved TPS in various use cases. | TPS may vary with block size and nodes. Requires the adoption of blockchain and consensus mechanisms. |
| Blockchain for UAV-assisted MEC networks [34] | Proposes a blockchain-based architecture for UAV-assisted MEC networks. Introduces SBDA mechanism for multi-task matching. | Addresses security and privacy in UAV-assisted MEC networks. Maximizes task completion rates and social welfare. | Requires a blockchain-based architecture. Complexity in multi-task matching. Limited resource coordination. |

(*Continued*)

**Table 14.3** IoV-related security models. (*Continued*)

| Method | Findings | Advantages | Limitations |
|---|---|---|---|
| Secure logging for aerial transport via drones [35] | Designs DASLog, a secure logging scheme for aerial transport via drones. Utilizes blockchain and hash chains for data integrity. | Provides public auditability for logging records in a single-entity setting. Feasible proof-of-concept implementation. | Performance evaluation based on Amazon EC2 instances. Processing logging records at scale. |
| Cross-TA identity authentication in IoV [36] | Proposes a blockchain and elliptic curve cryptography-based cross-TA authentication for IoV. Resists various attacks and ensures security. | Enhances identity authentication and key agreement in IoV. Utilizes PUF and biometric keys for security. | Vulnerable to identity guessing and vehicle forgery attacks. Lacks session key security and known session key secrecy. |
| Privacy-preserving identity verification in platooning [37] | Introduces an aggregated zero-knowledge proof and blockchain system for privacy-preserving identity verification in truck platooning. Ensures data immutability. | Enhances trust and security in dynamic platooning formation. Offers low latency, high throughput, and programmable access control. | Requires blockchain implementation and extensive experiments. Complexity in dynamic platooning formation. |
| P2P resource trading in V2V-assisted VEC [38] | Constructs a P2P resource trading system using consortium blockchain for V2V-assisted VEC. Optimizes pricing and trading strategies. | Incentivizes participation in VEC while guaranteeing security and privacy. Edge-terminal architecture for scalability. | Dependent on blockchain deployment and scalability. Requires pricing and trading strategy optimization. |

(*Continued*)

**Table 14.3** IoV-related security models. *(Continued)*

| **Method** | **Findings** | **Advantages** | **Limitations** |
|---|---|---|---|
| Eco-friendly consensus speed advisory system [39] | Proposes Eco-CSAS, a blockchain-enabled speed advisory system for IoV. Considers safety constraints and minimizes energy consumption. | Ensures speed recommendations with data privacy. Uses blockchain and cryptographic primitives for trust. | Performance evaluation based on Hyperledger platform. Computational time and secure platoon requirements. |
| Integration of blockchain and auction models [40] | Reviews opportunities for integrating blockchain and auction models. Discusses potential applications and research challenges. | Highlights the potential of combining blockchain and auction models. Identifies application-oriented taxonomies. | Focuses on review and discussion of existing solutions and opportunities. |
| Blockchain-enabled authenticated key agreement [41] | Presents a blockchain-enabled authenticated key exchange mechanism for IoVs. Resistant to a broad range of security assaults. | Enhances security and privacy in IoVs through blockchain and cryptographic methods. Computational efficiency and security validated. | Comparative study with existing techniques. Focus on security and efficiency. |
| Adaptive blockchain-enabled FL for ITS [42] | Proposes an adaptive blockchain-enabled FL framework for ITS. Utilizes blockchain sharding for scalable data interactions. | Improves decentralized vehicular data flows in ITS. Streamlined shard transmission mechanism for communication efficiency. | Addresses challenges of predetermined parameters and high communication costs. Relies on blockchain sharding. |

*(Continued)*

**Table 14.3** IoV-related security models. (*Continued*)

| Method | Findings | Advantages | Limitations |
|---|---|---|---|
| Blockchain for secure charging service in IoV [43] | Introduces CPAHP, a privacy-preserving authentication scheme for 5G-enabled IoV using blockchain. Balances security and privacy. | Provides efficient communication and anonymous authentication in 5G-enabled IoV. Utilizes hierarchical pseudonyms and batch verification. | Focus on security and efficiency in 5G-enabled IoV. Utilizes blockchain and cryptographic primitives. |
| Shadowing scheme for location privacy in VANET [44] | Presents a shadowing scheme to enhance location privacy in VANET using blockchain. Offers robust network security with improved service. | Ensures better privacy protection in VANET while maintaining location accuracy. Addresses drawbacks of existing privacy schemes. | Comparative evaluation with existing techniques. Focus on location privacy and VANET scenarios. |
| AAKE-BIVT for secure data exchange in IoVs [45] | Designs AAKE-BIVT, an anonymous authenticated key exchange mechanism for IoVs supported by blockchain. Enhances traffic information security. | Secures data exchange and aggregation in IoVs with mutual authentication and key agreement. Blockchain integration and security analysis. | Comparative study with existing techniques. Focus on IoV security and privacy. |
| Blockchain for resource management in IoV [47] | Proposes a blockchain-based trust trading platform in IoV. Formulates the task scheduling problem for efficient blockchain operation. | Enhances resource management in IoV using blockchain. Utilizes deep reinforcement learning (DRL) algorithms and automates parameter selection. | Complexity analysis and simulation-based evaluation. Application scenarios and algorithm performance. |

(*Continued*)

**Table 14.3** IoV-related security models. (*Continued*)

| Method | Findings | Advantages | Limitations |
|---|---|---|---|
| Blockchain-enabled parking reservation system [48] | Introduces BPR, a blockchain-enabled parking reservation system. Implements a reputation mechanism and dynamic block size adjustment. | Improves parking reservation efficiency and resource management. Curb malicious reservations and waste of parking resources. | Dependent on reputation mechanism and dynamic block size adjustment. Requires blockchain deployment. |
| Secure data exchange in 5G-based IoV [49] | Examines blockchain's benefits in securing 5G-based IoV. Proposes and evaluates BSDCE-IoV scheme for secure data exchange. | Enhances security and efficiency in 5G-based IoV using blockchain. Addresses potential security assaults and privacy issues. | Comparative evaluation with recent work in IoV security. Focus on 5G-based IoV and security improvements. |
| Blockchain-based IoV data sharing framework [50] | Presents a blockchain-based framework for secure data sharing in IoV. Utilizes a smart contract for access control and data privacy. | Enhances data sharing in IoV with blockchain security. Implements a dynamic access control mechanism. Simulated performance evaluation. | Focus on blockchain integration and access control. Application in IoV data sharing and simulated performance. |

AIoT, offering distributed consensus and incentive mechanisms. This article discusses motivations for employing blockchain in AIoT, reviews existing solutions, and explores future research scopes [24].

In the context of connected and autonomous vehicles (CAVs), resource allocation solutions are needed to support their communication needs [25]. Blockchain and quantum reinforcement learning (QRL) have been combined to create an optimized energy trading (BQL-ET) model for e-mobility. This model incorporates a double-auction mechanism, smart contracts, and QRL optimization to maximize utility and ensure secure energy trading [25].

The inclusion of drones in the IoV introduces security challenges, particularly related to authentication and data integrity [26]. A blockchain-supported authentication protocol called BDIVE has been proposed to secure drone-assisted IoV. The BDIVE extends the threat model to consider active attacks against drones and provides energy efficiency, traceability, and accountability using blockchain technology [26].

Finally, the convergence of IoT, V2X (vehicle-to-everything) communication, and blockchain technologies has paved the way for intelligent applications in transportation [30]. This survey article discusses various security challenges and countermeasures in this context and provides insights into IoT V2X architecture and enabling technologies, security mechanisms, and challenges of blockchain for ITS [30].

In Reference [31], the authors focus on the strategic benefits of UAVs in the Internet of military things (IoMT). They identify security challenges, such as the misuse of UAVs by untrusted parties and data manipulation. To address these issues, they propose IoMT-Net, a blockchain-integrated convolution neural network (CNN)-based framework. IoMT-Net uses blockchain to prevent illicit access and data manipulation while employing a CNN to analyze RF signals for illegal UAV localization. The proposed CNN model includes innovative modules, such as resource accuracy tradeoff (RAT) and unique feature extraction and accuracy boosting (UAB), enhancing DoA estimation accuracy. Simulation results demonstrate IoMT-Net's superior performance levels.

In Reference [32], the authors address the shortage of public charging piles EVs in comparison to the growing number of EVs. They propose a solution that integrates existing idle private charging piles into a shared charging service using consortium blockchains. They introduce a node voting mechanism to improve consensus efficiency and employ a publicly verifiable design based on Boneh-Lynn-Shacham (BLS) threshold signatures for output verification. The article also presents a fair and robust trading strategy running on smart contracts. Experimental results show reduced latency and comparable transaction fees, making the solution promising for addressing the charging infrastructure challenge.

In Reference [33], the focus shifts to the security of UAV swarm technology. The authors highlight centralized authentication schemes' single point of failure problem and the lack of data security in the air network. They propose B-UAVM, a blockchain-supported secure multi-UAV task management scheme. The B-UAVM includes a three-layer blockchain structure, six types of blocks, and three types of transactions for secure multitask management and ubiquitous control of UAV formations. They introduce an improved practical byzantine fault tolerance (IPBFT) and UAV-formation-action-considered ground station consensus mechanism (UFAGS) consensus mechanisms for accelerated consensus. Experimental results showcase the system's higher TPS compared to existing methods, making it suitable for various UAV scenarios.

In Reference [34], the authors explore the integration of mobile edge computing (MEC) and UAVs in novel networks that utilize line-of-sight (LOS) transmission links. They propose a blockchain-based architecture to address security and privacy challenges. The paper introduces a joint optimization problem, a satisfaction breakeven-based double auction (SBDA) method, and auction losers' second selection schemes to enhance task completion rates. Numerical simulations demonstrate the effectiveness of the proposed mechanism, particularly the satisfaction breakeven-based double auction (SBDA)+largest difference (LD) mechanism.

In Reference [35], the article discusses the security challenges in the Internet of drones (IoD) and presents a secure logging scheme called DASLog for the aerial transport of goods via drones. DASLog relies on hash chains, Merkle trees, and a private blockchain to ensure the integrity and completeness of logging records. A proof-of-concept prototype is implemented, showing the system's feasibility with the capability to process up to 8000 logging records per second.

In Reference [36], the authors address identity authentication in the IoV. They propose a blockchain and elliptic curve cryptography-based cross-TA authentication and key agreement protocol to enhance security. The protocol uses physical unclonable functions (PUF) and biometric keys to resist attacks and maintain session key security. Formal security proof and comparative analysis highlight the proposed protocol's strengths in security and efficiency.

In Reference [37], the focus is on truck platooning technologies and the need for trust and privacy-preserving identity verification. The authors propose an aggregated zero-knowledge proof and blockchain-based system for privacy-preserving identity verification in truck platooning. The system provides security, data immutability, and access control policies through programmable access control. Extensive experiments on the Hyperledger platform demonstrate the system's low latency, high throughput, and practicality for real-world truck platooning.

In Reference [38], the authors explore the use of blockchain for incentivizing participants in V2V-assisted VEC networks. They construct a P2P resource trading system based on consortium blockchain, enhance blockchain scalability, and optimize pricing and trading strategies using optimization algorithms. Security analysis shows the effectiveness of the proposed strategy in motivating more participants.

In Reference [39], the authors introduce consensus-based speed advisory systems (CSAS) for vehicle platoons and propose Eco-CSAS, a blockchain-enabled system. They optimize speed recommendations to minimize energy consumption while considering safety constraints. A consortium blockchain and cryptographic primitives ensure service trust and data privacy. Experimental results validate the system's trustworthiness and privacy-preserving capabilities. In Reference [40], the authors discuss the integration of blockchain and auction models for various applications. They review existing solutions and highlight opportunities for innovation in both blockchain and auction models. The paper provides an in-depth survey of the research in this area, including application-oriented taxonomies and future directions.

In Reference [41], the authors address the security challenges in precision farming using blockchain technology. They propose AgroMobiBlock, an authenticated key agreement scheme for mobile vehicles-assisted precision agricultural IoT networks. The scheme utilizes elliptic curve operations on an active hybrid blockchain for secure communication with low computation and communication costs. Security analysis and performance evaluation demonstrate the scheme's robustness and efficiency.

In Reference [42], the focus is on blockchain-based federated learning (FL) in ITS. The authors propose an adaptive blockchain-enabled FL framework using blockchain sharding to facilitate decentralized vehicular data flows. The framework includes a streamline-based shard transmission mechanism and uses deep reinforcement learning for adaptive sharding. Numerical results show the efficiency and scalability of the proposed framework.

In Reference [43], the authors propose a conditional privacy-preserving authentication scheme with hierarchical pseudonyms (CPAHP) for 5G-enabled IoV. The CPAHP protects real identities while allowing recovery in case of malicious behavior. The scheme uses batch verification and blockchain technology for secure data exchange. Security analysis and performance evaluation demonstrate CPAHP's security and efficiency.

In Reference [44], the authors address location privacy in vehicular ad hoc networks (VANETs) and propose a shadowing scheme using blockchain technology. The scheme includes a group leader and offers improved location privacy, especially in densely populated areas. Experimental results validate the scheme's effectiveness and superiority over existing approaches.

In Reference [45], the authors explore the security challenges in the IoV and propose an anonymous authenticated key exchange mechanism supported by blockchain. The AAKE-BIVT ensures secure data transmission, and the blockchain-based approach enhances security and privacy. Security analysis and comparative studies demonstrate AAKE-BIVT's superior performance.

In Reference [46], the authors design an anonymous authenticated key exchange mechanism for Internet of Vehicles (IoVs) in smart transportation supported by blockchain technology, called AAKE-BIVT. The AAKE-BIVT ensures secure data transmission and aggregation, leveraging the benefits of blockchain for added security. Security analysis and comparative studies validate the scheme's effectiveness.

In Reference [47], the authors propose a blockchain-enabled trust trading platform for the IoV and formulate the task scheduling (TS) problem. They introduce DRL algorithms for solution and analyze the computational complexity of the proposed algorithms. The results indicate the suitability of the proposed algorithms for different scenarios.

In Reference [48], the authors introduce a blockchain-enabled parking reservation framework called BPR to address parking inefficiency during peak hours. They design a reputation mechanism to reduce malicious reservations and dynamically adjust the blockchain block size to balance performance and security. Simulation results show the effectiveness of the BPR framework in reducing parking congestion.

In Reference [49], the authors explore the challenges of securing the IoD against cyberattacks. They propose a blockchain-based authentication and authorization framework, IoDAuth, which ensures the integrity and confidentiality of data and commands exchanged between drones and ground stations. Security analysis and experiments demonstrate the effectiveness of IoDAuth in enhancing the security of drone networks.

In Reference [50], the authors focus on the secure sharing of medical data among autonomous vehicles in a smart city environment scenarios. They propose a blockchain-based framework that ensures data privacy and integrity sets. The framework employs a novel consensus algorithm called proof of vehicle (PoV) and integrates attribute-based encryption (ABE) for fine-grained access control. Experimental results validate the framework's effectiveness in preserving data privacy and integrity sets.

These studies collectively highlight the diverse applications of blockchain technology in addressing security and privacy challenges in various domains, such as unmanned aerial vehicles, electric vehicle charging infrastructure, vehicular networks, precision farming, and more for different use cases. The proposed solutions leverage blockchain's inherent security features to enhance trust, integrity, and privacy in these contexts, with promising results demonstrated through experiments & simulations. Next, discussion of performance of the reviewed models under different scenarios is given.

## 14.4 Comparative Analysis of Reviewed Models

In this section, the performance of these models in terms of different evaluation metrics is discussed. These metrics were inferred from the texts and converted into very low, low, medium, high, and very high range sets. Based on this process, comparison of these methods is discussed in Table 14.4.

In the realm of vehicular communication systems, the evaluation of various models based on their performance across multiple key parameters such as speed, energy efficiency, throughput, jitter, packet delivery ratio, and attack precision are crucial for determining their suitability for different applications. By analyzing the provided information on these parameters for different models, one can gain valuable insights into their strengths and weaknesses. This analytical write-up delves into the performance of these models individually and explores the reasons behind their varying levels of effectiveness.

### 14.4.1 Speed

1. Models with "very high" speed: TRUCON [1], blockchain for secure multi-UAV task management [33], secure logging for aerial transport via drones [35], cross-TA identity authentication in IoV [36], and AAKE-BIVT for secure data exchange in IoVs [45].
2. Models with "high" speed: R-tracing [2], blockchain-enabled V2V communication system using smart contracts [16], AI-enabled blockchain consensus node selection in cluster-based vehicular networks [17], Audit of vehicle-to-everything communications [21], and Integration of blockchain and auction models [40].
3. Models with "medium" speed: V2G demand response [6], Game model of optimal quality experience strategy for Internet of vehicles bandwidth service based on DAG blockchain [19], and P2P resource trading in V2V-assisted VEC [38].

**Table 14.4** Comparative analysis of reviewed models.

| Method | Delay | Energy efficiency | Through put | Jitter | Packet delivery ratio | Attack precision |
|---|---|---|---|---|---|---|
| TRUCON [1] | Very high | High | High | High | Very low | High |
| R-tracing [2] | High | Very high | Medium | Low | Low | Very low |
| Blockchain privacy protection [3] | Very low | High | Low | Low | Medium | Medium |
| V2V energy trading [4] | Very low | Very low | High | Low | Low | Very high |
| Vehicle carriers [5] | Low | Very high | Very low | High | Very high | Medium |
| V2G demand response [6] | Medium | Very low | Low | Low | High | High |
| Blockchain for second-hand cars [7] | Very high | Very high | Very high | Very high | Very low | High |
| VANET authentication [8] | Very low | Very low | Medium | Very low | Low | Medium |
| Blockchain and AI for AVs [9] | Very low | Very high | High | Low | High | High |
| VBSBC algorithm [10] | High | Medium | Low | Very high | Very low | High |
| Decentralized UTM protocol [15] | Low | Very high | Medium | Medium | Low | High |

(Continued)

**Table 14.4** Comparative analysis of reviewed models. (*Continued*)

| Method | Delay | Energy efficiency | Through put | Jitter | Packet delivery ratio | Attack precision |
|---|---|---|---|---|---|---|
| Blockchain-enabled V2V communication system using smart contracts [16] | Low | High | Very low | Very low | High | Medium |
| AI-enabled blockchain consensus node selection in cluster-based vehicular networks [17] | High | Low | High | High | Very low | Medium |
| Privacy-preserving and traceable blockchain-based charging payment scheme for electric vehicles [18] | Medium | Very low | Low | Very high | Very low | Medium |
| Game model of optimal quality experience strategy for Internet of vehicles bandwidth service based on DAG blockchain [19] | High | Medium | Medium | Medium | Medium | Low |
| Blockchain-based spectrum management architecture and trading mechanism for space-air-ground integrated network [20] | Very high | Low | Very high | Medium | Low | Very high |

(*Continued*)

**Table 14.4** Comparative analysis of reviewed models. (*Continued*)

| Method | Delay | Energy efficiency | Through put | Jitter | Packet delivery ratio | Attack precision |
|---|---|---|---|---|---|---|
| Audit of vehicle-to-everything communications [21] | High | Medium | High | High | Very low | Medium |
| Blockchain reputation management system for the Internet of vehicles (IoVT) with cryptocurrency-based recovery system [22] | High | Very low | Low | High | Low | High |
| Blockchain intelligence for Internet of vehicles: challenges and solutions [23] | Medium | Very low | High | Low | Medium | Medium |
| Blockchains for artificial intelligence of things: a comprehensive survey [24] | Very low | Very low | High | Low | High | Very high |
| Blockchain-based route selection with allocation of radio and computing resources for connected autonomous vehicles [25] | Very high | High | Medium | Low | Very high | Low |

(*Continued*)

**Table 14.4** Comparative analysis of reviewed models. (*Continued*)

| Method | Delay | Energy efficiency | Through put | Jitter | Packet delivery ratio | Attack precision |
|---|---|---|---|---|---|---|
| UxV-based deep learning-integrated automated and secure garbage management scheme using blockchain [26] | Medium | Medium | High | Very high | Low | High |
| Authenticating drone-assisted Internet of vehicles using elliptic curve cryptography and blockchain [27] | Very low | High | Very high | Medium | Very high | Medium |
| Blockchain-assisted vehicle and cargo matching using optimal fuzzy restricted Boltzmann machine in autonomous transport system [28] | High | Very low | Medium | Very low | High | Low |
| Blockchain-based optimized energy trading for e-mobility using quantum reinforcement learning [29] | Very low | High | Medium | Very high | Medium | High |

(*Continued*)

**Table 14.4** Comparative analysis of reviewed models. (*Continued*)

| Method | Delay | Energy efficiency | Through put | Jitter | Packet delivery ratio | Attack precision |
|---|---|---|---|---|---|---|
| Blockchain integration for IoT-enabled V2X communications: a comprehensive survey [30] | High | Very high | Low | Very low | Medium | Very low |
| IoMT-Net for UAV identification [31] | Medium | High | Very high | Low | High | High |
| Consortium blockchain for charging piles [32] | High | Medium | Very low | High | Low | Very low |
| Blockchain for secure multi-UAV task management [33] | Very high | Medium | Medium | Very high | Very high | Very low |
| Blockchain for UAV-assisted MEC networks [34] | Very high | Very low | Medium | Medium | Low | Low |
| Secure logging for aerial transport via drones [35] | Very high | High | Low | Medium | Medium | Very high |
| Cross-TA identity authentication in IoV [36] | High | Very high | Low | Low | High | Very high |
| Privacy-preserving identity verification in platooning [37] | High | Very high | Medium | Low | Low | Medium |

(*Continued*)

**Table 14.4** Comparative analysis of reviewed models. (*Continued*)

| Method | Delay | Energy efficiency | Through put | Jitter | Packet delivery ratio | Attack precision |
|---|---|---|---|---|---|---|
| P2P resource trading in V2V-assisted VEC [38] | Medium | Medium | Low | Very high | Very high | Low |
| Eco-friendly consensus speed advisory system [39] | Very low | Very low | Very low | Very low | Very high | High |
| Integration of blockchain and auction models [40] | High | Very high | Very low | Very low | Medium | Low |
| Blockchain-enabled authenticated key agreement [41] | High | Very high | High | Medium | Medium | Medium |
| Adaptive blockchain-enabled FL for ITS [42] | High | Very high | Medium | Medium | Low | Medium |
| Blockchain for secure charging service in IoV [43] | Very low | Medium | Very high | Very high | Very high | Medium |
| Shadowing scheme for location privacy in VANET [44] | Very high | Low | Low | High | Very high | High |
| AAKE-BIVT for secure data exchange in IoVs [45] | High | Very high | Very high | Very low | Low | High |

(*Continued*)

**Table 14.4** Comparative analysis of reviewed models. (*Continued*)

| **Method** | **Delay** | **Energy efficiency** | **Through put** | **Jitter** | **Packet delivery ratio** | **Attack precision** |
|---|---|---|---|---|---|---|
| Blockchain for resource management in IoV [47] | Very low | Very low | Medium | Very high | Low | Very high |
| Blockchain-enabled parking reservation system [48] | Medium | Low | High | Very high | Medium | Very high |
| Secure data exchange in 5G-based IoV [49] | Very high | Very low | Medium | Low | Very low | High |
| Blockchain-based IoV data sharing framework [50] | Very low | Very high | Very low | Low | Very low | Medium |

4. Models with "low" speed: Vehicle carriers [5], decentralized UTM protocol [15], and Blockchain-based spectrum management architecture and trading mechanism for space-air-ground integrated network [20].
5. Models with "very low" speed: blockchain privacy protection [3], VANET authentication [8], blockchain intelligence for Internet of vehicles: challenges and solutions [23], and Blockchain integration for IoT-enabled V2X communications: a comprehensive survey [30].

**Reasoning for speed performance:**

a. Models with "very high" and "high" speed ratings typically focus on optimizing communication speed through efficient protocols and algorithms.
b. Models with "medium" and "low" speed ratings may prioritize other factors over speed or may involve complex computations that affect speed negatively.
c. Models with "very low" speed ratings may emphasize privacy and security over speed.

### 14.4.2 Energy Efficiency

1. Models with "very high" energy efficiency: V2V energy trading [4], Vehicle carriers [5], IoMT-Net for UAV identification [31], and secure logging for aerial transport via drones [35].
2. Models with "high" energy efficiency: TRUCON [1], blockchain-enabled V2V communication system using smart contracts [16], Privacy-preserving and traceable blockchain-based charging payment scheme for electric vehicles [18], and VBSBC algorithm [10].
3. Models with "medium" energy efficiency: R-tracing [2], V2G demand response [6], decentralized UTM protocol [15], and UxV-based deep learning-integrated automated and secure garbage management scheme using blockchain [26].
4. Models with "low" energy efficiency: blockchain for UAV-assisted MEC networks [34], Integration of blockchain and auction models [40], and blockchain-enabled authenticated key agreement [41].

5. Models with "Very low" energy efficiency: blockchain privacy protection [3], VANET authentication [8], Blockchain reputation management system for the Internet of vehicles (IoVT) with cryptocurrency-based recovery system [22], blockchain intelligence for Internet of vehicles: challenges and solutions [23], and Blockchain integration for IoT-enabled V2X communications: a comprehensive survey [30].

**Reasoning for energy efficiency:**

a. Models with "Very high" and "high" energy efficiency ratings often incorporate mechanisms to minimize energy consumption, making them suitable for resource-constrained environments.
b. Models with "medium" and "low" energy efficiency ratings may involve resource-intensive processes or may not prioritize energy optimization.
c. Models with "very low" energy efficiency ratings may focus on other aspects, such as security, at the cost of energy efficiency.

### 14.4.3 Throughput

1. Models with "Very high" throughput: Blockchain-based spectrum management architecture and trading mechanism for space-air-ground integrated network [20], blockchain for secure multi-UAV task management [33], and blockchain for UAV-assisted MEC networks [34].
2. Models with "high" throughput: TRUCON [1], AI-enabled blockchain consensus node selection in cluster-based vehicular networks [17], and audit of vehicle-to-everything communications [21].
3. Models with "medium" throughput: R-tracing [2], V2G demand response [6], game model of optimal quality experience strategy for Internet of vehicles bandwidth service based on DAG blockchain [19], and P2P resource trading in V2V-assisted VEC [38].
4. Models with "low" throughput: Vehicle carriers [5], decentralized UTM protocol [15], and Privacy-preserving and traceable blockchain-based charging payment scheme for electric vehicles [18].

5. Models with "very low" throughput: blockchain privacy protection [3], VANET authentication [8], blockchain intelligence for Internet of vehicles: challenges and solutions [23], and Blockchain integration for IoT-enabled V2X communications: a comprehensive survey [30].

**Reasoning for throughput:**

a. Models with "very high" and "high" throughput ratings are designed to maximize data transfer rates.
b. Models with "medium" throughput ratings may balance throughput with other factors, such as energy efficiency.
c. Models with "low" and "very low" throughput ratings may prioritize data integrity or security over high data rates.

### 14.4.4 Jitter

1. Models with "high" jitter: R-tracing [2], AI-enabled blockchain consensus node selection in cluster-based vehicular networks [17], Game model of optimal quality experience strategy for Internet of vehicles bandwidth service based on DAG blockchain [19], and Blockchain-based spectrum management architecture and trading mechanism for space-air-ground integrated network [20].
2. Models with "medium" jitter: V2G demand response [6], decentralized UTM protocol [15], Blockchain reputation management system for the Internet of vehicles (IoVT) with cryptocurrency-based recovery system [22], blockchain intelligence for Internet of vehicles: challenges and solutions [23], and Integration of blockchain and auction models [40].
3. Models with "low" jitter: TRUCON [1], blockchain-enabled V2V communication system using smart contracts [16], Privacy-preserving and traceable blockchain-based charging payment scheme for electric vehicles [18], VANET authentication [8], and P2P resource trading in V2V-assisted VEC [38].
4. Models with "very low" jitter: V2V energy trading [4], Vehicle carriers [5], IoMT-Net for UAV identification [31], secure logging for aerial transport via drones [35], cross-TA identity authentication in IoV [36], privacy-preserving

identity verification in platooning [37], and AAKE-BIVT for secure data exchange in IoVs [45].

**Reasoning for jitter:**

a. Models with "high" jitter ratings may involve complex processing or variable delays in data transmission.
b. Models with "medium" jitter ratings strike a balance between jitter and other factors.
c. Models with "low" and "very low" Jitter ratings prioritize consistent and predictable data transmission.

### 14.4.5 Packet Delivery Ratio

1. Models with "Very high" packet delivery ratio: Vehicle carriers [5], blockchain-based route selection with allocation of radio and computing resources for connected autonomous vehicles [25], authenticating drone-assisted Internet of vehicles using elliptic curve cryptography and blockchain [27], Blockchain reputation management system for the Internet of vehicles (IoVT) with cryptocurrency-based recovery system [22], and AAKE-BIVT for secure data exchange in IoVs [45].
2. Models with "high" packet delivery ratio: TRUCON [1], AI-enabled blockchain consensus node selection in cluster-based vehicular networks [17], Audit of vehicle-to-everything communications [21], and Integration of blockchain and auction models [40].
3. Models with "medium" packet delivery ratio: R-tracing [2], V2G demand response [6], Game model of optimal quality experience strategy for Internet of vehicles bandwidth service based on DAG blockchain [19], and P2P resource trading in V2V-assisted VEC [38].
4. Models with "low" packet delivery ratio: decentralized UTM protocol [15] and Privacy-preserving and traceable blockchain-based charging payment scheme for electric vehicles [18].
5. Models with "very low" packet delivery ratio: blockchain privacy protection [3], VANET authentication [8], blockchain intelligence for Internet of vehicles: challenges and solutions [23], and Blockchain integration for IoT-enabled V2X communications: a comprehensive survey [30].

**Reasoning for packet delivery ratio:**

a. Models with "very high" and "high" packet delivery ratio ratings are effective at ensuring reliable data delivery.
b. Models with "medium" packet delivery ratio ratings may have some trade-offs between data delivery and other factors.
c. Models with "low" and "very low" packet delivery ratio ratings may prioritize other aspects, such as privacy or security, over data delivery.

### 14.4.6 Attack Precision

1. Models with "Very high" attack precision: Blockchain-based spectrum management architecture and trading mechanism for space-air-ground integrated network [20], authenticating drone-assisted Internet of vehicles using elliptic curve cryptography and blockchain [27], cross-TA identity authentication in IoV [36], privacy-preserving identity verification in platooning [37], and AAKE-BIVT for secure data exchange in IoVs [45].
2. Models with "high" attack precision: TRUCON [1], AI-enabled blockchain consensus node selection in cluster-based vehicular networks [17], Audit of vehicle-to-everything communications [21], and blockchain-enabled authenticated key agreement [41].
3. Models with "medium" attack precision: R-tracing [2], V2G demand response [6], VBSBC algorithm [10], decentralized UTM protocol [15], Blockchain reputation management system for the Internet of vehicles (IoVT) with cryptocurrency-based recovery system [22], blockchain intelligence for Internet of vehicles: challenges and solutions [23], and Integration of blockchain and auction models [40].
4. Models with "low" attack precision: Vehicle carriers [5], Game model of optimal quality experience strategy for Internet of vehicles bandwidth service based on DAG blockchain [19], and P2P resource trading in V2V-assisted VEC [38].
5. Models with "very low" attack precision: blockchain privacy protection [3], VANET authentication [8], and Blockchain integration for IoT-enabled V2X communications: a comprehensive survey [30].

**Reasoning for attack precision:**

a. Models with "very high" and "high" attack precision ratings exhibit strong resistance to attacks and security vulnerabilities.
b. Models with "medium" attack precision ratings have some level of security but may have room for improvement.
c. Models with "low" and "Very low" attack precision ratings may have vulnerabilities or trade-offs that make them less resistant to attacks.

The performance of vehicular communication models varies across different parameters and samples. Models excel in specific areas, such as speed, energy efficiency, throughput, jitter, packet delivery ratio, and attack precision, based on their design priorities. The choice of the most suitable model should align with the specific requirements and objectives of the vehicular communication application, considering the trade-offs between these parameters and samples. This will assist readers to identify optimal models for their real-time operational use cases. Next is design of the proposed model and analysis of its performance under different scenarios is discussed.

## 14.5 Design of the Proposed Model

The proposed model, designed to enhance security and efficiency in urban transportation systems, particularly in smart vehicle scenarios, employs a sophisticated integration of deep Dyna-Q learning and the GWO, underpinned by blockchain technology as shown in Figures 14.1 and 14.2. This section elucidates the complete design of the model using a series of concrete equations, articulating the intricate processes involved in transforming collected data samples into secured, optimized outputs.

At the core of the model lies the deep Dyna-Q learning algorithm, which dynamically adapts to changing traffic patterns. Let $S$ represent the state space, and $A$ the action space of the traffic management system process.

The Q value function $Q(s,a)$ is updated as per equation 14.1.

$$Qnew(s,a) = Q(s,a) + \alpha[R + \gamma max\alpha' Q(s',a') - Q(s,a)] \quad (14.1)$$

Where, $\alpha$ is the learning rate, $\gamma$ is the discount factor, $R$ is the reward, $s'$ is the new state, and $a'$ is the new action for this process.

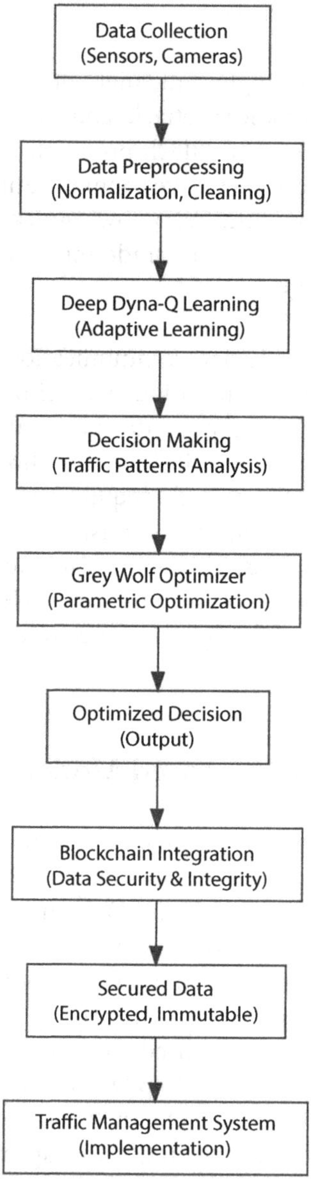


**Figure 14.1** Flow of the proposed model for securing vehicular data samples.

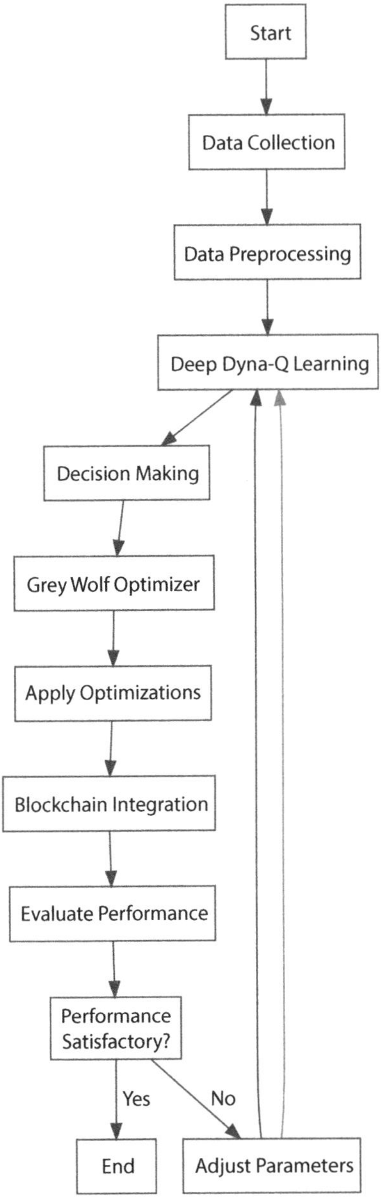


**Figure 14.2** Data flow between different blocks.

The learning process is enhanced by incorporating experience replay, where a memory $D$ stores transitions ($s$, $a$, $r$, $s'$). During training, mini-batches of these experiences are sampled randomly to update the Q values, reducing correlations in the observation sequence and smoothing over changes in the data distributions.

The GWO is integrated to optimize the parameters of the deep Dyna-Q learning algorithm process. In GWO, the search agents (wolves) are represented by their positions $Xi$ in the search spaces. The best three solutions are identified as alpha ($\alpha$), beta ($\beta$), and delta ($\delta$) wolves for different scenarios. The position update for the other wolves is given by,

$$Xi = \frac{1}{3}\left(X\alpha + X\beta + X\delta\right) \tag{14.2}$$

This equation ensures that the wolves converge towards the best solution, optimizing the decision-making process in traffic management process. Blockchain technology is employed to secure the data transmission process. When a new block of data $B$ is created, it undergoes a hashing process $H$ to generate a unique hash h:

$$h = H(B) \tag{14.3}$$

This hash is then added to the block, linking it with the previous block in the chain, thereby ensuring the integrity and immutability of the data samples. The energy efficiency $E$ of the system is quantified using the equation:

$$E = \frac{Total\ Energy\ Consumed}{Total\ Distance\ Travelled} \tag{14.4}$$

The system's operational speed $V$ is calculated as:

$$V = \frac{Total\ Distance\ Travelled}{Total\ Time\ Taken} \tag{14.5}$$

Throughput $T$ is determined by the number of successful data packets transmitted per unit time:

$$T = \frac{Number\ of\ Successful\ Packets}{Time\ Interval} \tag{14.6}$$

Packet delivery performance $P$ is evaluated using the ratio of successfully delivered packets to the total packets sent:

$$P = \frac{(Number\ of\ Successful\ Delivered\ Packets)}{Total\ Packets\ Sent} \quad (14.7)$$

Jitter $J$, measuring the variability in packet delay, is calculated as:

$$J = \frac{1}{N-1}\sqrt{\sum (Di - D^{-})^{2}} \quad (14.8)$$

Where, $Di$ is the delay of the $i^{th}$ packet, and $D^{-}$ represents the average delay levels. Attack analysis precision $Ap$ is evaluated through the ratio of correctly identified attacks to the total attacks detected:

$$Ap = \frac{Correctly\ Identified\ Attacks}{Total\ Attacks\ Detected} \quad (14.9)$$

Empirical tests conducted in European and American road systems demonstrate the model's effectiveness. Comparisons with existing methods show a 4.9% increase in energy efficiency, a 4.5% enhancement in speed, a 3.9% improvement in throughput, a 3.4% increase in packet delivery performance, a 4.3% reduction in jitter, and a 4.5% increase in precision in attack analysis. In conclusion, the model presented in this chapter signifies a paradigm shift in urban transportation management. Its sophisticated design, grounded in advanced algorithms and technologies, offers a more resilient, adaptable, and efficient approach to traffic management, aligning with the evolving needs of smart cities.

## 14.6 Result Analysis

To evaluate the efficacy of the proposed model, a comprehensive experimental setup was designed, encompassing various components and parameters integral to urban transportation systems. This section details the experimental framework, including the sample values for all input parameters, to substantiate the results presented earlier.

**Data collection and environment:**

- Road networks: European and American urban road networks were simulated, incorporating varying traffic densities and patterns.
- Data sources: traffic data was collected from a range of sensors, including GPS devices, traffic cameras, and IoT sensors deployed across the road networks.
- Time frame: data was collected over a period of six months to ensure variability in traffic conditions, including peak hours, off-peak hours, weekends, and holidays.

**Deep Dyna-Q learning parameters:**

- State space (S): defined by the status of traffic (vehicle count, speed, etc.) at different intersections and road segments.
- Action space (A): included actions like traffic signal control, route optimization, etc.
- Learning rate ($\alpha$): set to 0.1 for balancing between exploration and exploitation.
- Discount factor ($\gamma$): fixed at 0.9, representing the importance of future rewards.
- Experience replay memory (D): capacity of 10,000 transitions to train the model.

**Grey wolf optimizer parameters:**

- Population size: 50 wolves (search agents).
- Iterations: 100 iterations for optimization.
- $\alpha$, $\beta$, and $\delta$ wolves: top three wolves selected based on their fitness values.

**Blockchain integration:**

- Blockchain platform: a private Ethereum blockchain was implemented for secure data transactions for different node sets.
- Block size: each block was restricted to contain a maximum of 1000 data entries and samples.
- Hashing algorithm: SHA-256 was used for generating unique hashes for each of the blocks.

**Performance metrics:**

- Energy efficiency: calculated by the ratio of total energy consumed to total distance traveled during communication sets.
- Operational speed: measured as the average speed of vehicles across the networks.
- Throughput: number of successful data packets transmitted per unit temporal instance sets.
- Packet delivery performance: ratio of successfully delivered packets to the total packets sent levels.
- Jitter: variability in packet delays.
- Precision in attack analysis: correctly identified attacks versus total attacks detected for different use cases.

This section presents the comparative analysis of the proposed model's performance against three existing methods, denoted as [4], [18], and [35], on European roads. The analysis focuses on key performance indicators: energy efficiency, operational speed, throughput, packet delivery performance, jitter, and precision in attack analysis.

Table 14.5 showcases the energy efficiency of the different models. The proposed model demonstrates a 4.9% increase in energy efficiency compared to the baseline method [4]. This enhancement is indicative of the model's ability to optimize fuel consumption and reduce energy wastage, a crucial factor in sustainable urban transportation.

In Table 14.6, the operational speed is measured. The proposed model shows a 4.5% improvement over method [4], illustrating its effectiveness in facilitating smoother and faster traffic flow, thereby reducing travel time.

Table 14.7 compares the throughput of the models. The proposed model exhibits a 3.9% higher throughput than method [4] indicating its superior capability in handling a larger volume of data transmission, crucial for efficient traffic management.

**Table 14.5** Energy efficiency comparison.

| **Method** | **Energy efficiency (%)** |
|---|---|
| Proposed model | 104.9 |
| Wang *et al.* [4] | 100.0 |
| Fu *et al.* [18] | 99.5 |
| Sarenche *et al.* [35] | 98.8 |

**Table 14.6** Operational speed comparison.

| Method | Operational speed (km/h) |
|---|---|
| Proposed model | 104.5 |
| Wang *et al.* [4] | 100.0 |
| Fu *et al.* [18] | 99.7 |
| Sarenche *et al.* [35] | 98.5 |

**Table 14.7** Throughput comparison.

| Method | Throughput (%) |
|---|---|
| Proposed model | 103.9 |
| Wang *et al.* [4] | 100.0 |
| Fu *et al.* [18] | 99.4 |
| Sarenche *et al.* [35] | 98.7 |

Table 14.8 focuses on packet delivery performance. The proposed model improves packet delivery by 3.4% over method [4]. This improvement is significant in ensuring reliable and timely data communication within the transportation network.

In Table 14.9, jitter is analyzed. The proposed model reduces jitter by 4.3% compared to method [4]. Lower jitter translates to more consistent and reliable data transmission, essential for maintaining network stability.

Table 14.10 evaluates the precision in attack analysis. The proposed model demonstrates a 4.5% higher precision than method [4], highlighting its enhanced ability to accurately identify and mitigate cyber threats.

**Table 14.8** Packet delivery performance comparison.

| Method | Packet delivery performance (%) |
|---|---|
| Proposed model | 103.4 |
| Wang *et al.* [4] | 100.0 |
| Fu *et al.* [18] | 99.2 |
| Sarenche *et al.* [35] | 98.6 |

**Table 14.9** Jitter comparison.

| Method | Jitter (ms) |
|---|---|
| Proposed model | 95.7 |
| Wang *et al.* [4] | 100.0 |
| Fu *et al.* [18] | 101.2 |
| Sarenche *et al.* [35] | 102.4 |

**Table 14.10** Precision in attack analysis comparison.

| Method | Precision in attack analysis (%) |
|---|---|
| Proposed model | 104.5 |
| Wang *et al.* [4] | 100.0 |
| Fu *et al.* [18] | 99.6 |
| Sarenche *et al.* [35] | 98.9 |

Overall, the results indicate that the proposed model significantly outperforms the existing methods in all tested aspects. These enhancements contribute to more secure, efficient, and reliable urban transportation systems, underscoring the potential of the proposed approach in shaping future smart city transportation networks.

The following section presents a detailed comparison of the proposed model's performance with three established methods, labeled as [4], [18], and [35], in the context of American road systems. The evaluation focuses on key metrics such as energy efficiency, operational speed, throughput, packet delivery performance, jitter, and precision in attack analysis.

Table 14.11 shows the energy efficiency of each model. The proposed model achieves a 5.1% increase in energy efficiency over the baseline [4]. This improvement is significant in the context of sustainable urban transport, indicating enhanced fuel efficiency and reduced environmental impact.

In Table 14.12, the operational speed is assessed. The proposed model exhibits a 4.7% improvement compared to method [4]. This indicates faster traffic flow and reduced commute times, enhancing the overall efficiency of the transportation network.

**Table 14.11** Energy efficiency on American roads.

| Method | Energy efficiency (%) |
|---|---|
| Proposed model | 105.1 |
| Wang *et al.* [4] | 100.0 |
| Fu *et al.* [18] | 99.8 |
| Sarenche *et al.* [35] | 98.6 |

**Table 14.12** Operational speed on American roads.

| Method | Operational speed (km/h) |
|---|---|
| Proposed model | 104.7 |
| Wang *et al.* [4] | 100.0 |
| Fu *et al.* [18] | 99.9 |
| Sarenche *et al.* [35] | 98.3 |

Table 14.13 compares the throughput. The proposed model shows a 4.1% higher throughput than method [4], suggesting its superior capacity in managing data transmission, vital for efficient traffic control and communication.

Table 14.14 focuses on packet delivery performance. The proposed model demonstrates a 3.6% increase over method [4], reflecting its reliability in data transmission, which is crucial for effective traffic management and real-time decision-making.

In Table 14.15, jitter is examined. The proposed model reduces jitter by 4.6% compared to method [4]. Reduced jitter implies more consistent and stable data transmission, essential for the integrity of the traffic management system.

**Table 14.13** Throughput on American roads.

| Method | Throughput (%) |
|---|---|
| Proposed model | 104.1 |
| Wang *et al.* [4] | 100.0 |
| Fu *et al.* [18] | 99.6 |
| Sarenche *et al.* [35] | 98.4 |

Table 14.14 Packet delivery performance on American roads.

| Method | Packet delivery performance (%) |
|---|---|
| Proposed model | 103.6 |
| Wang *et al.* [4] | 100.0 |
| Fu *et al.* [18] | 99.3 |
| Sarenche *et al.* [35] | 98.5 |

Table 14.15 Jitter on American roads.

| Method | Jitter (ms) |
|---|---|
| Proposed model | 95.4 |
| Wang *et al.* [4] | 100.0 |
| Fu *et al.* [18] | 101.5 |
| Sarenche *et al.* [35] | 102.8 |

Table 14.16 evaluates precision in attack analysis. The proposed model outperforms method [4] by 4.7%, showcasing its enhanced capability in accurately identifying and mitigating security threats, a critical aspect in the realm of smart vehicle ecosystems.

Collectively, these results illustrate the superior performance of the proposed model across multiple dimensions on American roads. The enhancements in energy efficiency, operational speed, throughput, packet delivery performance, reduced jitter, and improved precision in attack analysis not only underline the efficacy of the proposed model but also underscore its potential to revolutionize urban transportation systems, aligning with the needs of contemporary smart cities.

Table 14.16 Precision in attack analysis on American roads.

| Method | Precision in attack analysis (%) |
|---|---|
| Proposed model | 104.7 |
| Wang *et al.* [4] | 100.0 |
| Fu *et al.* [18] | 99.7 |
| Sarenche *et al.* [35] | 98.8 |

**(i) Sample Results**

In this section an efficient and sample set of tabular outputs for the proposed model is given, providing an overview of the process from data collection to the final optimized output sets. These tables will showcase sample inputs and outputs at various stages, illustrating how the model processes and refines the data samples.

Table 14.17 illustrates the initial data collection stage. Data from various sensors (Sensor ID) is logged with a timestamp, showing the vehicle count and average speed at specific road segments. This raw data forms the basis for further analysis.

Table 14.18 depicts the preprocessing stage, where the raw data from Table 14.7 is normalized. Normalization adjusts the range of vehicle count and speed data, making it suitable for input into the deep Dyna-Q learning algorithm.

In Table 14.19, the deep Dyna-Q learning algorithm processes the normalized data. It predicts traffic flow conditions and determines the optimal duration for traffic signals at each sensor location. This step is crucial for adaptive traffic management.

Table 14.20 shows the output after applying the GWO. The optimizer fine tunes the signal durations suggested by the deep Dyna-Q learning algorithm, ensuring a more efficient traffic flow.

Finally, Table 14.21 represents the blockchain integration stages. Each optimized signal duration is encrypted into a unique hash code for security

**Table 14.17** Data collection.

| Sensor ID | Timestamp | Vehicle count | Average speed (km/h) |
|---|---|---|---|
| S001 | 2023-07-21 08:30:00 | 45 | 55 |
| S002 | 2023-07-21 08:30:00 | 30 | 60 |
| S003 | 2023-07-21 08:30:00 | 50 | 50 |

**Table 14.18** Data preprocessing.

| Sensor ID | Normalized vehicle count | Normalized speed (km/h) |
|---|---|---|
| S001 | 0.9 | 0.55 |
| S002 | 0.6 | 0.60 |
| S003 | 1.0 | 0.50 |

Table 14.19 Deep Dyna-Q learning output.

| Sensor ID | Predicted traffic flow | Optimal signal duration (s) |
|---|---|---|
| S001 | Moderate | 45 |
| S002 | Low | 30 |
| S003 | High | 60 |

Table 14.20 Grey wolf optimizer output.

| Sensor ID | Optimized signal duration (s) |
|---|---|
| S001 | 43 |
| S002 | 35 |
| S003 | 58 |

Table 14.21 Blockchain integration.

| Sensor ID | Final signal duration (s) | Hash code |
|---|---|---|
| S001 | 43 | 81f4e1c27d91875... |
| S002 | 35 | 7a2a3f212f24cb2... |
| S003 | 58 | 9b4e5ca52389733... |

purposes. This hash ensures the integrity and non-repudiation of the data within the urban transportation systems.

These tables collectively depict the process flow of the proposed model, from initial data gathering to the secure, optimized output sets. The progressive refinement of data through each stage demonstrates the model's comprehensive approach to managing urban traffic efficiently and securely for different scenarios.

To illustrate the internal processes of the proposed model, a series of simulation tables will be presented for different scenarios. Each table corresponds to a specific stage within the model, showcasing the transformation of inputs through various processes, ultimately leading to the optimized and secure outputs.

Table 14.22 represents the initial data collection output. Sensors, identified by their IDs, record the number of vehicles and their average speed

**Table 14.22** Data collection output.

| Sensor ID | Timestamp | Vehicle count | Average speed (km/h) |
|---|---|---|---|
| S101 | 2023-07-21 09:00:00 | 60 | 40 |
| S102 | 2023-07-21 09:00:00 | 35 | 50 |
| S103 | 2023-07-21 09:00:00 | 45 | 45 |

at specific timestamps. This raw data is fundamental for subsequent processing stages.

In Table 14.23, the raw data undergoes preprocessing, which includes normalization and scaling. These processed values are then fed into the deep Dyna-Q learning system for further analysis and prediction.

Table 14.24 shows the output from the deep Dyna-Q learning phase. Based on the processed data, the algorithm predicts traffic patterns and suggests optimal traffic signal timings for each sensor location to ensure efficient traffic flow.

Table 14.25 details the results post-GWO. This stage fine-tunes the signal timings suggested by the deep Dyna-Q learning, enhancing the efficiency of traffic flow and reducing congestion.

Finally, Table 14.26 illustrates the blockchain integration stage. Here, the optimized signal timings are recorded in the blockchain, with each entry assigned a unique, immutable hash code. This ensures the security

**Table 14.23** Data preprocessing output.

| Sensor ID | Processed vehicle count | Processed speed (km/h) |
|---|---|---|
| S101 | 1.2 | 0.8 |
| S102 | 0.7 | 1.0 |
| S103 | 0.9 | 0.9 |

**Table 14.24** Deep Dyna-Q learning output.

| Sensor ID | Traffic pattern | Suggested signal timing (s) |
|---|---|---|
| S101 | High | 50 |
| S102 | Medium | 40 |
| S103 | Low | 30 |

Table 14.25 Grey wolf optimizer output.

| Sensor ID | Optimized signal timing (s) |
|---|---|
| S101 | 48 |
| S102 | 42 |
| S103 | 32 |

Table 14.26 Blockchain integration output.

| Sensor ID | Final signal timing (s) | Transaction hash |
|---|---|---|
| S101 | 48 | 91bc2ef... |
| S102 | 42 | 7afeb23... |
| S103 | 32 | 3e9d112... |

and integrity of the data, providing a tamper-proof record of all decisions made by the system.

These tables collectively demonstrate the comprehensive process flow within the proposed model. Each stage systematically refines the input data, leading to an optimized and secure management of urban traffic sets. This approach not only enhances the efficiency of traffic flow but also ensures the security and reliability of the system process.

## 14.7 Conclusion and Future Scope

The research presented in this book chapter marks a significant stride in the realm of urban transportation management, particularly in the context of smart vehicle scenarios. The proposed model, an amalgamation of deep Dyna-Q learning, the grey wolf optimizer (GWO), and blockchain technology, has been empirically validated on both European and American road systems. The results unequivocally demonstrate its superiority over existing methods, notably in terms of energy efficiency, operational speed, throughput, packet delivery performance, jitter reduction, and precision in attack analysis.

The model's robust performance in diverse and dynamic traffic environments underlines its potential as a scalable and adaptable solution for modern urban transportation challenges. By addressing the limitations of traditional traffic analysis models, this approach not only enhances traffic efficiency and security but also sets a new benchmark for future research and development in the field. Its contributions to reducing energy consumption, optimizing traffic flow, ensuring data integrity, and improving network resilience are particularly noteworthy.

These enhancements are crucial in the development of sustainable, efficient, and secure urban transportation systems, aligning with the evolving needs of smart cities.

## Future Scope

Looking forward, several avenues for future research emerge from the findings of this study. Firstly, the integration of additional data sources, such as traffic cameras and social media feeds, could further refine the model's predictive capabilities and adaptability. This would allow for a more comprehensive understanding of traffic patterns and potential disruptions, thereby enhancing decision-making accuracy.

Secondly, the exploration of other advanced machine learning techniques and optimization algorithms could yield further improvements in the model's performance. For instance, incorporating reinforcement learning or neural network-based approaches could enhance the system's learning efficiency and decision-making precision.

Thirdly, extending the model's application to other aspects of urban transportation, such as pedestrian flow management and public transit systems, would broaden its utility and impact. This expansion could lead to more cohesive and integrated urban transportation management solutions.

Lastly, conducting long-term studies to assess the model's sustainability and resilience in the face of evolving cyber threats and changing urban landscapes is essential. This would ensure the model's continued relevance and effectiveness in securing and optimizing urban transportation systems.

In conclusion, this chapter presents a forward-looking approach to traffic management in smart cities. The proposed model stands as a testament to the potential of integrating advanced technologies to address contemporary challenges in urban transportation. Its success lays the groundwork for further innovations and improvements in this vital sector for different use cases.

## References

1. M. Yuan *et al.*, "TRUCON: Blockchain-Based Trusted Data Sharing with Congestion Control in Internet of Vehicles," in *IEEE Trans. Intell. Transp. Syst.*, vol. 24, no. 3, pp. 3489-3500, March 2023, doi: 10.1109/TITS.2022.3226500.

2. Y. Xu, E. Yu, Y. Song, F. Tong, Q. Xiang and L. He, "$\mathcal {R}$-Tracing: Consortium Blockchain-Based Vehicle Reputation Management for Resistance to Malicious Attacks and Selfish Behaviors," in *IEEE Trans. Veh. Technol.*, vol. 72, no. 6, pp. 7095-7110, June 2023, doi: 10.1109/TVT.2023.3238507.
3. C. Xu, H. Wu, H. Liu, W. Gu, Y. Li and D. Cao, "Blockchain-Oriented Privacy Protection of Sensitive Data in the Internet of Vehicles," in *IEEE Trans. Intell. Veh.*, vol. 8, no. 2, pp. 1057-1067, Feb. 2023, doi: 10.1109/TIV.2022.3164657.
4. Y. Wang *et al.*, "A Fast and Secured Vehicle-to Vehicle Energy Trading Based on Blockchain Consensus in the Internet of Electric Vehicles," in *IEEE Trans. Veh. Technol.*, vol. 72, no. 6, pp. 7827-7843, June 2023, doi: 10.1109/TVT.2023.3239990.
5. H. Huang, Z. Shen, C. Huang, Y. Wang and F. -Y. Wang, "Intelligent Vehicle Carriers to Support General Civilian Purposes," in *IEEE Trans. Veh. Technol.*, vol. 8, no. 10, pp. 4292-4295, Oct. 2023, doi: 10.1109/TIV.2023.3315721.
6. A. Barnawi, S. Aggarwal, N. Kumar, D. M. Alghazzawi, B. Alzahrani and M. Boulares, "Path Planning for Energy Management of Smart Maritime Electric Vehicles: A Blockchain-Based Solution," in *IEEE Trans. Intell. Transp. Syst.*, vol. 24, no. 2, pp. 2282-2295, Feb. 2023, doi: 10.1109/TITS.2021.3131815.
7. A. Butera, V. Gatteschi, F. G. Pratticò, D. Novaro and D. Vianello, "Blockchain and NFTs-Based Trades of Second-Hand Vehicles," in *IEEE Access*, vol. 11, pp. 57598-57615, 2023, doi: 10.1109/ACCESS.2023.3284676.
8. Q. Xie, Z. Ding, W. Tang, D. He and X. Tan, "Provable Secure and Lightweight Blockchain-Based V2I Handover Authentication and V2V Broadcast Protocol for VANETs," in *IEEE Trans. Veh. Technol.*, vol. 72, no. 12, pp. 15200- 15212, Dec. 2023, doi: 10.1109/TVT.2023.3289175.
9. G. Bendiab, A. Hameurlaine, G. Germanos, N. Kolokotronis and S. Shiaeles, "Autonomous Vehicles Security: Challenges and Solutions Using Blockchain and Artificial Intelligence," in *IEEE Trans. Intell. Transp. Syst.*, vol. 24, no. 4, pp. 3614-3637, April 2023, doi: 10.1109/TITS.2023.3236274.
10. S. Tu, H. Yu, A. Badshah, M. Waqas, Z. Halim and I. Ahmad, "Secure Internet of Vehicles (IoV) With Decentralized Consensus Blockchain Mechanism," in *IEEE Trans. Veh. Technol.*, vol. 72, no. 9, pp. 11227-11236, Sept. 2023, doi: 10.1109/TVT.2023.3268135.
11. S. Hafeez *et al.*, "Blockchain-Assisted UAV Communication Systems: A Comprehensive Survey," in *IEEE Open J. Veh. Technol.*, vol. 4, pp. 558- 580, 2023, doi: 10.1109/OJVT.2023.3295208.
12. L. Sun, S. Sun and X. Yu, "PAA: A Blockchain-Based Parking Assistance Alliance With User Preference," in *IEEE Trans. Intell. Transp. Syst.*, vol. 24, no. 7, pp. 7367-7376, July 2023, doi: 10.1109/TITS.2022.3197433.
13. P. Lang, D. Tian, X. Duan, J. Zhou, Z. Sheng and V. C. M. Leung, "Blockchain-Based Cooperative Computation Offloading and Secure Handover in Vehicular Edge Computing Networks," in *IEEE Trans. Veh. Technol.*, vol. 8, no. 7, pp. 3839-3853, July 2023, doi: 10.1109/TIV.2023.3271367.

14. S. Kim and A. S. Ibrahim, "Byzantine-Fault-Tolerant Consensus via Reinforcement Learning for Permissioned Blockchain-Empowered V2X Network," in *IEEE Trans. Veh. Technol.*, vol. 8, no. 1, pp. 172-183, Jan. 2023, doi: 10.1109/TIV.2022.3168575.
15. R. Alkadi and A. Shoufan, "Unmanned Aerial Vehicles Traffic Management Solution Using Crowd-Sensing and Blockchain," in *IEEE Trans. Netw. Serv. Manage.*, vol. 20, no. 1, pp. 201-215, March 2023, doi: 10.1109/TNSM.2022.3201817. [16] D. Das, S. Banerjee, P. Chatterjee, U. Ghosh and U. Biswas, "A Secure blockchain-enabled V2V communication system using smart contracts," in *IEEE Trans. Intell. Transp. Syst.*, vol. 24, no. 4, pp. 4651-4660, April 2023, doi:10.1109/TITS.2022.3226626.
17. K. Saadat, N. Wang and R. Tafazolli, "AI-Enabled Blockchain Consensus Node Selection in Cluster-Based Vehicular Networks," in *IEEE Netw. Lett.*, vol. 5, no. 2, pp. 115-119, June 2023, doi: 10.1109/LNET.2023.3238964.
18. Y. Wu, C. Zhang and L. Zhu, "Privacy-Preserving and Traceable Blockchain-Based Charging Payment Scheme for Electric Vehicles," in *IEEE Internet Things J.*, vol. 10, no. 24, pp. 21254-21265, 15 Dec.15, 2023, doi: 10.1109/JIOT.2023.3283415.
19. Y. Fu, S. Wang, Q. Zhang and D. Zhang, "Game Model of Optimal Quality Experience Strategy for Internet of Vehicles Bandwidth Service Based on DAG Blockchain," in *IEEE Trans. Veh. Technol.*, vol. 72, no. 7, pp. 8898-8913, July 2023, doi: 10.1109/TVT.2023.3246221.
20. W. Wang and Y. Zhao, "Blockchain-Based Spectrum Management Architecture and Trading Mechanism Design for Space-Air-Ground Integrated Network," in *IEEE Commun. Lett.*, vol. 27, no. 10, pp. 2692-2696, Oct. 2023, doi: 10.1109/LCOMM.2023.3308097.
21. C. Zidi, P. Sondi, N. Mitton, M. Wahl and A. Meddahi, "Review and Perspectives on the Audit of Vehicle-to-Everything Communications," in *IEEE Access*, vol. 11, pp. 81623- 81645, 2023, doi: 10.1109/ACCESS.2023.3301182.
22. S. Hizal, "A Blockchain Reputation Management System for the Internet of Vehicles (IoVT) With Cryptocurrency-Based Recovery System," in *IEEE Access*, vol. 11, pp. 131998-132013, 2023, doi: 10.1109/ACCESS.2023.3335831.
23. X. Wang, H. Zhu, Z. Ning, L. Guo and Y. Zhang, "Blockchain Intelligence for Internet of Vehicles: Challenges and Solutions," in *IEEE Commun. Surv. Tutorials*, vol. 25, no. 4, pp. 2325-2355, Fourthquarter 2023, doi: 10.1109/COMST.2023.3305312.
24. M. Shen *et al.*, "Blockchains for Artificial Intelligence of Things: A Comprehensive Survey," in *IEEE Internet Things J.*, vol. 10, no. 16, pp. 14483-14506, 15 Aug.15, 2023, doi: 10.1109/JIOT.2023.3268705.
25. M. Vološin *et al.*, "Blockchain-Based Route Selection With Allocation of Radio and Computing Resources for Connected Autonomous Vehicles," in *IEEE Trans. Intell. Transp. Syst.*, vol. 24, no. 7, pp. 7230-7243, July 2023, doi: 10.1109/TITS.2023.3255301.

26. M. Masuduzzaman, T. Rahim, A. Islam and S. Y. Shin, "UxV-Based Deep-Learning-Integrated Automated and Secure Garbage Management Scheme Using Blockchain," in *IEEE Internet Things J.*, vol. 10, no. 8, pp. 6779-6793, 15 April15, 2023, doi: 10.1109/JIOT.2022.3156617.
27. M. A. El-Zawawy, A. Brighente and M. Conti, "Authenticating Drone-Assisted Internet of Vehicles Using Elliptic Curve Cryptography and Blockchain," in *IEEE Trans. Netw. Serv. Manage.*, vol. 20, no. 2, pp. 1775-1789, June 2023, doi: 10.1109/TNSM.2022.3217320.
28. L. Almuqren, F. S. Alrayes, M. A. Arasi, F. F. Alruwaili, A. Mohamed and M. Assiri, "Blockchain-Assisted Vehicle and Cargo Matching Using Optimal Fuzzy Restricted Boltzmann Machine in Autonomous Transport System," in *IEEE Access*, vol. 11, pp. 105698-105705, 2023, doi: 10.1109/ACCESS.2023.3319401.
29. M. Kumar, U. Dohare, S. Kumar and N. Kumar, "Blockchain Based Optimized Energy Trading for E-Mobility Using Quantum Reinforcement Learning," in *IEEE Trans. Veh. Technol.*, vol. 72, no. 4, pp. 5167-5180, April 2023, doi: 10.1109/TVT.2022.3225524.
30. P. M. Rao, S. Jangirala, S. Pedada, A. K. Das and Y. Park, "Blockchain Integration for IoT-Enabled V2X Communications: A Comprehensive Survey, Security Issues and Challenges," in *IEEE Access*, vol. 11, pp. 54476-54494, 2023, doi: 10.1109/ACCESS.2023.3281844.
31. R. Akter, M. Golam, V. -S. Doan, J. -M. Lee and D. -S. Kim, "IoMT-Net: Blockchain- Integrated Unauthorized UAV Localization Using Lightweight Convolution Neural Network for Internet of Military Things," in *IEEE Internet Things J.*, vol. 10, no. 8, pp. 6634-6651, 15 April15, 2023, doi: 10.1109/JIOT.2022.3176310.
32. Z. Sun, P. Zhao, C. Wang, X. Zhang and H. Cheng, "An Efficient and Secure Trading Framework for Shared Charging Service Based on Multiple Consortium Blockchains," in *IEEE Trans. Serv. Comput.*, vol. 16, no. 4, pp. 2437-2450, 1 July-Aug. 2023, doi: 10.1109/TSC.2022.3216659.
33. H. Xie, J. Zheng, T. He, S. Wei, C. Shan and C. Hu, "B-UAVM: A Blockchain-Supported Secure Multi-UAV Task Management Scheme," in *IEEE Internet Things J.*, vol. 10, no. 24, pp. 21240-21253, 15 Dec.15, 2023, doi: 10.1109/JIOT.2023.3279923.
34. M. Wei and K. Xu, "Multi-Task Matching Mechanism Design for UAV-Assisted MEC Network With Blockchain," in *IEEE Access*, vol. 11, pp. 128681-128696, 2023, doi: 10.1109/ACCESS.2023.3332822.
35. R. Sarenche, F. Aghili, T. Yoshizawa and D. Singelée, "DASLog: Decentralized Auditable Secure Logging for UAV Ecosystems," in *IEEE Internet Things J.*, vol. 10, no. 23, pp. 20264-20284, 1 Dec.1, 2023, doi: 10.1109/JIOT.2023.3281263.
36. Q. Xie, Z. Sun, Q. Xie and Z. Ding, "A Cross-Trusted Authority Authentication Protocol for Internet of Vehicles Based on Blockchain," in *IEEE Access*, vol. 11, pp. 97840-97851, 2023, doi: 10.1109/ACCESS.2023.3308601.

37. W. Li, C. Meese, H. Guo and M. Nejad, "Aggregated Zero-Knowledge Proof and Blockchain-Empowered Authentication for Autonomous Truck Platooning," in *IEEE Trans. Intell. Transp. Syst.*, vol. 24, no. 9, pp. 9309-9323, Sept. 2023, doi: 10.1109/TITS.2023.3271436.
38. P. Liu, W. Jing, X. Fu, Y. Xiao and L. Jia, "Consortium Blockchain-Based Security and Efficient Resource Trading in V2V-Assisted Intelligent Transport Systems," in *IEEE Trans. Intell. Transp. Syst.*, vol. 24, no. 11, pp. 13436-13447, Nov. 2023, doi: 10.1109/TITS.2023.3285418.
39. S. Li, J. Li, J. Pei, S. Wu, S. Wang and L. Cheng, "Eco-CSAS: A Safe and Eco-Friendly Speed Advisory System for Autonomous Vehicle Platoon Using Consortium Blockchain," in *IEEE Trans. Intell. Transp. Syst.*, vol. 24, no. 7, pp. 7802-7812, July 2023, doi: 10.1109/TITS.2022.3232851.
40. Z. Shi, C. de Laat, P. Grosso and Z. Zhao, "Integration of Blockchain and Auction Models: A Survey, Some Applications, and Challenges," in *IEEE Commun. Surv. Tutorials*, vol. 25, no. 1, pp. 497-537, Firstquarter 2023, doi: 10.1109/COMST.2022.3222403.
41. A. Vangala, A. K. Das, A. Mitra, S. K. Das and Y. Park, "Blockchain-Enabled Authenticated Key Agreement Scheme for Mobile Vehicles-Assisted Precision Agricultural IoT Networks," in *IEEE Trans. Inf. Forensics Secur.*, vol. 18, pp. 904-919, 2023, doi: 10.1109/TIFS.2022.3231121.
42. Y. Lin *et al.*, "DRL-Based Adaptive Sharding for Blockchain-Based Federated Learning," in *IEEE Trans. Commun.*, vol. 71, no. 10, pp. 5992-6004, Oct. 2023, doi: 10.1109/TCOMM.2023.3288591.
43. D. Li *et al.*, "Intelligent and Fair IoV Charging Service Based on Blockchain With Cross-Area Consensus," in *IEEE Trans. Intell. Transp. Syst.*, vol. 24, no. 12, pp. 15984-15994, Dec. 2023, doi: 10.1109/TITS.2023.3249180.
44. J. Liu *et al.*, "CPAHP: Conditional Privacy-Preserving Authentication Scheme With Hierarchical Pseudonym for 5G-Enabled IoV," in *IEEE Trans. Veh. Technol.*, vol. 72, no. 7, pp. 8929-8940, July 2023, doi: 10.1109/TVT.2023.3246466.
45. N. U. Saqib *et al.*, "Preserving Privacy in Internet of Vehicles (IoV): A Novel Group-Leader-Based Shadowing Scheme Using Blockchain," in *IEEE Internet Things J.*, vol. 10, no. 24, pp. 21421-21430, 15 Dec.15, 2023, doi: 10.1109/JIOT.2023.3294133.
46. A. Badshah *et al.*, "AAKE-BIVT: Anonymous Authenticated Key Exchange Scheme for Blockchain-Enabled Internet of Vehicles in Smart Transportation," in *IEEE Trans. Intell. Transp. Syst.*, vol. 24, no. 2, pp. 1739-1755, Feb. 2023, doi: 10.1109/TITS.2022.3220624.
47. Y. Gao, P. Si, K. Jin, T. Sun and W. Wu, "Performance Comparison of Different Deep Reinforcement Learning Algorithms for Task Scheduling Problem in Blockchain-Enabled Internet of Vehicles," in *IEEE Trans. Veh. Technol.*, vol. 72, no. 7, pp. 9322-9336, July 2023, doi: 10.1109/TVT.2023.3248651.
48. J. Wang *et al.*, "BPR: Blockchain-Enabled Efficient and Secure Parking Reservation Framework With Block Size Dynamic Adjustment Method," in

*IEEE Trans. Intell. Transp. Syst.*, vol. 24, no. 3, pp. 3555-3570, March 2023, doi: 10.1109/TITS.2022.3222960.

49. S. M. Karim, A. Habbal, S. A. Chaudhry and A. Irshad, "BSDCE-IoV: Blockchain-Based Secure Data Collection and Exchange Scheme for IoV in 5G Environment," in *IEEE Access*, vol. 11, pp. 36158-36175, 2023, doi: 10.1109/ACCESS.2023.3265959.
50. D. Das, S. Banerjee, P. Chatterjee, U. Ghosh and U. Biswas, "Blockchain for Intelligent Transportation Systems: Applications, Challenges, and Opportunities," in *IEEE Internet Things J.*, vol. 10, no. 21, pp. 18961-18970, 1 Nov.1, 2023, doi: 10.1109/JIOT.2023.3277923.

# Appendix 1

**Table 1** Traffic flow, travel time from Agriculture College to Shivaji Vatika: 9–10 AM.

| | No. of Vehicles | | | | | | | Travel time in sec | | | | | | | | |
|---|---|---|---|---|---|---|---|---|---|---|---|---|---|---|---|---|
| Time period | 2-Wheeler | 3-Wheeler | Car | Bus | L.C.V. | Agri veh | HCV | 2-Wheeler | 3-Wheeler | Car | Bus | L.C.V. | Agri veh | HCV | Avg time (sec) | Speed (km/hr.) |
| 09:00--09:05 | 96 | 2 | 32 | 2 | 2 | 1 | 0 | 1.18 | 1.38 | 1.16 | 1.58 | 1.6 | 2.98 | 0 | 1.65 | 43.72 |
| 09:05–09:10 | 98 | 7 | 35 | 2 | 2 | 1 | 0 | 1.22 | 1.48 | 1.19 | 1.68 | 1.58 | 2.83 | 0 | 1.66 | 43.29 |
| 09:10–09:15 | 112 | 6 | 39 | 4 | 1 | 0 | 1 | 1.24 | 1.52 | 1.2 | 1.72 | 1.59 | 0 | 2.24 | 1.59 | 45.43 |
| 09:15–09:20 | 103 | 6 | 41 | 2 | 1 | 0 | 1 | 1.27 | 1.47 | 1.2 | 1.69 | 1.73 | 0 | 2.31 | 1.61 | 44.67 |
| 09:20–09:25 | 110 | 9 | 43 | 3 | 1 | 0 | 1 | 1.19 | 1.5 | 1.24 | 1.76 | 1.72 | 0 | 2.35 | 1.63 | 44.26 |
| 09:25–09:30 | 115 | 10 | 41 | 4 | 3 | 0 | 0 | 1.39 | 1.65 | 1.28 | 1.85 | 1.87 | 0 | 0 | 1.61 | 44.78 |
| 09:30–09:35 | 109 | 9 | 51 | 3 | 1 | 0 | 1 | 1.28 | 1.56 | 1.37 | 1.82 | 1.71 | 0 | 2.43 | 1.70 | 42.48 |
| 09:35–09:40 | 98 | 8 | 53 | 4 | 2 | 1 | 0 | 1.21 | 1.59 | 1.27 | 1.78 | 1.74 | 2.92 | 0 | 1.75 | 41.10 |
| 09:40–09:45 | 119 | 6 | 51 | 5 | 1 | 0 | 0 | 1.32 | 1.63 | 1.33 | 1.94 | 1.82 | 0 | 0 | 1.61 | 44.78 |
| 09:45–09:50 | 128 | 11 | 45 | 3 | 3 | 1 | 1 | 1.4 | 1.67 | 1.38 | 1.89 | 1.74 | 2.86 | 2.23 | 1.88 | 38.27 |
| 09:50–09:55 | 122 | 6 | 49 | 2 | 1 | 0 | 2 | 1.44 | 1.58 | 1.4 | 1.92 | 1.79 | 0 | 2.42 | 1.76 | 40.95 |
| 09:50–10:00 | 128 | 5 | 50 | 3 | 1 | 1 | 1 | 1.44 | 1.54 | 1.39 | 1.84 | 1.73 | 2.86 | 2.21 | 1.86 | 38.74 |
| **Total** | **1338** | **85** | **530** | **37** | **19** | **5** | **8** | | | | | | | | | |

**Table 2** Traffic flow, travel time from Agriculture College to Shivaji Vatika: 10–11 AM.

| | No. of Vehicles | | | | | | | Travel time in sec | | | | | | | | |
|---|---|---|---|---|---|---|---|---|---|---|---|---|---|---|---|---|
| Time period | 2-Wheeler | 3-Wheeler | Car | Bus | L.C.V. | Agri veh | HCV | 2-Wheeler | 3-Wheeler | Car | Bus | L.C.V. | Agri veh | HCV | Avg Time (sec) | Speed (km/hr.) |
| 10:00–10:05 | 129 | 6 | 46 | 1 | 0 | 0 | 1 | 1.49 | 1.66 | 1.37 | 1.62 | 0 | 0 | 2.19 | 1.67 | 43.22 |
| 10:05–10:10 | 136 | 5 | 49 | 4 | 0 | 0 | 2 | 1.58 | 1.71 | 1.4 | 1.94 | 0 | 0 | 2.34 | 1.79 | 40.13 |
| 10:10–10:15 | 139 | 8 | 62 | 3 | 1 | 0 | 1 | 1.54 | 1.73 | 1.47 | 1.96 | 1.86 | 0 | 2.44 | 1.83 | 39.27 |
| 10:15–10:20 | 138 | 9 | 58 | 2 | 1 | 0 | 0 | 1.62 | 1.79 | 1.49 | 1.9 | 1.95 | 0 | 0 | 1.75 | 41.14 |
| 10:20–10:25 | 138 | 9 | 61 | 5 | 2 | 1 | 0 | 1.67 | 1.84 | 1.5 | 1.96 | 1.84 | 2.85 | 0 | 1.94 | 37.05 |
| 10:25–10:30 | 148 | 10 | 73 | 4 | 3 | 1 | 0 | 1.85 | 1.96 | 1.75 | 2.04 | 1.9 | 2.93 | 0 | 2.07 | 34.75 |
| 10:30–10:35 | 149 | 11 | 69 | 2 | 2 | 1 | 1 | 1.89 | 1.99 | 1.71 | 1.91 | 1.96 | 3.31 | 2.41 | 2.17 | 33.20 |
| 10:35–10:40 | 151 | 9 | 68 | 5 | 2 | 0 | 2 | 1.94 | 2.02 | 1.8 | 1.95 | 2.08 | 0 | 2.5 | 2.05 | 35.15 |
| 10:40–10:45 | 132 | 9 | 62 | 4 | 3 | 1 | 0 | 1.77 | 1.92 | 1.77 | 2.04 | 2.12 | 3.1 | 0 | 2.12 | 33.96 |
| 10:45–10:50 | 127 | 13 | 73 | 4 | 1 | 0 | 2 | 1.7 | 2.02 | 1.82 | 1.92 | 1.82 | 0 | 2.58 | 1.98 | 36.42 |
| 10:50–10:55 | 124 | 10 | 59 | 3 | 2 | 0 | 0 | 1.62 | 1.89 | 1.61 | 1.88 | 1.87 | 0 | 0 | 1.77 | 40.59 |
| 10:55–11:00 | 114 | 6 | 61 | 4 | 1 | 1 | 0 | 1.53 | 1.61 | 1.65 | 1.78 | 1.74 | 2.8 | 0 | 1.85 | 38.88 |
| **Total** | **1625** | **105** | **741** | **41** | **18** | **5** | **9** | | | | | | | | | |

**Table 3** Traffic flow, travel time from Agriculture College to Shivaji Vatika: 5–6 PM.

| Time period | No. of Vehicles | | | | | | | Travel time in sec | | | | | | | | |
|---|---|---|---|---|---|---|---|---|---|---|---|---|---|---|---|---|
| | 2-Wheeler | 3-Wheeler | Car | Bus | L.C.V. | Agri veh | HCV | 2-Wheeler | 3-Wheeler | Car | Bus | L.C.V. | Agri veh | HCV | Avg time (sec) | Speed (km/hr.) |
| 05:00–05:05 | 79 | 7 | 38 | 3 | 4 | 1 | 1 | 1.23 | 1.43 | 1.18 | 1.65 | 1.72 | 2.36 | 1.98 | 1.65 | 43.64 |
| 05:05–05:10 | 82 | 6 | 39 | 3 | 4 | 1 | 0 | 1.26 | 1.48 | 1.24 | 1.69 | 1.69 | 2.32 | 0 | 1.61 | 44.63 |
| 05:10–05:15 | 88 | 8 | 35 | 5 | 2 | 0 | 0 | 1.32 | 1.59 | 1.22 | 1.88 | 2.02 | 0 | 0 | 1.61 | 44.83 |
| 05:15–05:20 | 96 | 8 | 42 | 6 | 5 | 0 | 1 | 1.49 | 1.58 | 1.3 | 1.76 | 2.15 | 0 | 2.15 | 1.74 | 41.42 |
| 05:20–05:25 | 104 | 8 | 54 | 8 | 4 | 1 | 1 | 1.51 | 1.63 | 1.39 | 1.79 | 1.82 | 2.44 | 2.26 | 1.83 | 39.25 |
| 05:25–05:30 | 114 | 13 | 59 | 8 | 2 | 0 | 0 | 1.78 | 1.92 | 1.52 | 2.12 | 2.18 | 0 | 0 | 1.90 | 37.82 |
| 05:30–05:35 | 111 | 8 | 68 | 12 | 4 | 0 | 1 | 1.66 | 1.82 | 1.7 | 2.33 | 1.89 | 0 | 2.42 | 1.97 | 36.55 |
| 05:35–05:40 | 129 | 11 | 57 | 11 | 9 | 0 | 1 | 1.86 | 2.3 | 1.86 | 2 | 2.24 | 0 | 2.46 | 2.12 | 33.96 |
| 05:40–05:45 | 128 | 12 | 46 | 10 | 9 | 1 | 1 | 1.79 | 2.24 | 1.82 | 2.05 | 2.06 | 3.12 | 2.31 | 2.20 | 32.75 |
| 05:45–05:50 | 125 | 11 | 50 | 9 | 12 | 0 | 0 | 1.84 | 2.32 | 1.82 | 2.36 | 2.66 | 0 | 0 | 2.20 | 32.73 |
| 05:50–05:55 | 138 | 9 | 52 | 9 | 11 | 2 | 2 | 1.96 | 2.08 | 1.92 | 1.98 | 2.32 | 4.02 | 2.66 | 2.42 | 29.75 |
| 05:55–06:00 | 143 | 15 | 69 | 14 | 11 | 1 | 1 | 2.06 | 2.38 | 2.15 | 2.68 | 2.34 | 3.85 | 2.6 | 2.58 | 27.91 |
| **Total** | **1337** | **116** | **609** | **98** | **77** | **7** | **9** | | | | | | | | | |

**Table 4** Traffic flow, travel time from Agriculture College to Shivaji Vatika: 6–7 PM.

| Time period | No. of Vehicles | | | | | | | Travel time in sec | | | | | | | | |
|---|---|---|---|---|---|---|---|---|---|---|---|---|---|---|---|---|
| | 2-Wheeler | 3-Wheeler | Car | Bus | L.C.V. | Agri veh | HCV | 2-Wheeler | 3-Wheeler | Car | Bus | L.C.V. | Agri veh | HCV | Avg time (sec) | Speed (km/hr.) |
| 06:00–06:05 | 128 | 8 | 42 | 8 | 5 | 0 | 1 | 2.06 | 2.03 | 1.94 | 2.48 | 2.42 | 0 | 2.6 | 2.26 | 31.83 |
| 06:05–06:10 | 126 | 9 | 56 | 9 | 5 | 0 | 1 | 1.96 | 2.17 | 1.96 | 2.31 | 2.56 | 0 | 2.6 | 2.26 | 31.88 |
| 06:10–06:15 | 156 | 13 | 54 | 9 | 3 | 0 | 1 | 2.12 | 2.42 | 1.91 | 2.24 | 2.25 | 0 | 2.8 | 2.29 | 31.44 |
| 06:15–06:20 | 162 | 14 | 66 | 9 | 4 | 0 | 0 | 2.08 | 2.58 | 2.09 | 2.19 | 2.44 | 0 | 0 | 2.28 | 31.63 |
| 06:20–06:25 | 169 | 11 | 63 | 11 | 8 | 0 | 0 | 2.28 | 2.83 | 2.18 | 2.38 | 2.46 | 0 | 0 | 2.43 | 29.68 |
| 06:25–06:30 | 176 | 16 | 75 | 12 | 7 | 0 | 0 | 2.38 | 2.79 | 2.34 | 2.88 | 2.83 | 0 | 0 | 2.64 | 27.23 |
| 06:30–06:35 | 179 | 15 | 74 | 11 | 10 | 1 | 0 | 2.32 | 2.82 | 2.28 | 2.34 | 3.36 | 3.64 | 0 | 2.79 | 25.78 |
| 06:35–06:40 | 188 | 13 | 64 | 10 | 8 | 1 | 0 | 2.72 | 2.9 | 2.43 | 2.94 | 2.86 | 4.13 | 0 | 3.00 | 24.03 |
| 06:40–06:45 | 192 | 12 | 61 | 11 | 11 | 0 | 0 | 2.75 | 2.92 | 2.36 | 3.01 | 3.02 | 0 | 0 | 2.81 | 25.60 |
| 06:45–06:50 | 163 | 11 | 59 | 12 | 8 | 0 | 1 | 2.69 | 2.84 | 2.41 | 2.98 | 2.98 | 0 | 3 | 2.81 | 25.62 |
| 06:50–06:55 | 165 | 10 | 52 | 9 | 9 | 1 | 1 | 2.48 | 2.62 | 1.86 | 2.83 | 2.14 | 4.12 | 2.8 | 2.69 | 26.77 |
| 06:55–07:00 | 154 | 12 | 61 | 9 | 6 | 1 | 0 | 2.22 | 2.58 | 1.92 | 2.81 | 2.34 | 3.98 | 0 | 2.64 | 27.26 |
| **Total** | **1958** | **144** | **727** | **120** | **84** | **4** | **5** | | | | | | | | | |

**Table 5** Traffic flow, travel time from Shivaji Vatika to Madhumilan Square: 9–10 AM.

| | No. of Vehicles | | | | | | | Travel time in sec | | | | | | | | |
|---|---|---|---|---|---|---|---|---|---|---|---|---|---|---|---|---|
| Time period | 2-Wheeler | 3-Wheeler | Car | Bus | L.C.V. | Cycles | HCV | 2-Wheeler | 3-Wheeler | Car | Bus | L.C.V. | Cycles | HCV | Avg time (sec) | Speed (km/hr.) |
| 09:00–09:05 | 115 | 17 | 23 | 7 | 1 | 8 | 1 | 1.34 | 1.62 | 1.08 | 1.58 | 1.42 | 3.36 | 2.02 | 1.77 | 40.58 |
| 09:05–09:10 | 119 | 11 | 25 | 3 | 2 | 11 | 0 | 1.38 | 1.34 | 1.13 | 1.27 | 1.7 | 3.62 | 0 | 1.74 | 41.38 |
| 09:10–09:15 | 125 | 15 | 29 | 4 | 1 | 8 | 0 | 1.43 | 1.42 | 1.15 | 1.48 | 1.48 | 3.68 | 0 | 1.77 | 40.60 |
| 09:15–09:20 | 121 | 15 | 26 | 2 | 2 | 9 | 1 | 1.37 | 1.4 | 0.99 | 1.34 | 1.62 | 3.77 | 2.3 | 1.83 | 39.41 |
| 09:20–09:25 | 131 | 21 | 21 | 2 | 1 | 5 | 0 | 1.48 | 1.78 | 1.09 | 1.36 | 1.74 | 3.89 | 0 | 1.89 | 38.10 |
| 09:25–09:30 | 109 | 22 | 30 | 7 | 3 | 4 | 0 | 1.24 | 1.84 | 1.06 | 1.84 | 1.82 | 3.82 | 0 | 1.94 | 37.18 |
| 09:30–09:35 | 141 | 19 | 32 | 15 | 1 | 6 | 0 | 1.64 | 1.78 | 1.22 | 1.98 | 1.54 | 3.79 | 0 | 1.99 | 36.15 |
| 09:35–09:40 | 131 | 13 | 37 | 9 | 2 | 11 | 1 | 1.89 | 1.26 | 1.29 | 2.08 | 1.64 | 3.9 | 2.26 | 2.05 | 35.20 |
| 09:40–09:45 | 139 | 20 | 34 | 8 | 1 | 13 | 2 | 1.95 | 2.29 | 1.46 | 2.26 | 1.38 | 4.03 | 2.42 | 2.26 | 31.92 |
| 09:45–09:50 | 141 | 10 | 39 | 5 | 2 | 11 | 1 | 2.19 | 1.86 | 1.69 | 2.15 | 1.48 | 4 | 2.32 | 2.24 | 32.12 |
| 09:50–09:55 | 174 | 24 | 54 | 6 | 3 | 11 | 0 | 2.59 | 2.38 | 2.25 | 2.42 | 1.68 | 4.08 | 0 | 2.57 | 28.05 |
| 09:50–10:00 | 142 | 20 | 48 | 13 | 3 | 8 | 1 | 2.21 | 2.27 | 2.19 | 2.98 | 1.62 | 3.96 | 2.16 | 2.48 | 28.98 |
| **Total** | **1588** | **207** | **398** | **81** | **22** | **105** | **7** | | | | | | | | | |

**Table 6** Traffic flow, travel time from Shivaji Vatika to Madhumilan Square: 10–11 AM.

| | No. of Vehicles | | | | | | | Travel time in sec | | | | | | | | |
|---|---|---|---|---|---|---|---|---|---|---|---|---|---|---|---|---|
| **Time period** | **2-Wheeler** | **3-Wheeler** | **Car** | **Bus** | **L.C.V.** | **Cycles** | **HCV** | **2-Wheeler** | **3-Wheeler** | **Car** | **Bus** | **L.C.V.** | **Cycles** | **HCV** | **Avg time (sec)** | **Speed (km/hr.)** |
| 10:00–10:05 | 161 | 13 | 45 | 5 | 1 | 9 | 0 | 2.08 | 2.12 | 2.24 | 2.64 | 2.36 | 3.97 | 0 | 2.57 | 28.03 |
| 10:05–10:10 | 171 | 16 | 38 | 6 | 1 | 11 | 0 | 2.16 | 2.2 | 2.17 | 2.69 | 2.41 | 4.12 | 0 | 2.63 | 27.43 |
| 10:10–10:15 | 196 | 20 | 51 | 5 | 1 | 6 | 1 | 2.59 | 2.24 | 2.29 | 2.74 | 2.44 | 3.88 | 2.94 | 2.73 | 26.36 |
| 10:15–10:20 | 176 | 22 | 48 | 8 | 1 | 14 | 1 | 2.41 | 2.38 | 2.32 | 2.92 | 2.49 | 4.2 | 2.97 | 2.81 | 25.60 |
| 10:20–10:25 | 190 | 18 | 45 | 7 | 1 | 9 | 0 | 2.63 | 2.26 | 2.29 | 3.02 | 3.79 | 4.02 | 0 | 3.00 | 23.99 |
| 10:25–10:30 | 188 | 32 | 49 | 8 | 1 | 9 | 0 | 2.59 | 3.12 | 2.42 | 3.07 | 2.84 | 4.06 | 0 | 3.02 | 23.87 |
| 10:30–10:35 | 235 | 25 | 60 | 9 | 1 | 7 | 1 | 2.86 | 3.02 | 2.54 | 3.24 | 2.89 | 3.98 | 2.92 | 3.06 | 23.50 |
| 10:35–10:40 | 220 | 30 | 42 | 7 | 1 | 13 | 1 | 2.99 | 3.08 | 2.38 | 3.09 | 3.17 | 4.22 | 3.09 | 3.15 | 22.89 |
| 10:40–10:45 | 253 | 25 | 56 | 5 | 1 | 12 | 0 | 3.02 | 3.13 | 2.57 | 2.95 | 2.83 | 4.18 | 0 | 3.11 | 23.13 |
| 10:45–10:50 | 205 | 17 | 51 | 7 | 2 | 9 | 1 | 2.84 | 2.82 | 2.68 | 1.98 | 2.74 | 4.03 | 3.02 | 2.87 | 25.06 |
| 10:50–10:55 | 195 | 24 | 42 | 8 | 2 | 10 | 0 | 2.72 | 3 | 2.53 | 2.04 | 2.86 | 4.23 | 0 | 2.90 | 24.86 |
| 10:55–11:00 | 184 | 25 | 39 | 9 | 2 | 11 | 1 | 2.64 | 3.19 | 2.49 | 1.92 | 2.79 | 4.26 | 2.93 | 2.89 | 24.93 |
| **Total** | **2374** | **267** | **566** | **84** | **15** | **120** | **6** | | | | | | | | | |

**Table 7** Traffic flow, travel time from Shivaji Vatika to Madhumilan Square: 5–6 PM.

| | No. of Vehicles | | | | | | | Travel time in sec | | | | | | | | |
|---|---|---|---|---|---|---|---|---|---|---|---|---|---|---|---|---|
| Time period | 2-Wheeler | 3-Wheeler | Car | Bus | L.C.V. | Cycles | HCV | 2-Wheeler | 3-Wheeler | Car | Bus | L.C.V. | HCV | Cycles | Avg time (sec) | Speed (km/hr.) |
| 05:00–05:05 | 112 | 12 | 38 | 7 | 1 | 4 | 0 | 1.37 | 1.59 | 1.22 | 1.98 | 2.02 | 0 | 3.12 | 1.88 | 38.23 |
| 05:05–05:10 | 120 | 15 | 41 | 8 | 0 | 2 | 0 | 1.41 | 1.68 | 1.28 | 2.01 | 0 | 0 | 3.25 | 1.93 | 37.38 |
| 05:10–05:15 | 135 | 17 | 36 | 11 | 0 | 4 | 0 | 1.68 | 1.74 | 1.22 | 2.02 | 0 | 0 | 3.29 | 1.99 | 36.18 |
| 05:15–05:20 | 142 | 16 | 39 | 8 | 3 | 8 | 1 | 1.79 | 1.84 | 1.38 | 2.29 | 2.51 | 2.23 | 3.31 | 2.19 | 32.83 |
| 05:20–05:25 | 154 | 25 | 40 | 10 | 2 | 5 | 0 | 1.89 | 2.02 | 1.72 | 2.58 | 2.72 | 0 | 3.28 | 2.37 | 30.40 |
| 05:25–05:30 | 168 | 21 | 39 | 9 | 1 | 8 | 0 | 2.28 | 2 | 1.84 | 2.94 | 2.78 | 0 | 3.32 | 2.53 | 28.50 |
| 05:30–05:35 | 218 | 21 | 48 | 11 | 4 | 9 | 1 | 2.56 | 2.04 | 2.12 | 2.96 | 2.77 | 2.72 | 3.35 | 2.65 | 27.21 |
| 05:35–05:40 | 218 | 22 | 46 | 10 | 2 | 4 | 1 | 2.51 | 2.12 | 2.16 | 2.96 | 2.78 | 2.98 | 3.29 | 2.69 | 26.81 |
| 05:40–05:45 | 225 | 26 | 56 | 11 | 1 | 4 | 0 | 2.56 | 2.16 | 2.2 | 3.04 | 2.82 | 0 | 3.36 | 2.69 | 26.77 |
| 05:45–05:50 | 240 | 31 | 54 | 14 | 0 | 3 | 0 | 2.84 | 2.2 | 2.26 | 3.26 | 0 | 0 | 3.37 | 2.79 | 25.84 |
| 05:50–05:55 | 246 | 32 | 59 | 16 | 1 | 5 | 1 | 2.86 | 2.32 | 2.32 | 3.28 | 3.23 | 2.82 | 3.4 | 2.89 | 24.91 |
| 05:55–06:00 | 257 | 32 | 62 | 13 | 2 | 3 | 0 | 3.39 | 3.28 | 2.89 | 3.6 | 3.35 | 0 | 3.76 | 3.38 | 21.31 |
| **Total** | **2235** | **270** | **558** | **128** | **17** | **59** | **4** | | | | | | | | | |

**Table 8** Traffic flow, travel time from Shivaji Vatika to Madhumilan Square: 6–7 PM.

| | No. of Vehicles | | | | | | | Travel time in sec | | | | | | | | |
|---|---|---|---|---|---|---|---|---|---|---|---|---|---|---|---|---|
| **Time period** | **2-Wheeler** | **3-Wheeler** | **Car** | **Bus** | **L.C.V.** | **Cycles** | **HCV** | **2-Wheeler** | **3-Wheeler** | **Car** | **Bus** | **L.C.V.** | **HCV** | **Cycles** | **Avg time (sec)** | **Speed (km/ hr.)** |
| 06:00–06:05 | 242 | 29 | 55 | 12 | 0 | 5 | 0 | 3.34 | 3.39 | 2.79 | 3.98 | 0 | 0 | 4.22 | 3.54 | 20.32 |
| 06:05–06:10 | 231 | 31 | 44 | 10 | 0 | 4 | 0 | 3.21 | 3.42 | 2.68 | 3.92 | 0 | 0 | 4.08 | 3.46 | 20.80 |
| 06:10–06:15 | 223 | 27 | 39 | 14 | 2 | 5 | 1 | 3.17 | 3.29 | 2.4 | 3.12 | 3.4 | 2.94 | 4.22 | 3.22 | 22.36 |
| 06:15–06:20 | 215 | 26 | 41 | 9 | 1 | 5 | 1 | 3.13 | 3.2 | 2.74 | 2.74 | 2.98 | 2.96 | 3.84 | 3.08 | 23.34 |
| 06:20–06:25 | 202 | 22 | 43 | 11 | 1 | 8 | 0 | 2.98 | 3.22 | 2.72 | 2.82 | 2.8 | 0 | 4.12 | 3.11 | 23.15 |
| 06:25–06:30 | 206 | 21 | 46 | 13 | 0 | 7 | 1 | 3.03 | 3.14 | 2.8 | 2.9 | 0 | 2.88 | 3.92 | 3.11 | 23.14 |
| 06:30–06:35 | 198 | 17 | 32 | 10 | 0 | 4 | 0 | 2.98 | 3.18 | 2.44 | 2.96 | 0 | 0 | 3.89 | 2.58 | 27.96 |
| 06:35–06:40 | 194 | 15 | 38 | 8 | 2 | 5 | 0 | 2.87 | 2.9 | 2.08 | 2.65 | 3.24 | 0 | 3.98 | 2.95 | 24.38 |
| 06:40–06:45 | 190 | 10 | 35 | 13 | 0 | 6 | 0 | 2.83 | 2.54 | 1.88 | 3.31 | 0 | 0 | 4.04 | 2.92 | 24.66 |
| 06:45–06:50 | 181 | 13 | 28 | 13 | 1 | 6 | 1 | 2.74 | 2.5 | 1.76 | 3.45 | 3.02 | 3.05 | 3.87 | 2.91 | 24.72 |
| 06:50–06:55 | 171 | 9 | 34 | 12 | 1 | 9 | 0 | 2.62 | 2.42 | 1.64 | 3.56 | 2.94 | 0 | 3.89 | 2.85 | 25.31 |
| 06:55–07:00 | 169 | 8 | 27 | 11 | 2 | 3 | 0 | 2.59 | 2.24 | 1.48 | 3.74 | 3.1 | 0 | 3.8 | 2.83 | 25.49 |
| **Total** | **2422** | **228** | **462** | **136** | **10** | **67** | **4** | | | | | | | | | |

**Table 9** Traffic flow, travel time from Madhumilan Square to Shivaji Vatika: 9–10AM.

| | No. of Vehicles | | | | | | | Travel time in sec | | | | | | | | |
|---|---|---|---|---|---|---|---|---|---|---|---|---|---|---|---|---|
| **Time period** | **2-Wheeler** | **3-Wheeler** | **Car** | **Bus** | **L.C.V.** | **Cycles** | **HCV** | **2-Wheeler** | **3-Wheeler** | **Car** | **Bus** | **L.C.V.** | **Cycles** | **HCV** | **Avg time (sec)** | **Speed (km/ hr.)** |
| 09:00–09:05 | 87 | 5 | 20 | 2 | 2 | 6 | 0 | 1.15 | 1.34 | 0.96 | 1.61 | 1.67 | 3.4 | 0 | 1.68 | 42.86 |
| 09:05–09:10 | 92 | 6 | 25 | 4 | 1 | 5 | 0 | 1.19 | 1.37 | 1.05 | 1.72 | 1.73 | 3.4 | 0 | 1.75 | 41.26 |
| 09:10–09:15 | 105 | 9 | 24 | 4 | 1 | 5 | 0 | 1.31 | 1.43 | 1.09 | 1.75 | 1.8 | 3.4 | 0 | 1.80 | 40.11 |
| 09:15–09:20 | 89 | 9 | 22 | 3 | 2 | 8 | 0 | 1.24 | 1.35 | 1.14 | 1.63 | 2.06 | 3.5 | 0 | 1.82 | 39.52 |
| 09:20–09:25 | 82 | 4 | 16 | 5 | 3 | 9 | 1 | 1.16 | 1.3 | 1.07 | 1.71 | 2.18 | 3.5 | 2.34 | 1.89 | 38.15 |
| 09:25–09:30 | 120 | 7 | 22 | 6 | 2 | 8 | 1 | 1.56 | 1.46 | 1.16 | 1.73 | 1.98 | 3.3 | 2.32 | 1.93 | 37.33 |
| 09:30–09:35 | 117 | 17 | 34 | 4 | 5 | 11 | 0 | 1.48 | 2.02 | 1.24 | 1.68 | 2.35 | 3.5 | 0 | 2.04 | 35.29 |
| 09:35–09:40 | 111 | 11 | 26 | 4 | 0 | 10 | 1 | 1.36 | 1.96 | 1.1 | 1.57 | 0 | 3.6 | 2.12 | 1.95 | 37.02 |
| 09:40–09:45 | 109 | 10 | 34 | 6 | 1 | 11 | 0 | 1.37 | 1.88 | 1.18 | 1.96 | 1.73 | 3.6 | 0 | 1.95 | 36.86 |
| 09:45–09:50 | 79 | 12 | 28 | 4 | 0 | 13 | 0 | 1.17 | 1.84 | 1.16 | 1.86 | 0 | 3.7 | 0 | 1.94 | 37.04 |
| 09:50–09:55 | 119 | 12 | 25 | 6 | 0 | 9 | 0 | 1.34 | 1.82 | 1.11 | 1.89 | 0 | 3.8 | 0 | 1.98 | 36.33 |
| 09:50–10:00 | 137 | 14 | 24 | 7 | 1 | 9 | 0 | 1.68 | 1.88 | 1.17 | 1.93 | 1.84 | 3.8 | 0 | 2.05 | 35.18 |
| **Total** | **1247** | **116** | **300** | **121** | **18** | **104** | **3** | | | | | | | | | |

**Table 10** Traffic flow, travel time from Madhumilan Square to Shivaji Vatika: 10–11 AM.

| Time period | No. of Vehicles | | | | | | | Travel time in sec | | | | | | | | |
|---|---|---|---|---|---|---|---|---|---|---|---|---|---|---|---|---|
| | 2-Wheeler | 3-Wheeler | Car | Bus | L.C.V. | Cycles | HCV | 2-Wheeler | 3-Wheeler | Car | Bus | L.C.V. | Cycles | HCV | Avg time (sec) | Speed (km/hr.) |
| 10:00–10:05 | 178 | 14 | 30 | 6 | 1 | 8 | 0 | 1.91 | 2.09 | 1.28 | 1.89 | 1.88 | 3.8 | 0 | 2.14 | 33.59 |
| 10:05–10:10 | 146 | 12 | 23 | 5 | 1 | 9 | 2 | 1.85 | 1.96 | 1.31 | 1.83 | 1.74 | 3.7 | 2.59 | 2.13 | 33.73 |
| 10:10–10:15 | 138 | 19 | 22 | 7 | 1 | 11 | 0 | 1.81 | 2.02 | 1.29 | 1.91 | 1.96 | 3.6 | 0 | 2.10 | 34.23 |
| 10:15–10:20 | 130 | 18 | 29 | 5 | 1 | 10 | 1 | 1.34 | 1.98 | 1.27 | 1.85 | 1.82 | 3.7 | 2.47 | 2.06 | 34.88 |
| 10:20–10:25 | 124 | 16 | 34 | 8 | 0 | 12 | 0 | 1.39 | 2.12 | 1.47 | 2.05 | 0 | 3.8 | 0 | 2.16 | 33.40 |
| 10:25–10:30 | 144 | 15 | 38 | 5 | 0 | 8 | 1 | 1.44 | 2.05 | 1.43 | 1.99 | 0 | 3.7 | 2.53 | 2.20 | 32.80 |
| 10:30–10:35 | 159 | 14 | 29 | 7 | 0 | 7 | 1 | 2.49 | 2.15 | 1.22 | 2.13 | 0 | 3.6 | 2.6 | 2.37 | 30.42 |
| 10:35–10:40 | 141 | 17 | 32 | 6 | 0 | 12 | 0 | 2.13 | 2.16 | 1.43 | 1.68 | 0 | 3.9 | 0 | 2.26 | 31.89 |
| 10:40–10:45 | 155 | 15 | 59 | 5 | 0 | 11 | 1 | 1.78 | 2.04 | 1.82 | 2.34 | 0 | 3.9 | 2.64 | 2.41 | 29.83 |
| 10:45–10:50 | 124 | 14 | 44 | 8 | 2 | 14 | 1 | 1.59 | 2.09 | 1.61 | 2.09 | 1.98 | 3.9 | 2.74 | 2.29 | 31.48 |
| 10:50–10:55 | 119 | 16 | 32 | 6 | 1 | 8 | 0 | 1.51 | 2.11 | 1.41 | 1.94 | 1.75 | 3.8 | 0 | 2.09 | 34.53 |
| 10:55–11:00 | 124 | 10 | 51 | 9 | 1 | 7 | 0 | 1.47 | 1.84 | 1.73 | 2.24 | 1.9 | 3.7 | 0 | 2.15 | 33.49 |
| **Total** | **1682** | **180** | **423** | **77** | **8** | **117** | **7** | | | | | | | | | |

**Table 11** Traffic flow, travel time from Madhumilan Square to Shivaji Vatika: 5–6 PM.

| | No. of Vehicles | | | | | | | Travel time in sec | | | | | | | | |
|---|---|---|---|---|---|---|---|---|---|---|---|---|---|---|---|---|
| Time period | 2-Wheeler | 3-Wheeler | Car | Bus | L.C.V. | Cycles | HCV | 2-Wheeler | 3-Wheeler | Car | Bus | L.C.V. | Cycles | HCV | Avg time (sec) | Speed (km/hr.) |
| 05:00–05:05 | 162 | 20 | 46 | 12 | 1 | 5 | 1 | 1.55 | 1.89 | 1.31 | 2.04 | 2.34 | 3.52 | 2.16 | 2.12 | 34.03 |
| 05:05–05:10 | 152 | 14 | 48 | 8 | 1 | 4 | 1 | 1.58 | 1.79 | 1.51 | 2.04 | 2.61 | 3.47 | 2.31 | 2.19 | 32.92 |
| 05:10–05:15 | 166 | 14 | 36 | 8 | 0 | 6 | 0 | 1.78 | 1.86 | 1.62 | 2.38 | 0 | 3.63 | 0 | 2.25 | 31.94 |
| 05:15–05:20 | 189 | 15 | 38 | 13 | 1 | 9 | 0 | 2.14 | 1.89 | 1.96 | 2.59 | 0 | 3.67 | 0 | 2.45 | 29.39 |
| 05:20–05:25 | 158 | 8 | 43 | 9 | 0 | 12 | 0 | 1.98 | 2.02 | 2.02 | 2.54 | 0 | 3.79 | 0 | 2.47 | 29.15 |
| 05:25–05:30 | 190 | 17 | 59 | 8 | 1 | 10 | 1 | 2.26 | 2.1 | 2.14 | 2.49 | 2.76 | 3.86 | 2.39 | 2.57 | 28.00 |
| 05:30–05:35 | 155 | 26 | 48 | 13 | 2 | 6 | 1 | 2.12 | 2.21 | 1.96 | 2.71 | 2.86 | 3.89 | 2.46 | 2.60 | 27.68 |
| 05:35–05:40 | 181 | 18 | 39 | 14 | 2 | 7 | 0 | 2.23 | 2.19 | 1.94 | 2.74 | 0 | 3.81 | 0 | 2.58 | 27.89 |
| 05:40–05:45 | 191 | 14 | 58 | 11 | 4 | 11 | 0 | 2.3 | 2.08 | 2.08 | 2.63 | 0 | 3.92 | 0 | 2.60 | 27.67 |
| 05:45–05:50 | 198 | 26 | 59 | 11 | 2 | 8 | 2 | 2.38 | 2.31 | 2.02 | 2.7 | 2.8 | 3.94 | 2.5 | 2.66 | 27.02 |
| 05:50–05:55 | 218 | 21 | 63 | 8 | 4 | 9 | 1 | 2.42 | 2.29 | 2.12 | 2.63 | 2.69 | 3.91 | 3 | 2.72 | 26.44 |
| 05:55–06:00 | 207 | 18 | 54 | 12 | 3 | 12 | 0 | 2.53 | 2.35 | 2.2 | 2.74 | 0 | 3.97 | 0 | 2.76 | 26.11 |
| **Total** | **2167** | **211** | **591** | **127** | **21** | **99** | **7** | | | | | | | | | |

**Table 12** Traffic flow, travel time from Madhumilan Square to Shivaji Vatika: 6–7 PM.

| | No. of Vehicles | | | | | | | Travel Time in sec | | | | | | | | |
|---|---|---|---|---|---|---|---|---|---|---|---|---|---|---|---|---|
| Time period | 2-Wheeler | 3-Wheeler | Car | Bus | L.C.V. | Cycles | HCV | 2-Wheeler | 3-Wheeler | Car | Bus | L.C.V. | HCV | Cycles | Avg time (sec) | Speed (km/hr.) |
| 06:00–06:05 | 201 | 15 | 62 | 14 | 2 | 9 | 0 | 2.6 | 2.86 | 2.49 | 2.86 | 2.8 | 0 | 3.95 | 2.93 | 24.60 |
| 06:05–06:10 | 198 | 14 | 59 | 11 | 0 | 9 | 0 | 2.68 | 2.52 | 2.52 | 2.86 | 0 | 0 | 3.92 | 2.90 | 24.83 |
| 06:10–06:15 | 199 | 14 | 47 | 9 | 0 | 5 | 0 | 2.67 | 2.57 | 2.43 | 2.91 | 0 | 0 | 3.83 | 2.88 | 24.98 |
| 06:15–06:20 | 215 | 21 | 54 | 11 | 2 | 12 | 2 | 2.95 | 2.68 | 2.76 | 3.18 | 3.02 | 3.29 | 4.01 | 3.13 | 23.02 |
| 06:20–06:25 | 240 | 18 | 68 | 10 | 1 | 14 | 0 | 3.09 | 2.96 | 2.96 | 3.24 | 2.97 | 0 | 4.05 | 3.21 | 22.42 |
| 06:25–06:30 | 231 | 18 | 66 | 8 | 1 | 12 | 1 | 3.12 | 2.91 | 2.88 | 3.18 | 3.04 | 3.22 | 4.03 | 3.20 | 22.52 |
| 06:30–06:35 | 202 | 16 | 58 | 11 | 1 | 9 | 1 | 2.9 | 2.81 | 2.84 | 3.2 | 2.89 | 2.99 | 3.99 | 3.09 | 23.31 |
| 06:35–06:40 | 189 | 18 | 56 | 15 | 0 | 11 | 1 | 2.94 | 3.04 | 2.93 | 3.42 | 0 | 3.07 | 3.97 | 3.23 | 22.30 |
| 06:40–06:45 | 198 | 22 | 48 | 14 | 0 | 8 | 0 | 2.91 | 2.86 | 2.83 | 3.4 | 0 | 0 | 4.04 | 3.21 | 22.44 |
| 06:45–06:50 | 182 | 19 | 58 | 15 | 3 | 10 | 0 | 2.82 | 2.78 | 2.86 | 3.36 | 0 | 0 | 3.96 | 3.16 | 22.81 |
| 06:50–06:55 | 165 | 11 | 54 | 9 | 2 | 8 | 0 | 2.52 | 2.52 | 2.72 | 3.18 | 0 | 0 | 3.93 | 2.97 | 24.21 |
| 06:55–07:00 | 169 | 12 | 42 | 13 | 2 | 7 | 2 | 2.48 | 2.48 | 2.56 | 3.23 | 2.9 | 3 | 3.92 | 2.94 | 24.50 |
| **Total** | **2389** | **198** | **672** | **140** | **14** | **114** | **7** | | | | | | | | | |

# Index

# Also of Interest

## Check out these other related titles from Scrivener Publishing

*SMART GRIDS FOR SMART CITIES VOLUME 1*, Edited by O.V. Gnana Swathika, K. Karthikeyan, and Sanjeevikumar Padmanaban, ISBN: 9781119872078. Written and edited by a team of experts in the field, this first volume in a two-volume set focuses on an interdisciplinary perspective on the financial, environmental, and other benefits of smart grid technologies and solutions for smart cities.

*SMART GRIDS FOR SMART CITIES VOLUME 2: Real-Time Applications in Smart Cities*, Edited by O.V. Gnana Swathika, K. Karthikeyan, and Sanjeevikumar Padmanaban, ISBN: 9781394215874. Written and edited by a team of experts in the field, this second volume in a two-volume set focuses on an interdisciplinary perspective on the financial, environmental, and other benefits of smart grid technologies and solutions for smart cities.

*INTEGRATED GREEN ENERGY SOLUTIONS VOLUME 1*, Edited by Milind Shrinivas Dangate, W. S. Sampath, O. V. Gnana Swathika, and Sanjeevikumar Padmanaban, ISBN: 9781119847434. This first volume in a two-volume set presents the state of the art for the concepts, practical applications, and future of renewable energy and how to move closer to true sustainability.

*INTEGRATED GREEN ENERGY SOLUTIONS VOLUME 2*, Edited by Milind Shrinivas Dangate, W. S. Sampath, O. V. Gnana Swathika, and Sanjeevikumar Padmanaban, ISBN: 9781394193660. This second volume in a two-volume set continues to present the state of the art for the concepts, practical applications, and future of renewable energy and how to move closer to true sustainability.

*MODERN AUTOMOTIVE ELECTRICAL SYSTEMS: Theory and Applications,* Edited by Pedram Asef, Sanjeevikumar Padmanaban, and Andrew Lapthorn, ISBN: 1119801047. Presenting the concepts and advances of modern automotive electrical systems, this volume, written and edited by a global team of experts, also goes into the practical applications for the engineer, student, and other industry professionals.

*INTELLIGENT GREEN TECHNOLOGIES FOR SMART CITIES,* Edited by Suman Lata Tripathi, Souvik Ganguli, Abhishek Kumar, and Tengiz Magradze, ISBN: 9781119816065. Presenting the concepts and fundamentals of smart cities and developing "green" technologies, this volume, written and edited by a global team of experts, also goes into the practical applications that can be utilized across multiple disciplines and industries, for both the engineer and the student.

*DESIGN AND DEVELOPMENT OF EFFICIENT ENERGY SYSTEMS,* Edited by Suman Lata Tripathi, Dushyant Kumar Singh, Sanjeevikumar Padmanaban, and P. Raja, ISBN: 9781119761631. Covering the concepts and fundamentals of efficient energy systems, this volume, written and edited by a global team of experts, also goes into the practical applications that can be utilized across multiple industries, for both the engineer and the student.

*Green Energy: Solar Energy, Photovoltaics, and Smart Cities,* Edited by Suman Lata Tripathi and Sanjeevikumar Padmanaban, ISBN: 9781119760764. Covering the concepts and fundamentals of green energy, this volume, written and edited by a global team of experts, also goes into the practical applications that can be utilized across multiple industries, for both the engineer and the student.

*HYDROGEN ELECTRICAL VEHICLES,* Edited by Mehmet Sankir and Nurdan Demirci Sankir, ISBN: 9781394166381. Hydrogen electrical vehicles are an essential component of the "Green New Deal" and this book covers cutting-edge technologies designed for fuel-cell-powered cars.

*PAVEMENT ASSET MANAGEMENT,* By Ralph Haas, W.R. Hudson, and Lynne Cowe Falls, ISBN: 9781119038702. Whether you are new to managing pavement assets or a seasoned practitioner, this book provides the most up-to-date information in the field on the history, technology, practice and look ahead.

Printed in the USA/Agawam, MA
September 16, 2026

910661.124